S0-BLS-201

Jork, Funk, Fischer, Wimmer

Thin-Layer Chromatography

Volume 1a

Hellmut Jork, Werner Funk,
Walter Fischer, Hans Wimmer

Thin-Layer Chromatography

Reagents and Detection Methods

Volume 1
Physical and Chemical Detection Methods
(in several parts, parts 1b and 1c in preparation)

Volume 2
Radiometric Detection Methods
(in preparation)

Volume 3
Biochemical and Biological Detection Methods
(in preparation)

Distribution:

VCH Verlagsgesellschaft, P.O. Box 101161, D-6940 Weinheim (Federal Republic of Germany)

Switzerland: VCH Verlags-AG, P.O. Box, CH-4020 Basel (Switzerland)

United Kingdom and Ireland: VCH Publishers (UK) Ltd., 8 Wellington Court,
Wellington Street, Cambridge CB1 1HW (England)

USA and Canada: VCH Publishers, Suite 909, 220 East 23rd Street, New York,
NY 10010-4606 (USA)

ISBN 3-527-27834-6 (VCH, Weinheim) ISBN 0-89573-876-7 (VCH, New York)

Hellmut Jork, Werner Funk,
Walter Fischer, Hans Wimmer

Thin-Layer Chromatography

Reagents and Detection Methods

Volume 1a

Physical and Chemical Detection Methods:
Fundamentals, Reagents I

Translated by Frank and Jennifer A. Hampson

Prof. Dr. H. Jork
Universität des Saarlandes
Fachbereich 14
Stadtwald
D-6600 Saarbrücken

Prof. W. Funk
Fachbereich Technisches Gesundheitswesen
der Fachhochschule Gießen-Friedberg
Wiesenstraße 14
D-6300 Gießen

Dr. W. Fischer
c/o E. Merck
Abteilung V Reag SPA
Frankfurter Straße 250
D-6100 Darmstadt

Hans Wimmer
Eckhardt-Straße 23
D-6100 Darmstadt

Editorial Director: Dr. Hans F. Ebel
Production Manager: Dipl.-Ing. (FH) Hans Jörg Maier

Library of Congress Card No.:

89-16558

British Library Cataloguing-in-Publication Data:

Thinlayer chromatography : reagents and detection methods.
Vol. 1a : physical and chemical detection methods.
1. Thin layer chromatography
I. Jork, Hellmut
543'.08956
ISBN 3-527-27834-6

Deutsche Bibliothek Cataloguing-in-Publication Data:

Thin-Layer chromatography : reagents and detection methods /
Hellmut Jork... – Weinheim ; Basel (Switzerland) ; Cambridge
; New York, NY : VCH.
NE: Jork, Hellmut [Mitverf.]

Vol. 1. Physical and chemical detection methods.
a. Fundaments, reagents I. – 1990
ISBN 3-527-27834-6 (Weinheim...)
ISBN 0-89573-876-7 (New York...)

Printed on acid-free paper

Composition and Printing: Wiesbadener Graphische Betriebe GmbH, D-6200 Wiesbaden
Bookbinding: Georg Kränkl, D-6148 Heppenheim
Printed in the Federal Republic of Germany

Foreword

Thin-layer chromatography as practiced today seems to exist in two forms. Some scientists consider TLC to be a qualitative separation tool for simple mixtures where speed, low cost, and simplicity are its virtues. Others regard TLC as a powerful separation tool for the quantitative analysis of complex mixtures with a high sample throughput because of parallel sample processing, and as a technique that can tolerate cruder samples than column methods because the stationary phase is disposable, and which provides flexibility in the method and choice of detection since at the time of detection the separation is static and the layer open to inspection. Both groups of scientists use the same approaches and employ the same physical principles to achieve a separation, but only the second group does so in an optimized way. There remains an information gap which good books can fill to re-educate the scientific community of the current standing of TLC. I am delighted to affirm that the present book takes a needed step in this direction. As more scientists become acquainted with the modern practice of TLC they will need reliable and unbiased sources of information on the myriad of factors that influence quantitation of TLC chromatograms to avoid common pitfalls that follow in the wake of any technological advance. From physical principles, to working instruments, to the methodological requirements of an analytical protocol the reader will find such information, and just as importantly, experience, distilled into this book.

It seems to be a fact of life that real samples are too dilute or too complex for direct analyses no matter what new technology is available simply because the demand for analytical information is being continuously raised to a higher level. At this interface chemical intuition has always played an important role. Selective chemical reactions provide the methods to manipulate a sample to reveal the information desired. They provide the means to increase the response of an analyte to a particular detector and increase the selectivity of an analysis by targeting certain components of the sample to respond to a selected detector. Micropreparative chemistry and TLC have a long history of association because of the convenience of these reactions when performed with a static sample and because in TLC the separation and detection processes can be treated as separate steps and optimized independently of each other. A further important character of this book is the practical way it marries chemical and instrumental principles together providing an integrated source to the most important chemical reactions available and the details of their application to particular sample types. Until new detection principles are available, these reactions represent the most practical and, in many cases, an elegant solution to difficult analytical problems. This book should serve to revive interest in this area and to provide a methodological source for their practice.

The power of TLC is in its flexibility as a problem solving tool. As the problems in analysis become more complicated the sophistication by which we approach those problems is ever increasing. However, it behooves us as analytical chemists not to forget our fundamental training in chemistry and to apply those principles to today's problems. It is just this feature that the reader will find instilled into this book.

C. F. Poole
Department of Chemistry
Wayne State University
Detroit, MI 48202
USA

Preface

This book is the result of cooperation between four colleagues, who have been working in the field of thin-layer chromatography for many years and, in particular, took an active part in the development from hand-coated TLC plates to commercially available precoated plates and instrumental thin-layer chromatography. This development was accompanied by improvements in the field of detection of the separated zones. In particular, it became necessary to be able to deal with ever decreasing quantities of substance, so that the compilation "Anfärbereagenzien" by E. Merck, that had been available as a brochure for many, many years, no longer represented the state of the art of thin-layer chromatography.

It was against this background and in view of the fact that there is at present no contemporary monograph on thin-layer chromatography that this book was produced. It is intended as an introduction to the method, a reference book, and a laboratory handbook in one, i.e., far more than just a "Reagent Book".

The first part of the book consists of a detailed treatment of the fundamentals of thin-layer chromatography, and of measurement techniques and apparatus for the qualitative and quantitative evaluation of thin-layer chromatograms. In situ prechromatographic derivatization techniques used to improve the selectivity of the separation, to increase the sensitivity of detection, and to enhance the precision of the subsequent quantitative analysis are summarized in numerous tables.

Particular attention has been devoted to the fluorescence methods, which are now of such topicality, and to methods of increasing and stabilizing the fluorescence emissions. Nowhere else in the literature is there so much detailed information to be found as in the first part of this book, whose more than 600 literature references may serve to stimulate the reader to enlarge his or her own knowledge.

Nor has a general introduction to the microchemical postchromatographic reactions been omitted; it makes up the second part of the book.

This second part with its 80 worked-through and checked detection methods forms the foundation of a collection of reagent reports (monographs), which will be extended to several volumes and which is also sure to be welcomed by workers who carry out derivatizations in the fields of electrophoresis and high-pressure liquid chromatography. Alongside details of the reagents required and their handling and storage, the individual reports also contain details about the reaction concerned.

Wherever possible, dipping reagents have been employed instead of the spray reagents that were formerly commonplace. These make it easier to avoid contaminating the laboratory, because the coating of the chromatogram with the reagent takes place with less environmental pollution and lower health risks; furthermore, it is more homogeneous, which results in higher precision in quantitative analyses.

It is possible that the solvents suggested will not be compatible with all the substances detectable with a particular reagent, for instance, because the chromatographically separated substances or their reaction products are too soluble. Therefore, it should be checked in each case whether it is possible to employ the conditions suggested without modification. We have done this in each report for one chosen class of substance by working through an example for ourselves and have documented the results in the "Procedure Tested"; this includes not only the exact chromatographic conditions but also details concerning quantitation and the detection limits actually found. Other observations are included as "Notes". Various types of adsorbent have been included in these investigations and their applicability is also reported. If an adsorbent is not mentioned it only means that we did not check the application of the reagent to that type of layer and not that the reagent cannot be employed on that layer.

Since, in general, the reagent report includes at least one reference covering each substance or class of substances, it is possible to use Part II of this book with its ca. 750 references as a source for TLC applications. Only rarely are earlier references (prior to 1960), which were of importance for the development of the reagent, cited here.

There is no need to emphasize that many helpful hands are required in the compilation of such a review. Our particular thanks are due to Mrs. E. Kany, Mrs. I. Klein and Mrs. S. Netz together with Dipl.-Ing. M. Heiligenthal for their conscientious execution of the practical work.

We would also like to thank the graduate and postgraduate students who helped to check the derivatization reactions and Mrs. U. Enderlein, Mrs. E. Otto, and Mrs. H. Roth, whose capable hands took care of the technical preparations for the book and the production of the manuscript. We would particularly like to thank Dr. Kalinowski (Univ. Giessen) for his magnificent help in the formulation of the reaction paths for the reagent reports. Our thanks are also due to Dr. F. Hampson and Mrs. J. A. Hampson for translating the German edition of the book into English.

We thank the Baron, J. T. Baker, Camag, Desaga, Macherey-Nagel and E. Merck companies for their generous support of the experimental work.

Our particular thanks are also due to Dr. H. F. Ebel and his colleagues at VCH Verlagsgesellschaft for the realization of our concepts and for the design and presentation of the book and for the fact that this work has appeared in such a short time.

In spite of all our care and efforts we are bound to have made mistakes. For this reason we would like to ask TLC specialists to communicate to us any errors and any suggestions they may have for improving later volumes.

Saarbrücken, Giessen and Darmstadt,
October 1989

Hellmut Jork
Werner Funk
Walter Fischer
Hans Wimmer

Contents

Part I
Methods of Detection

Part II
Reagents in Alphabetical Order

Part I

Methods of Detection

1 Introduction

The *separation methods* routinely employed in the laboratory include the various chromatographic and electrophoretic techniques, whose selectivity is continually being increased by the introduction of new adsorbents, e.g. with chemically modified surfaces (Fig. 1).

Detection methods, which provide real information concerning the separated substances, are necessary in order to be able to analyze the separation result and separation performance achieved by such a system.

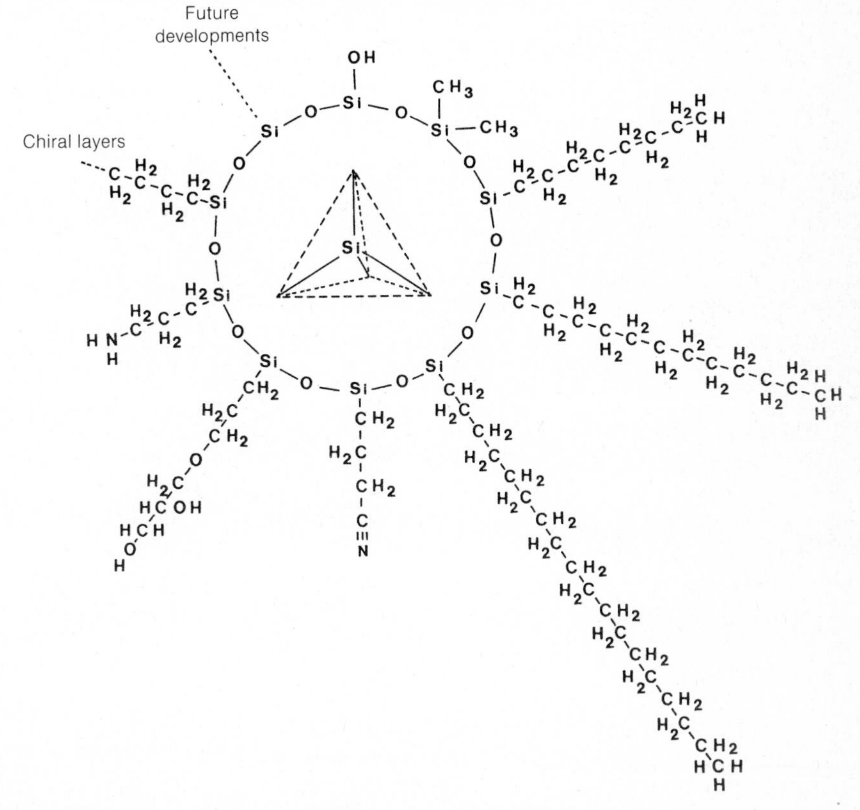

Fig. 1: Schematic view of the surface-modified silica gels at present commercially available.

In spite of numerous advances in the field of detection there are not and never have been any genuinely *substance-specific* chemical detection reactions. This means that, unlike the spectrometric methods, the methods of detection normally employed in chromatography cannot be employed for an unequivocal identification of compounds, they can only provide more or less definite indications for the characterization of the separated substances. Universal reagents are usually employed for a first analysis of the separation of samples of unknowns. This is then followed by the use of group-specific reagents. The more individual the pieces of information that can be provided from various sources for a presumed substance the more certainly is its presence indicated. However, all this evidence remains indicative; it is not a confirmation of identity.

The detection methods also serve especially to increase sensitivity and selectivity in addition to providing evidence concerning the quality of the separation. In the case of thin layer chromatography the *selectivity of the separation* which is achieved by the various techniques employed (e.g. multiple development, gradient elution, sequence TLC, AMD, HPPLC or OPLC techniques) (Fig. 2) is accompanied by *specificity of detection* [1–3]; the selectivity of detection can also be increased by the combination of several detection methods as shown in Figure 2. After chromatographic development the chromatogram may be regarded as being rather like a "diskette" with the individual pieces of information stored on it [4].

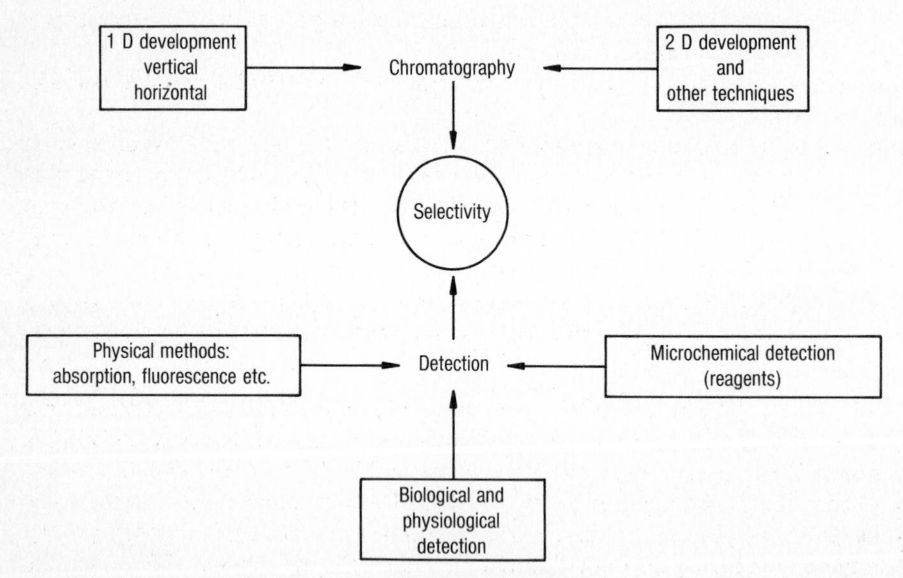

Fig. 2: Increasing the selectivity by combinations of methods.

The information can be read off as required at any time separately from the development process both in space and time. This possibility is *only* available in thin layer chromatography, because there is no direct on-line coupling between chromatographic development and detection. Thus, the analysis of a thin layer chromatogram can be repeated at any time as desired or carried out again according to other criteria than those employed in the first analysis. It follows that the detection technique does not restrict the choice of mobile phase. The chromatographic conditions can always be chosen to give the best separation for the particular sample, since the chromatogram is freed from mobile phase before detection is undertaken.

In addition to these advantages TLC also possesses other merits which ensure that it occupies a firm place in the arsenal of analytical techniques as a method for the separation of micro-, nano- and picogram quantities [5].

The method

- is easy to learn technically;
- is rapidly carried out, especially as HPTLC technique;
- is always available for use, since the precoated layers can usually be employed without pretreatment;
- can readily be monitored because the whole chromatogram (including the substances remaining at the start) can be taken in at a glance and it is not necessary to elute the individual components;
- does not require a regeneration step, since TLC and HPTLC plates are disposable items;
- can be economically employed for routine use because the consumption of mobile phase is low and, hence, there are scarcely any disposal problems (for instance up to 70 samples can be analyzed alongside each other and with authentic standard substances in a linear chamber with a very few milliliters of mobile phase [6]);
- is very adaptable: acidic, basic or purely aqueous mobile phases can be employed as can neutral lipophilic solvents, thus lending the whole thin layer chromatographic system a high degree of flexibility.

As a result of these merits thin layer chromatography finds application all over the world. The frequency of its application is documented in Figure 3. This CA search only includes those publications where TLC/HPTLC are included as key words. The actual application of the method is very much more frequent. The method is employed as a matter of course in many areas of quality control and routine monitoring of product purity. This was also true in the 1970s when the rapid development of high performance liquid chromatography (HPLC) led to a

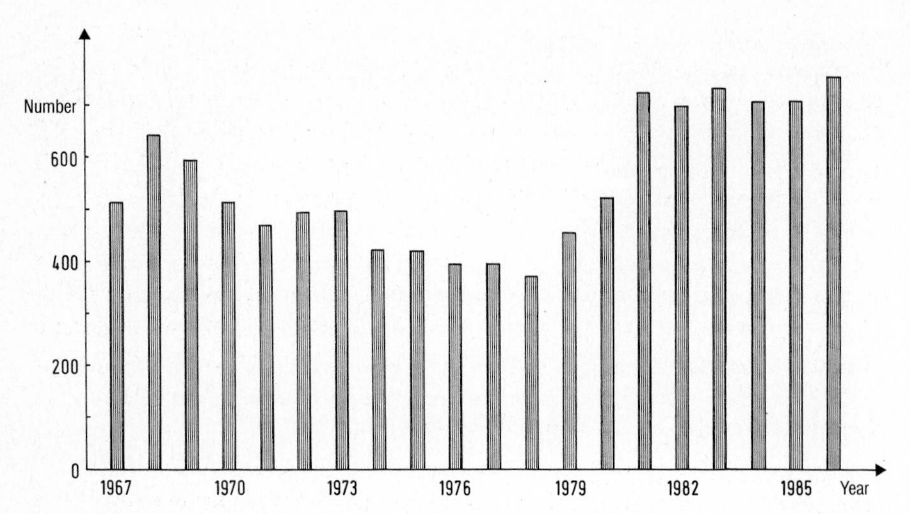

Fig. 3: Frequency distribution of TLC/HPTLC publications over the years 1967 – 1986 (search made from Chemical Abstracts).

temporary reduction in the frequency of academic TLC publications. Today both methods stand side by side as separation techniques in their own right. The analyst employs them where their respective advantages are evident. Economic considerations naturally have to be taken into account too [7] and the balance here is probably tipped in favor of thin-layer chromatography.

The information produced in a TLC separation can be recorded for storage or documentation. This can take the form of a manual-graphical reproduction, a photocopy, photograph or autoradiograph [8–10]. Increasingly, however, the chromatogram is scanned (fluorescence or absorption scanning curves) or the raw data are stored in data storage systems. The instrumentation required for these processes will be discussed in the subsequent chapters. A general discussion of documentation will follow in Chapter 4.

In principle it is possible to employ physical, microchemical and biological-physiological methods for detection in TLC (Fig. 2).

Physical methods: Physical methods include photometric absorption and fluorescence and phosphorescence inhibition, which is wrongly referred to as fluorescence quenching [1], and the detection of radioactively labelled substances by means of autoradiographic techniques, scintillation procedures or other radiometric methods. These methods are *nondestructive* (Chapt. 2).

Microchemical reactions: These can be carried out either with universal reagents [11] or with such substances which react with particular functional groups (group-characterizing reagents). If the separation process ensures that only *one* component occurs at a particular spot on the chromatogram, then this can be detected "substance-specifically". But specificity in an unequivocal sense can only be produced by a combination of the separation and the detection process. (The same is true of other forms of detection.)

Biological-physiological detection: The methods involved here take account of the biological activity of the separated components independent of their physical or chemical properties [12].

Their use is to be recommended [13] because

- such methods are highly specific (independent of the separation process);
- ineffective accompanying substances do not interfere with the investigation so that previous clean-up can often be omitted;
- the detection limits are comparable with those of classical detection methods.

These methods are employed for the detection and determination of antibiotics and substances with similar effects, like alkaloids, insecticides, fungicides, mycotoxins, vitamins, bitter principles and saponins [14].
We intend to devote separate volumes to each method of detection in the order discussed above.

References

[1] Jork, H.: *Qualitative und quantitative Auswertung von Dünnschicht-Chromatogrammen unter besonderer Berücksichtigung photoelektrischer Verfahren.* Professorial thesis, Universität des Saarlandes, Saarbrücken 1969.
[2] Jork, H.: More than 50 GDCh-training courses since 1972, Universität des Saarlandes, Saarbrücken.
[3] Funk, W.: *Fresenius Z. Anal. Chem.* **1984,** *318*, 206–219.
[4] Jork, H.: *Schnellmethoden in der Lebensmittel-Analytik.* Behr's Verlag, Hamburg 1987.
[5] Jork, H.: *Fresenius Z. Anal. Chem.* **1984,** *318*, 177–178.
[6] Jänchen, D.: *Proc. Int. Symp. Instrum. High Perform. Thin-Layer Chromatogr. (HPTLC), 1st, Bad Dürkheim,* 1980.

[7] Kelker, H.: *Nachr. Chem. Techn. Lab.* **1983**, *31*, 786.
[8] Stahl, E.: *Dünnschicht-Chromatographie, ein Laboratoriumshandbuch.* 2nd Ed., Springer, Berlin 1967.
[9] Kirchner, J. G.: *Thin-Layer Chromatography,* 2nd Ed., Wiley, New York 1978.
[10] Randerath, K.: *Dünnschicht-Chromatographie.* 2nd Ed., Verlag Chemie, Weinheim 1965.
[11] E. Merck, Company brochure "Dyeing Reagents for Thin Layer and Paper Chromatography", Darmstadt 1980.
[12] Wallhäuser, K. H.: in [8].
[13] Jork, H.: *GIT Fachz. Lab. Supplement 3 „Chromatographie"* **1986**, *30*, 79-87.
[14] Jork, H., Wimmer, H.: *Quantitative Auswertung von Dünnschicht-Chromatogrammen* **1986**, Arbeitsblatt I/1-7 und 1-8. GIT-Verlag, Darmstadt 1986.

2 Physical Methods of Detection

2.1 General

Physical detection methods are based on inclusion of substance-specific properties. The most commonly employed are the absorption or emission of electromagnetic radiation, which is detected by suitable detectors (the eye, photomultiplier). The β-radiation of radioactively labelled substances can also be detected directly. These *nondestructive* detection methods allow subsequent micropreparative manipulation of the substances concerned. They can also be followed by microchemical and/or biological-physiological detection methods.

A distinction is normally made between the visible and ultraviolet regions of the spectrum when detecting absorbing substances. Detection in the visible part of the spectrum can be carried out with the eye or with a photomultiplier.

2.2 Detection of Absorbing Substances

2.2.1 Visual Detection

The success of separation of *colored* compounds is usually monitored visually. Such compounds absorb a particular portion of the polychromatic (white) light in the visible wavelength range. The remaining radiation (complementary radiation) is reflected and detected by the eye; it determines the color of the substance zone. Table 1 correlates the wavelengths, colors and complementary colors.

Table 1. Correlation of wavelength, color and complementary color [1].

Wavelength [nm]	Color of radiation	Complementary color
620...700	red	bluish-green
590...620	orange	greenish-blue
570...590	yellow	blue
500...570	green	red/purple
450...500	blue	yellow
400...450	violet	yellowish green

Colorless substances absorb at wavelengths shorter than those of the visible range (the UV range normally amenable to analysis $\lambda = 400...200$ nm). Such compounds can be detected by the use of UV-sensitive detectors (photomultipliers, Sec. 2.2.3.1). Substances that absorb in the UV range and are stimulated to fluorescence or phosphorescence (luminescence) can be detected visually if they are irradiated with UV light.

2.2.2 Fluorescence and Phosphorescence Indicators

Fluorescent and phosphorescent substances are excited into an unstable energy state by UV light. When they return to the ground state they release a part of the energy taken up in the form of radiation. The emitted radiation is *less energetic* than the light absorbed and usually lies in the visible part of the spectrum. Since absorption (excitation) and emission obey a linear relationship over a certain range a reduction in absorption leads to a reduction in the luminescence, too.

This property can be applied to the detection of substances that absorb in the UV region: For on layers containing a fluorescent indicator or impregnated with a fluorescent substance the emission is reduced in regions where UV-active compounds partially absorb the UV light with which they are irradiated. Such substances, therefore, appear as *dark* zones on a fluorescent background (Fig. 4A).

This effect, which can also be produced if fluorescent substances are applied to the chromatogram by spraying or dipping after development, is an absorption effect and not a quenching process in the true sense of the word. It is correct to refer to fluorescence or phosphorescence diminishing. The more absorbant sample molecules there are present in the zone the darker this will appear (Fig. 4B).

This method of detection is at its most sensitive if the absorption maximum (λ_{max}) of the sample molecule is exactly at the wavelength of the UV light employed for irradiation. The further λ_{max} lies from this the less radiation is absorbed and the lower the sensitivity of detection. If the compound does not absorb at the wavelength of radiation or if it possesses an absorption minimum just there then such components are not detected by this method. Figure 4C illustrates this with the sweeteners saccharin and dulcin as examples.

Fluorescence and phosphorescence are both forms of *luminescence* [3]. If the emission of radiation has decayed within 10^{-8} s after the exciting radiation is cut off it is known as *fluorescence* [4], if the decay phase lasts longer (because the electrons return to the ground state from a forbidden triplet state (Fig. 5), then the phenomenon is known as *phosphorescence*. A distinction is also made between

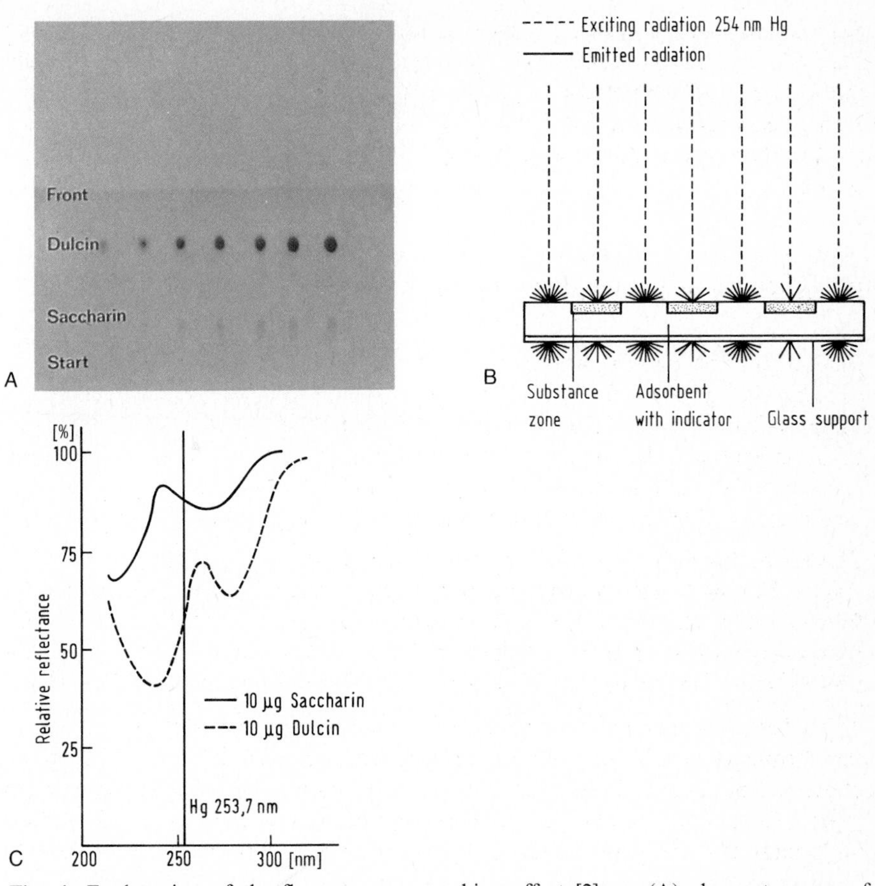

Fig. 4: Explanation of the fluorescence-quenching effect [2]. – (A) chromatograms of the same quantities of saccharin and dulcin observed under UV 254 light, (B) schematic representation of fluorescence quenching, (C) spectral reflectance curves of saccharin and dulcin.

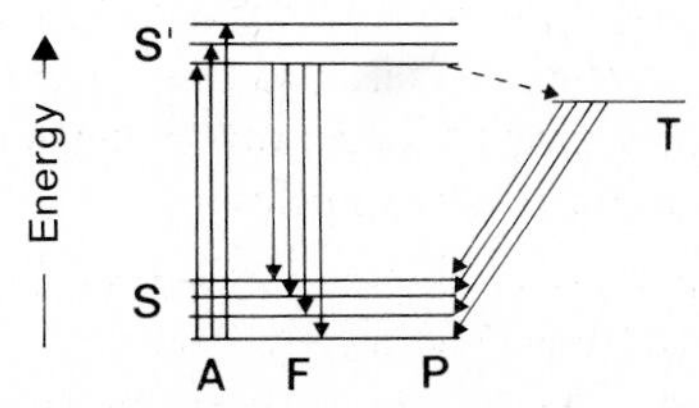

Fig. 5: Schematic representation of the electronic transitions during luminescence phenomena [5]. – A absorbed energy, F fluorescence emission, P phosphorescence, S ground state, S' excited singlet state, T "forbidden" triplet transition.

fluorescence and phosphorescence indicators. The former can be incorporated in the adsorbent layers or applied afterwards by spraying or dipping, the latter are always incorporated as homogeneously as possible into the stationary phase. Fluorescence occurs primarily in organic substances and phosphorescence, on the other hand, in inorganic compounds.

Organic fluorescence indicators for aluminium oxide, silica gel and cellulose layers (code F_{366}, UV_{366}) include:

- the sodium salt of 3-hydroxypyrene-5,8,10-trisulfonic acid [6],
- the sodium salt of 3,5-dihydroxypyrene-8,10-disulfonic acid [7],
- sodium fluorescein [8–11] and fluorescein [12] or 2′,7′-dichlorofluorescein [13–19],
- rhodamine B [12, 20–23] and rhodamine 6G [24–26],
- morin [11, 24, 27–29],
- cyanine dyestuffs [30, 31],
- stilbene derivatives (e.g. diaminostilbenetriazine) [12, 32] and
- optical brighteners (Ultraphor WT BASF [12, 33], Calcofluor R-white [34], Leukophor).

Oxytetracycline can also be employed at low pH on calcium-containing layers [35].

The scintillators are a special type of fluorescence indicators; they are employed for the fluorimetric detection of radioactively labelled substances. They are stimulated by β-radiation to the emission of electromagnetic radiation and will be discussed in Volume 2.

The substances employed as *inorganic phosphorescence indicators* (incorrectly referred to as fluorescence indicators) include blue (tin-activated strontium compounds), yellow (uranyl acetate [36]) and yellow-green (manganese-activated zinc silicate [37] or zinc cadmium sulfide [38]) emitting substances (code F_{254}, UV_{254}). Pigment ZS-super (Riedel de Haën) has also been employed [81]. Since these are not acid-stable they are replaced by substances such as alkaline earth metal tungstates in RP phases (code F_{254s}). These possess a pale blue emission [39].

The advantages of these inorganic indicators are:

- Such indicators do not migrate during chromatography to the solvent front under the influence of either polar or nonpolar organic solvents (uranyl acetate is an exception).

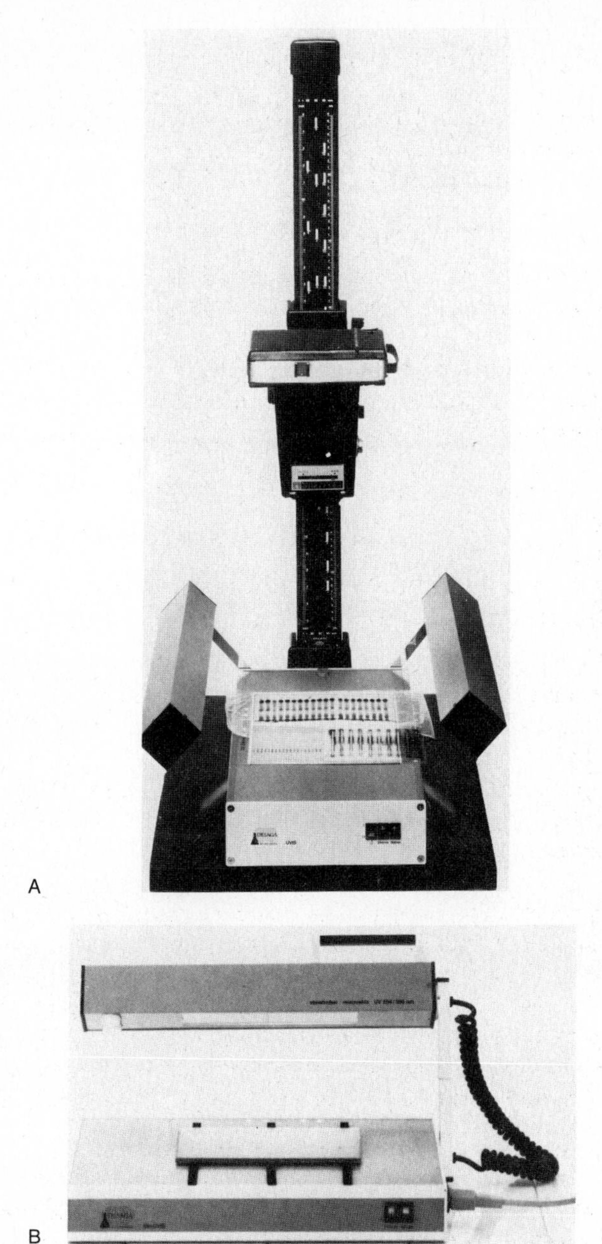

Fig. 6: UVIS and MinUVIS analysis lamps (DESAGA).

A

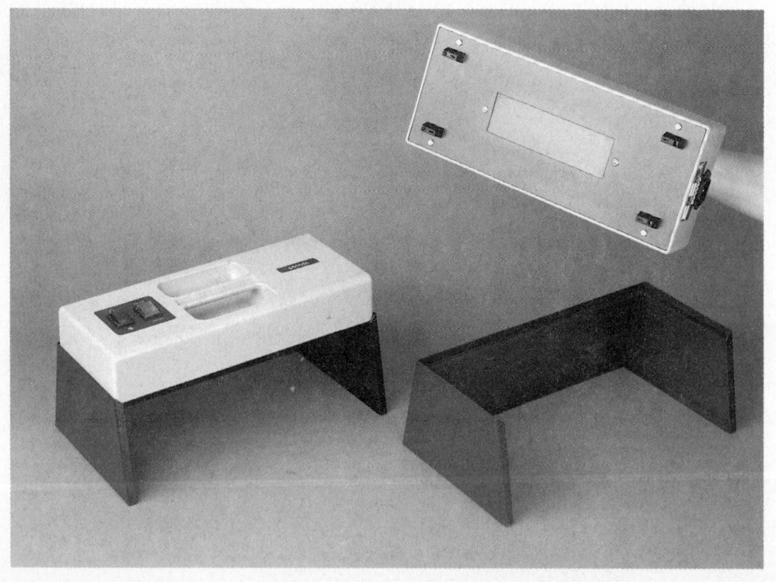

B

Fig. 7: UV hand lamps (CAMAG). (A) battery-powered UV lamp, (B) hand lamp with stand closed on three sides.

- Short-wavelength UV radiation ($\lambda = 254$ nm) is employed for excitation. This allows aromatic organic compounds, in particular, to be detected by fluorescence quenching. Uranylacetate may also be excited at $\lambda = 366$ nm.
- In the most favorable cases the detection limits are from 0.1 to 0.3 µg substance per chromatogram zone.

Energetic radiation sources are required to excite phosphorescence or fluorescence. Mercury line radiators are normally employed; these are readily available as relatively cheap mercury vapor lamps. The short-wavelength line at $\lambda = 254$ nm is

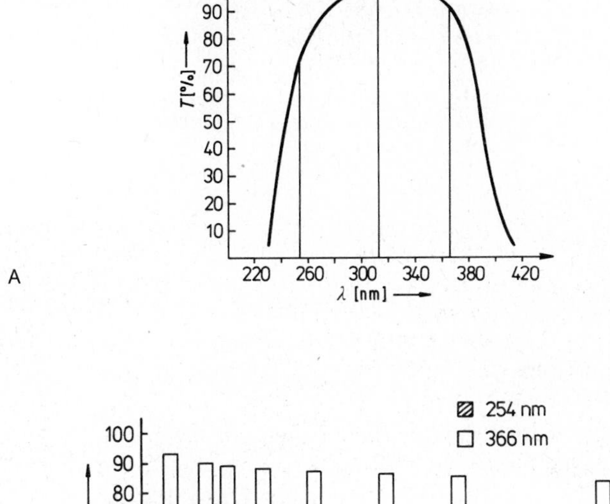

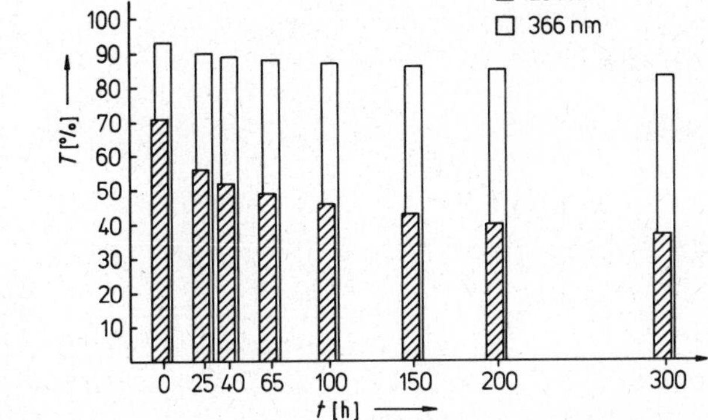

Fig. 8: Transmittance of black light filter as a function of wavelength (A) and as a function of the length of operation at $\lambda = 254$ and 365 nm (B).

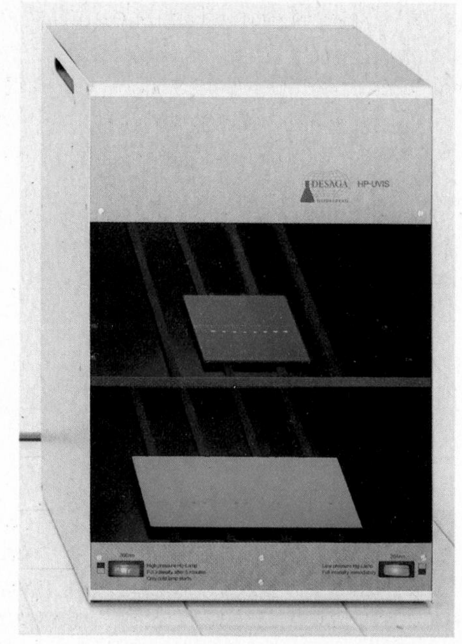

Fig. 9: HP-UVIS cabinet for UV inspection without a dark room (DESAGA).

mainly employed (although the absorption maxima of the indicators lie at $\lambda =$ 260 nm or 280 nm [37]) together with the long-wavelength double line at $\lambda = 365/366$ nm. While low pressure lamps deliver short-wavelength light almost exclusively, the proportion of long-wavelength UV radiation is much higher in the case of high pressure lamps. But these have the disadvantage that they require a starting up time of 2 to 5 minutes. They also operate relatively hot and can only be re-ignited after they have cooled down.

Note: The lamp can crack if the hot bulb comes into contact with a cold TLC plate (protective housing!).

These restrictions do not apply to the less intense fluorescent tubes installed in the UVIS or MinUVIS (Fig. 6) or Universal UV lamps (Fig. 7). Black glass surrounds or screens serve as filters. Unfortunately account is often not taken of the fact that the transparency for short-wavelength UV light decreases appreciably with increasing duration of irradiation (Fig. 8). So it is advisable to change the filters of lamps intended for short-wavelength radiation at regular intervals. They can still be employed in the long-wavelength region. This is particularly true if color photographs are taken for documentation purposes.

Combined compact instruments, where it is possible to switch from "daylight" to long- or short-wavelength UV light, are frequently offered for the examination of thin-layer chromatograms (Fig. 9). These are often fitted with a camera holder.

Caution: When working with UV light protective goggles should always be worn in order to avoid damage to the eyes.

2.2.3 Photometric Measurement of Absorption

2.2.3.1 Apparatus

Photomultipliers are appreciably more sensitive sensors than the eye in their response to line or continuum sources. Monochromators are fitted to the light beam in order to be able to operate as substance-specifically as possible [5]. Additional filter combinations (monochromatic and cut-off filters) are needed for the measurement of fluorescence. Appropriate instruments are not only suitable for the qualitative detection of separated substances (scanning absorption or fluorescence along the chromatogram) but also for characterization of the substance (recording of spectra in addition to hR_f) and for quantitative determinations.

Monochromators

Today's commercially available *chromatogram spectrometers* usually employ diffraction gratings for monochromation. These possess the following advantages over prism monochromators which are still employed in the SCHOEFFEL double-beam spectrodensitometer SD 3000 and in the ZEISS chromatogram spectrometer:

- The wavelength scale is approximately linear; this also means
- that the wavelength scan is also linear making automation easier using appropriate stepping motors;
- dispersion is almost constant and not wavelength-dependent, and
- the light transmission above $\lambda = 270$ nm is higher than is the case for prism monochromators.

However, the usable spectral region is limited by the wavelength-dependent efficiency of the gratings.

Fig. 10: TLC scanner II (CAMAG).

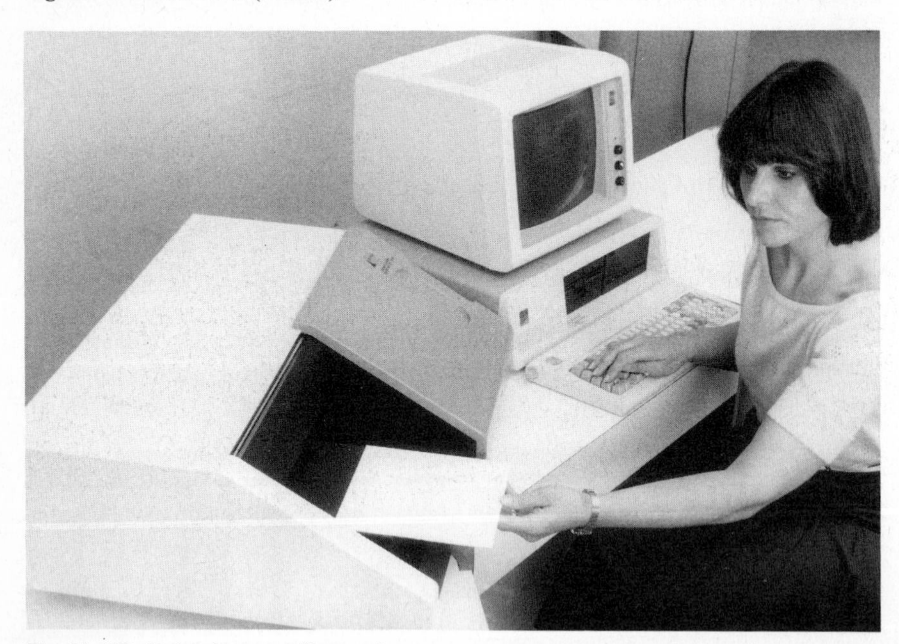

Fig. 11: CD-60 densitometer (DESAGA).

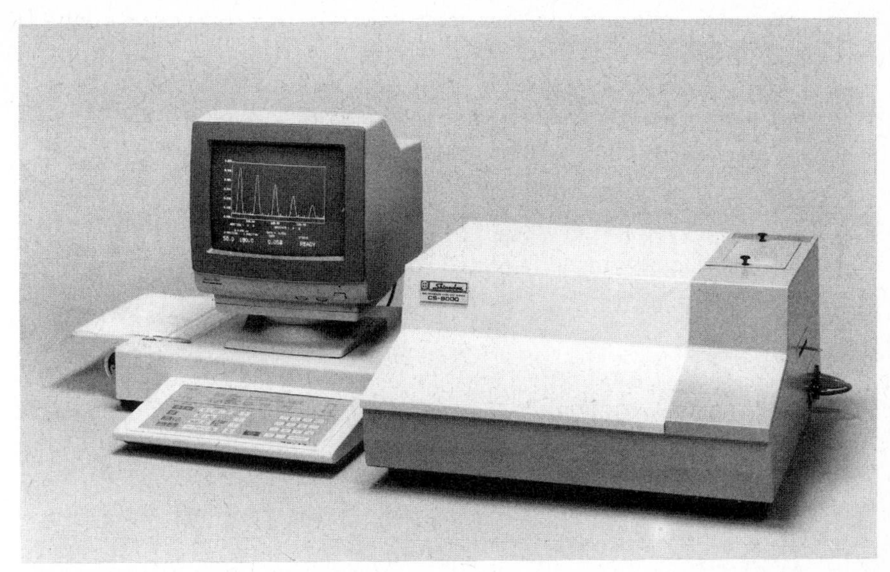

Fig. 12: Flying spot scanner CS 9000 (SHIMADZU).

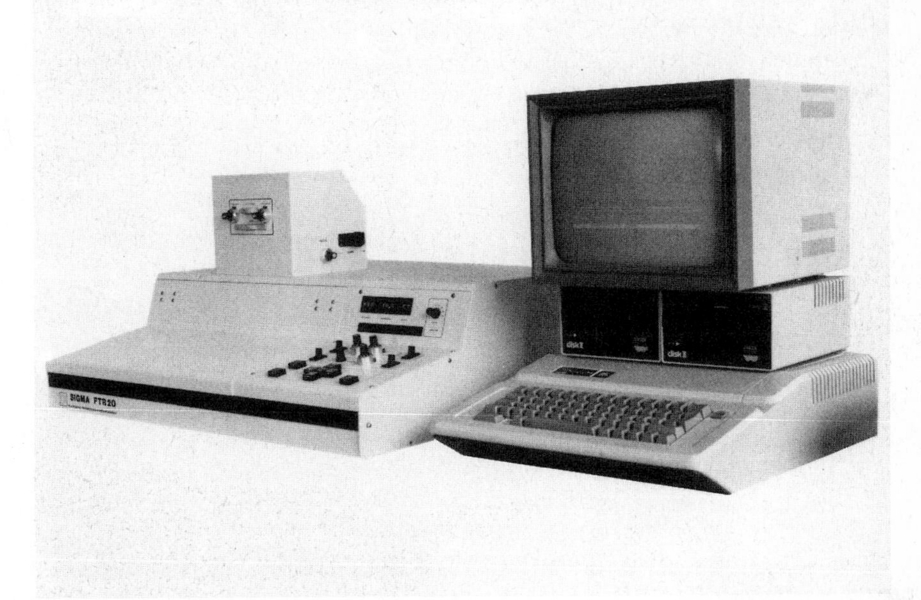

Fig. 13: FTR-20 scanner (SIGMA/BIOCHEM).

Note: Gratings should never be "polished" with the fingers or breathed on. This is also true of coated or "bloomed" gratings which have magnesium or lithium fluoride evaporated onto them.

Samples of spectrometers with grating monochromators

- TLC scanner II, CAMAG (Fig. 10)
- CD-60 densitometer, DESAGA (Fig. 11)
- CS 9000 Flying-spot scanner, SHIMADZU (Fig. 12)
- FTR-20 scanner, SIGMA/BIOCHEM (Fig. 13)

Light Sources

Lamps to be employed in photometry should

- produce radiation that is as constant as possible both in origin and intensity and
- be as good approximation as possible to a point source in order to facilitate the production of parallel beams [40].

A distinction must be made between *continuous sources* (hydrogen or deuterium lamps, incandescent tungsten lamps, high pressure xenon lamps) and *spectral line sources* (mercury lamps), which deliver spectrally purer light in the region of their emission lines.
A continuous source has to be employed to record absorption spectra. Fluorescence is usually excited with mercury vapor lamps; in the region of their major bands they radiate more powerfully than do xenon lamps (Fig. 14).

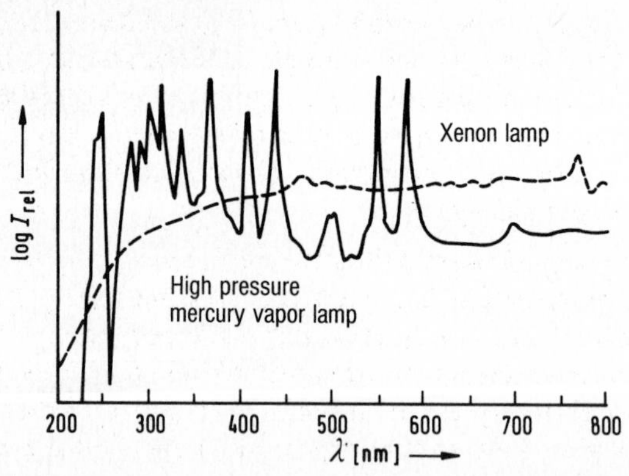

Fig. 14: Radiation characteristics of a high pressure Hg lamp (OSRAM HBO 100; continuous line) and of a xenon lamp (PEK 75; broken line) [4]. The intensity I is represented logarithmically in relative units.

Continuous sources: The sources of choice for measurements in the ultraviolet spectral region are *hydrogen* or *deuterium lamps* [1]. When the gas pressure is 30 to 60×10^{-3} Pa they yield a continuous emission spectrum. The maxima of their radiation emission occur at different wavelengths (H_2: λ = 280 nm; D_2: λ = 220 nm). This means that the deuterium lamp is superior for measurements in the lower UV region (Fig. 15).

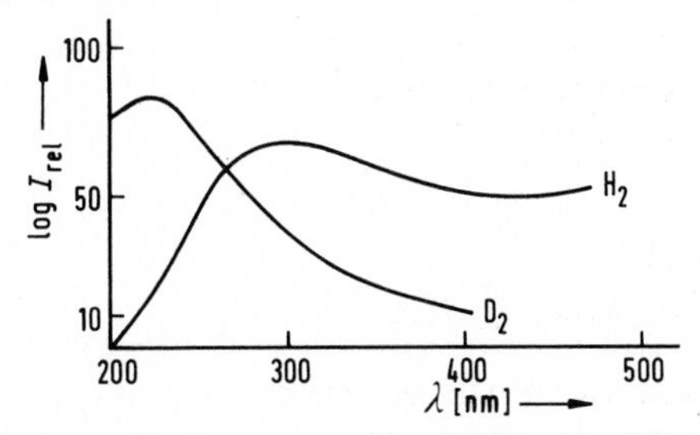

Fig. 15: Relative intensity distribution of the radiation produced by a hydrogen and a deuterium lamp.

Because of the high rate of diffusion of hydrogen the energy consumption resulting from thermal conduction is very large in the short-wave UV region and the radiation yield is relatively low. Deuterium diffuses more slowly, its thermal conductivity is lower and the radiation yield is ca. 30% higher than is the case for a hydrogen lamp.

The continuum produced by both of these lamps is accompanied by emission lines in the visible spectral region at $\lambda = 486.12$ nm (H_2) and $\lambda = 485.99$ nm (D_2); these can be employed for adjustment and calibration of the wavelength scale.
Hydrogen lamps are equipped with a rectangular slit for adjusting and centering the gas discharge; this ensures that the radiation intensity is particularly high in this region and the position of the radiation source is stable. But long-term drift cannot be excluded completely [1].

Tungsten incandescent lamps are primarily employed in the visible region (λ = 320...700 nm). They consist of an evacuated glass bulb [11] containing a thin, coiled tungsten wire which is heated to incandescence. Since tungsten melts at 3655 K the usual operating temperature is 2400 to 3450 K. The higher the temperature the higher is the vapor pressure of tungsten. The metal vapor is deposited on the relatively cool glass bulb so that the "transparency" of the glass is reduced, thus, reducing the operating life which is reported to be ca. 1000 hours at 2400 K. In order to reduce the rate of evaporation krypton or argon are often employed as protective gases, which means that 70 to 90% of the electrical energy is converted to radiation. The fact that a considerable proportion of the energy is radiated above $\lambda = 800$ nm is a disadvantage of the tungsten incandescent lamp.

Halogen lamps are tungsten lamps whose glass bulbs also contain iodine vapor [42]. When the coil is heated incandescent volatile tungsten iodine compounds are produced in the vapor phase and these are thermally decomposed at the glowing coil. This causes a reduction in the deposition of tungsten on the surface of the glass bulb so that such lamps can be operated at higher temperatures and generate a higher light yield. Since iodine vapor absorbs UV light these lamps have a purple tinge.

High pressure xenon lamps are also employed in some TLC scanners (e.g. the scanner of SCHOEFFEL and that of FARRAND). They produce higher intensity radiation than do hydrogen or tungsten lamps. The maximum intensity of the radiation emitted lies between $\lambda = 500$ and 700 nm. In addition to the continuum there are also weak emission lines below $\lambda = 495$ nm (Fig. 14). The intensity of the radiation drops appreciably below $\lambda = 300$ nm and the emission zone, which is stable for higher wavelengths, begins to move [43].

Spectral line radiators: In contrast to the lamps described above mercury vapor lamps are gas discharge lamps [1]. They are started by applying a higher voltage than the operating voltage. The power supply has to be well stabilized in order to achieve a constant rate of radiation and the radiator must always be allowed a few minutes to stabilize after it has been switched on. The waisting of the lamp body in the middle leads to a concentration of the incandescent region. This leads to stability of the arc. Its axis remains in the same position during operation and is readily optically imaged. However, it is impossible to ensure that the arc will not "jump". In addition the relative intensities of the individual lines can change with respect to each other, which can cause a short or long-term change in the recorded baseline. The physical data concerning the most frequently employed mercury lamps are listed in Table 2.

In contrast to the *low-pressure lamps* (1 – 130 Pa) which primarily emit at the resonance line at $\lambda = 254$ nm, *high-pressure lamps* ($10^4 - 10^6$ Pa) also produce numerous bands in the UV and VIS regions (Fig. 16). Table 3 lists the emission lines and the relative spectral energies of the most important mercury lamps (see also [44]). The addition of cadmium to a mercury vapor lamp increases the number of emission lines particularly in the visible region of the spectrum [45] so that it is also possible to work at $\lambda =$ 326, 468, 480, 509 and 644 nm [46].

Recently the Ar^+/He – Ne lasers have been employed for the analysis of thin-layer chromatograms [259 – 261]. However, instruments of this type have not yet come into general use.

Table 2. Summary of the most important technical data on the most frequently employed Hg lamps [47].

Parameter	High pressure Hg lamp			
	St 41	St 43	St 46	St 48
Current type	Direct current	Alternating current	Direct current	Direct current
Supply voltage [V]	220	220	220	220
Total length [mm]	120	90		85
Usable lit length [mm]	8	20		11.5
Gas pressure [Pa]	6×10^5	0.5×10^5		6×10^5
Emitter current [A]	0.6	1.0	0.6	0.6
Emitter power [W]	45	36	33	45
Luminous intensity [cd]	95	31	70	95
Luminous density [cd/cm^2]	500	25	375	500
Examples of scanners employing them	KM-3 chromato-gram spectro-photometer (C. ZEISS)	FTR-20 TLC scanner (SIGMA)	CD-60 densitometer (DESAGA)	CS-930 scanner (SHIMADZU) TLC scanner (CAMAG)

Table 3. List of emission lines and their relative intensities for the most important mercury lamps *).

Wavelength λ [nm]	Energy distribution of the emission bands		
	St 41	St 43	St 48
238 and 240	3	2	3
248	8	4	8
254	55	34	55
265	25	14	25
270	5	2	5
275	4	2	4
280	10	5	10
289	7	3	7
297	18	13	18
302	31	25	31
313	69	67	69
334	7	5	7
366	100	100	100
405 and 408	43	43	43
436	81	61	81
546	108	79	108
577 and 580	66	47	66

*) Exact intensities are given by KAASE et al. [48].

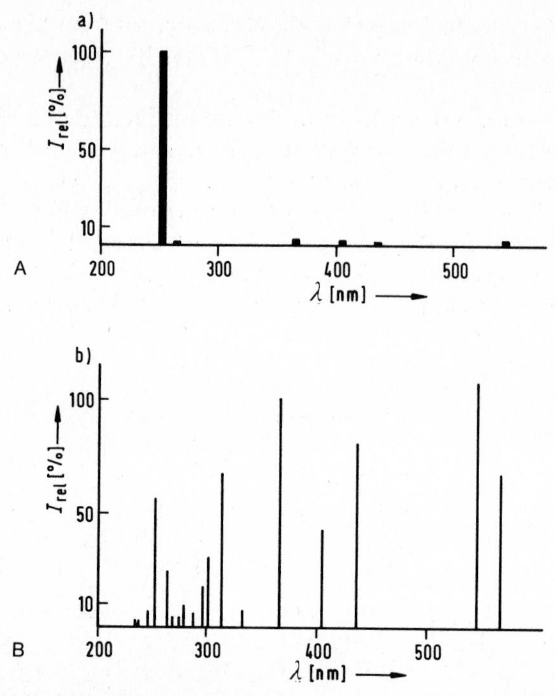

Fig. 16: Relative intensity distribution of a mercury NK 4/4 low pressure lamp (A) and of a mercury St 41 or St 48 lamp (B).

Detectors

The commercial instruments employ detectors of various types. Their utility depends fundamentally on

- the constancy with time of the photocurrent at constant radiation levels and constant external conditions,
- the proportionality of the photocurrent to the intensity of illumination and
- the signal to noise ratio of the photodetector.

A rôle is also played by the temperature and frequency dependence of the photocurrent, the variable surface sensitivity at various parts of the cathode and the vector effect of polarized radiation [40]. All the detectors discussed below are electronic components whose electrical properties vary on irradiation. The effects depend on external (photocells, photomultipliers) or internal photo effects (photoelements, photodiodes).

Photocells and photomultipliers (secondary electron multipliers, SEM) are mainly employed in photometry. These are detectors with an "external photo-effect".

Photocells: The basic construction of a photocell is illustrated in Figure 17. A photocurrent flows when the photocathode is illuminated, this is proportional to the intensity of illumination if the supply potential has been chosen to be higher than the saturation potential. A minimal potential is required between the photocathode and the anode in order to be able to "collect" the electrons that are emitted. The sensitivity is independent of frequency up to 10^7 Hz. The temperature sensitivity of evacuated photocells is very small. The dark current (see below) is ca. 10^{-11} A [1].

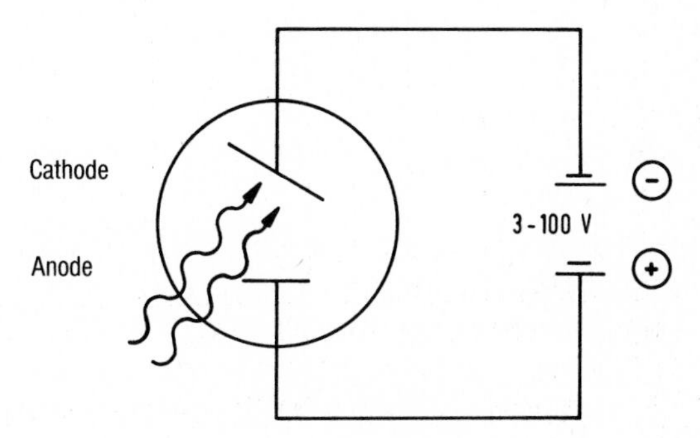

Fig. 17: The construction of a photocell, schematic [1].

Samples of analytical units with photocells:

- TURNER fluorimeter, model III (CAMAG)
- Quick Scan R & D densitometer (HELENA)
- Fiber optic densitometer, model 800 (KONTES)

Photomultipliers: Secondary electron multipliers, usually known as photomultipliers, are evacuated photocells incorporating an amplifier. The electrons emitted from the cathode are multiplied by 8 to 14 secondary electrodes (*dynodes*). A diagramatic representation for 9 dynodes is shown in Figure 18 [5]. Each electron impact results in the production of 2 to 4 and maximally 7 secondary electrons at each dynode. This results in an amplification of the photocurrent by a factor of 10^6 to 10^8. It is, however, still necessary to amplify the output of the photomultiplier.

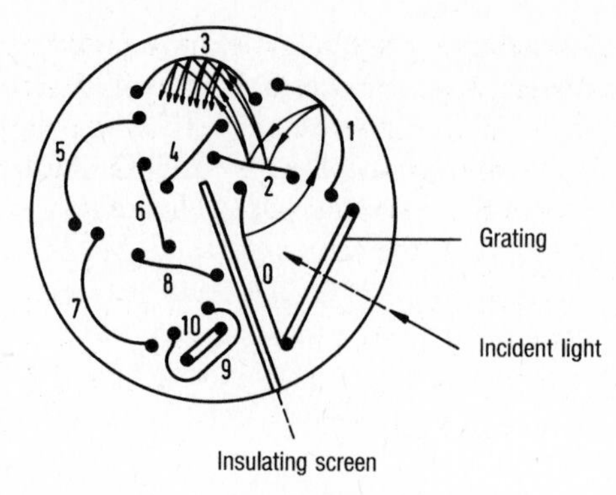

Fig. 18: Section through an RCA photomultiplier, schematic. – 0 photocathode, 1–9 dynodes, 10 anode.

The requirements for successful operation are a stable operating voltage of between 400 and 3000 V. The sensitivity of the photomultipliers is also dependent on this if a special compensation is not incorporated.

The absolute and spectral sensitivities can often vary by up to 100% within a few millimeters on the surface of the photocathode [49]. Figure 19 illustrates this effect for a sideways and vertical adjustment of a photomultiplier, in addition slight maladjustment of the light entrance can lead to "zero line runaway" as a result of thermal effects.

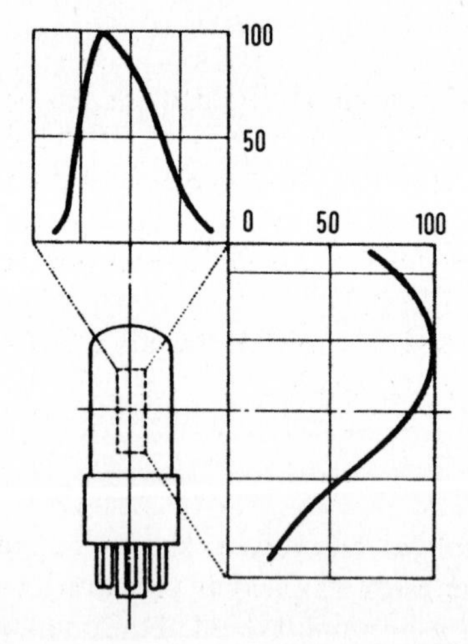

Fig. 19: Spacial dependence of the spectral sensitivity of a photocathode [50].

Depending on their positioning the dynodes are referred to as being "head-on" or "side-on". Commercial scanners mostly employ "side-on" secondary electron multipliers where, as the name implies, the radiation impinges from the side – as in Figure 19. Their reaction time is shorter than for head-on photomultipliers because the field strength between the dynodes is greater.

Head-on photomultipliers, on the other hand, possess a greater entry angle for the capturing photocathode (Fig. 20). A diffuse screen in front of the photocathode also allows the capture of light falling at an angle. These conditions are realized in the CAMAG TLC/HPTLC scanner I. The sensitivity of such head-on photomultipliers is independent of frequency up to 10^6 Hz.

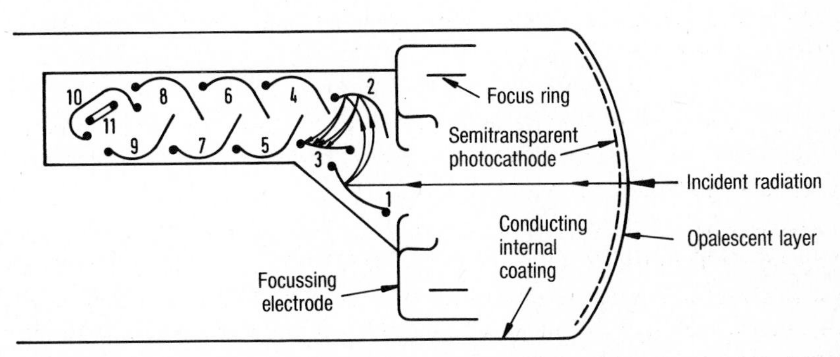

Fig. 20: Cross section through a "head on" photomultiplier [51, 52]. – 1–10 dynodes, 11 anode.

It is a disadvantage of all photomultipliers that the photocurrent is not completely proportional to the strength of illumination. Further, the photocurrent must not exceed 10^{-7} A or the photomultiplier becomes fatigued. Daylight switches are incorporated into some scanners for this reason in order to prevent over-illumination of the detector when the sample chamber is opened.

Detector noise: The detection limit for the recording of chromatographically separated substances is determined by

- the "chromatographic noise" and
- the "detector noise".

The latter mainly results from the thermal emission current. The dark current is apparent mainly in the long-wavelength range of the spectrum when the photocurrent is appropriately small [53, 54, 131]. It is relatively small for alloy cathodes (e.g. Sb-Cs cathodes), but not small enough to be negligible.

The emission of thermal electrons is subject to statistical fluctuations (lead shot effect). It is influenced by the current strength, the number of electric charges liberated and the frequency of the radiation [55] (see [40] for further details).

Range of application and spectral sensitivity: The photomultipliers most frequently employed in scanners possess antimony-caesium cathodes. These alloy cathodes are primarily sensitive to the short-wavelength part of visible light (Tab. 4).

The *long-wavelength limit* which depends on the cathode material is ca. $\lambda = 650$ nm in the red region of the spectrum. If this does not suffice for the determination an antimony-alkaline metal alloy is employed as the cathode material [56–58].

The range of application into the *short-wavelength region of the spectrum* depends on the window material employed in the photomultiplier. Borosilicate glass (KOVAR glass), for example, only transmits radiation down to about 280 nm, Suprasil down to 185 nm and fused quartz down to 160 nm. Hence fluorimeters which primarily detect long-wavelength radiation (fluorescent radiation) are often equipped with type S-4 detectors (Tab. 4), whose windows absorb a part of the short-wavelength radiations.

Table 4. Characteristics of the most important photomultipliers with Sb-Cs cathodes and 9 amplification steps [50].

Type	Origin			Spectral response [nm]	Wavelength maximum sensitivity	Window material
	RCA	HAMAMATSU	EMI			
S-4	1P21	1P21	9781A	300...650	400	borosilicate
	931B	R105	–	300...650	400	borosilicate
	–	R105UH	–	300...650	400	borosilicate
S-5	1P28	1P28	9661B	185...650	340	UV glass
	1P28/VI	R212	9781B	185...650	340	UV glass
	–	R212UH	9781B	185...650	340	UV glass
	1P28A	R454	9781R	185...650	450	UV glass
	1P28A/VI	R282	–	185...650	450	UV glass
S-19	4837	R106	9665A	160...650	340	fused silica
	–	R106UH	9783B	160...650	340	fused silica

Examples of scanners employing S-4 detectors:

- UV/VIS chromatogram analyzer for fluorescence measurements (FARRAND)
- Spectrofluorimeter SPF (AMERICAN INSTRUMENT CO.)
- TLD-100 scanner (VITATRON)

The majority of detectors, which are employed for the measurement of absorption, employ UV glass (e.g. Suprasil). All type S-5 photomultipliers possess sheaths of this material, so that they ought to be usable in the far UV region if N_2 purging is employed (to remove O_2) (Tab. 4).

Scanners with S-5 detectors

- KM 3 chromatogram spectrophotometer (C. ZEISS)
- SD 3000 spectrodensitometer (KRATOS/SCHOEFFEL)
- CD-60 densitometer (DESAGA)

Photomultipliers of type S-19 employ fused quartz instead of UV glass; this transmits down to $\lambda = 160$ nm although this far UV range is not normally employed in scanners.

Scanners with S-19 detectors

- CS 920 scanner (SHIMADZU)
- CS 930 scanner (SHIMADZU)

Photoelements and photodiodes: Both photoelements and photodiodes are photoelectric components depending on internal photoelectric effects.

In the case of *photoelements* incident quanta of light produce free charge carriers in the semiconductor layer; previously bound electrons become free. Thus, the nonconducting layer becomes conducting. In addition, the migrating electrons produce "holes" which increase the conductivity. The radiation energy is directly converted into electrical energy. The construction of a photoelement is illustrated in Figure 21.

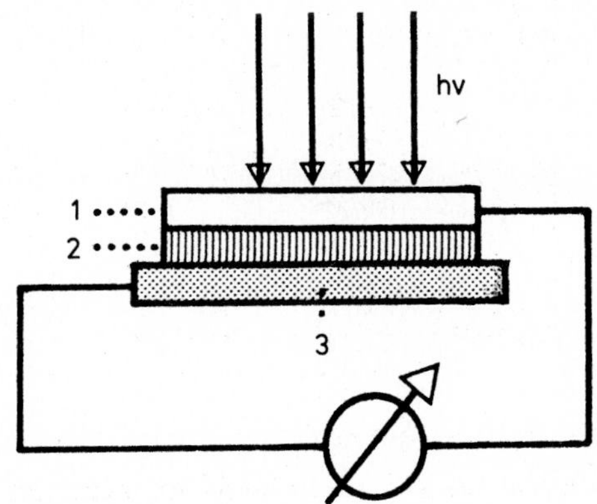

Fig. 21: Principle of construction of a photoelement [1]. – 1 light-transmitting metal layer, 2 semiconductor layer, 3 metal plate.

Photodiodes produce an electric field as a result of *pn* transitions. On illumination a photocurrent flows that is strictly proportional to the radiation intensity. Photodiodes are sensitive and free from inertia. They are, thus, suitable for rapid measurement [1, 59]; they have, therefore, been employed for the construction of diode array detectors.

2.2.3.2 Principles of Measurement

The scanners commercially available today operate on the basis of the optical train illustrated in Figure 22.

They can be employed to

- detect absorbing substances against a nonfluorescent plate background (Fig. 22A: recording a scanning curve, absorption spectra, quantitative analysis of absorbing substances);
- detect absorptions indirectly because of luminescence diminishing (Sec. 2.2.2). [Here, however, it is necessary to introduce a cut-off filter before the detector to absorb the shortwave excitation radiation (Fig. 22B)];

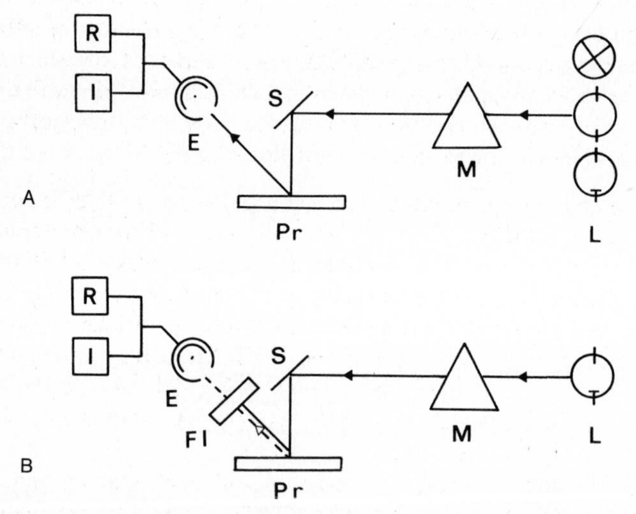

Fig. 22: Optical trains of the commercially available scanners. (A) absorption, (B) fluorescence quenching and true fluorescence.
R = recorder, I = integrator, E = detector, Pr = sample (TLC plate), S = mirror, M = monochromator, L = lamps (incandescent lamp ⊗, deuterium lamp ⏀ and mercury lamp ⏀), Fl = cutoff filter.

- detect and analyze quantitatively fluorescent substances against a nonfluorescent background without spectral analysis of the fluorescent light (Fig. 22B). [An additional filter is necessary here too (Sec. 2.3.3).]

On emitting phases it is not possible to determine directly (*in situ*) the fluorescence and absorption spectra of compounds that absorb in the excitation range of luminescence indicators without distorting the measurement signal.

Direct Determination of Absorbance

Determination of absorption spectra in reflectance: If there is no luminescence radiation *absorption spectra* can be determined using the light path sketched in Figure 22A. If absorbing substances, such as, for example, dyestuffs, caffein or PHB esters, are determined spectrophotometrically after chromatography, then, depending on the wavelength, these components absorb a proportion (I_{abs}) of the light irradiating them (I_0). The chromatographic zone remits a lower light intensity (I_{ref}) than the environment around it.

$$I_0 - I_{abs} = I_{ref}$$

Therefore it is possible to determine absorption spectra directly on the TLC plate by comparison with a substance-free portion of the layer. The wavelengths usually correspond to the spectra of the same substances in solution. However, adsorbents (silanols, amino and polyamide groups) and solvent traces (pH differences) can cause either bathochromic (ketones, aldehydes [60, 61], dyestuffs [62]) or hypsochromic (phenols, aniline derivatives [63]) shifts (Fig. 23).

However, these absorption spectra can be employed as an aid to characterization, particularly when authentic reference substances are chromatographed on a neighboring track. The use of differential spectrometry yields additional information [64]. Quantitative analysis is usually performed by scanning at the wavelength of greatest absorbance (λ_{max}). However, determinations at other wavelengths can sometimes be advantageous, e.g. when the result is a better baseline. An example is the determination of scopolamine at $\lambda = 220$ nm instead of at $\lambda_{max} = 205$ nm or of the fungicide vinclozolin at $\lambda = 245$ nm instead of at $\lambda_{max} = 220$ nm.

Absorbance (reflectance) scanning: The positions of the chromatographically separated substances are generally determined at λ_{max}. As the chromatogram is scanned the voltage differences produced at the detector are plotted as a function of position of measurement to yield an absorption scan (Fig. 24). Conclusions concerning the amount of substance chromatographed can be drawn from the areas or heights of the peaks [5].

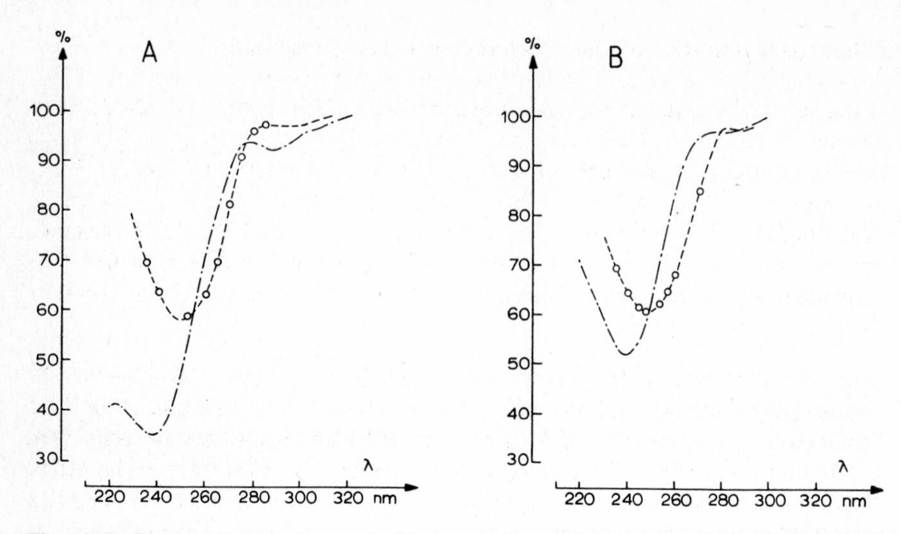

Fig. 23: Reflectance spectra (o – o –) of 3 μg testosterone (A) and 3 μg Δ^4-androstendione-(3,17) (B) taken up on a silica gel layer compared with the absorption spectra determined in methanolic solution (· – · –).

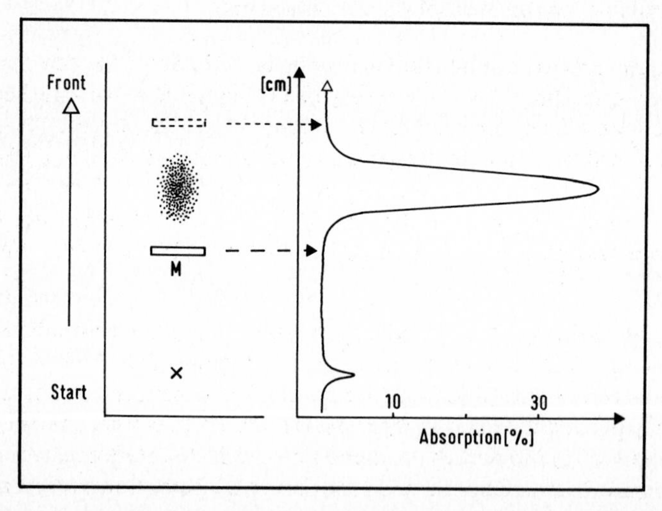

Fig. 24: Schematic representation of the recording of an absorbance scan. – M = measuring slit.

Indirect Determination of Absorption (Fluorescence Quenching)

If the substance under investigation absorbs with wavelengths between 250 and 300 nm, it should be checked whether it is possible to employ chromatographic layers containing luminescence indicators. For the inorganic indicators (Sec. 2.2.2) also absorb in this range and emit, for example, yellow-green long-wavelength radiation. Hence, it is the radiation that has not been absorbed by the substance plus the fluorescence/phosphorescence radiation emitted by the indicator that arrives at the detector. The signal produced is, therefore, a composite signal of these two radiation types.

When working with a deuterium lamp the radiation energy is so low that the luminescence radiation only makes up a few percent of the total radiation. This can be easily checked in the majority of scanners by setting the total radiation (e.g. $\lambda = 260$ nm) to 100% reflectance and then inserting a cut-off filter in the beam. This filter absorbs the short-wavelength radiation before it enters the detector (Fig. 22B). The energy that remains comes from the emission of the indicator or is produced by stray light. The remaining signal is almost always small. Hence, when a deuterium lamp is employed absorption determinations are only falsified to a small extent. This falsification is also reduced by the fact that at the site of a zone the absorbing substance also reduces the emission. The absorbing substances absorb energy in the excitation range of the luminescent indicator and, hence, less light is available for the stimulation of luminescence.

However, the optical train illustrated in Figure 22B allows the determination of fluorescence quenching. The "interfering effect" described above now becomes the major effect and determines the result obtained. For this purpose the deuterium lamp is replaced by a mercury vapor lamp, whose short-wavelength emission line ($\lambda = 254$ nm) excites the luminescence indicator in the layer. Since the radiation intensity is now much greater than was the case for the deuterium lamp, the fluorescence emitted by the indicator is also much more intense and is, thus, readily measured.

The emission of the indicator is reduced in places where there are substance zones that absorb at $\lambda = 254$ nm present in the chromatogram. This produces dark zones (Fig. 4A), whose intensity (or rather lack of it) is dependent on the amount of substance applied. If the plate background is set to 100% emission the phosphorescence is reduced appropriately in the region of the substance zones. When the chromatogram is scanned peaks are produced, whose position with respect to the start can be used to calculate R_f values and whose area or height can be used to construct calibration curves as a function of the amount applied (Fig. 25).

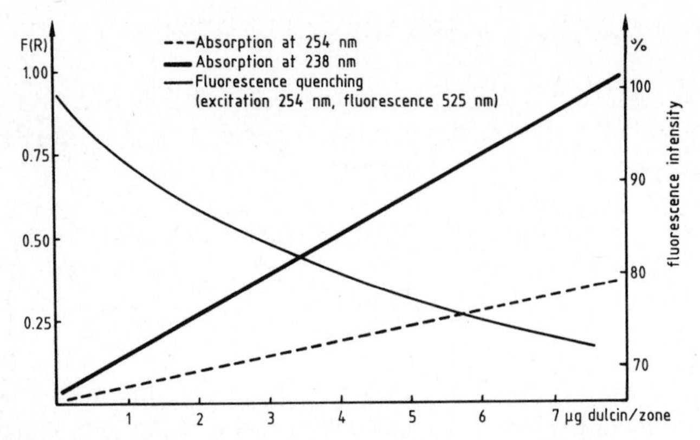

Fig. 25: Calibration curve for the determination of dulcin by fluorescence quenching and absorption [2].

However, the direct determination of absorption at the wavelength of maximum absorption is more sensitive (or in the worst case at least as sensitive) as the indirect measurement of absorption by fluorescence or phosphorescence quenching.

The fact that this type of analysis usually involves phosphorescence can be demonstrated by scanning a substance zone at various different rates. As can be seen in Figure 26 the rate of scanning of the TLC plate has an appreciable effect on the detector signal. The more rapidly the plate is moved the greater is the difference between the starting phosphorescence (chromatogram at rest) and the baseline during scanning (chromatogram in motion). Peak area and height also decrease appreciably; in addition, the baseline becomes more unsteady, which reduces the detection limit compared with that for absorption measurement at λ_{max}. So analysis of fluorescence quenching does not provide any advantages over the direct determination of absorption.

2.2.3.3 Quantitative Analysis

The determination of the spectrum reveals which wavelength is suitable for quantitative analysis. The wavelength of maximum absorption is normally chosen, because the difference from the background blank is greatest here. In this type of analysis the analyst "sacrifices" all the substance-characterizing information in the spectrum in favor of a single wavelength. When the chromatogram is scanned photometrically at this wavelength a plot is obtained of absorption versus position on chromatogram (Fig. 25); the peak heights and areas are a function of the

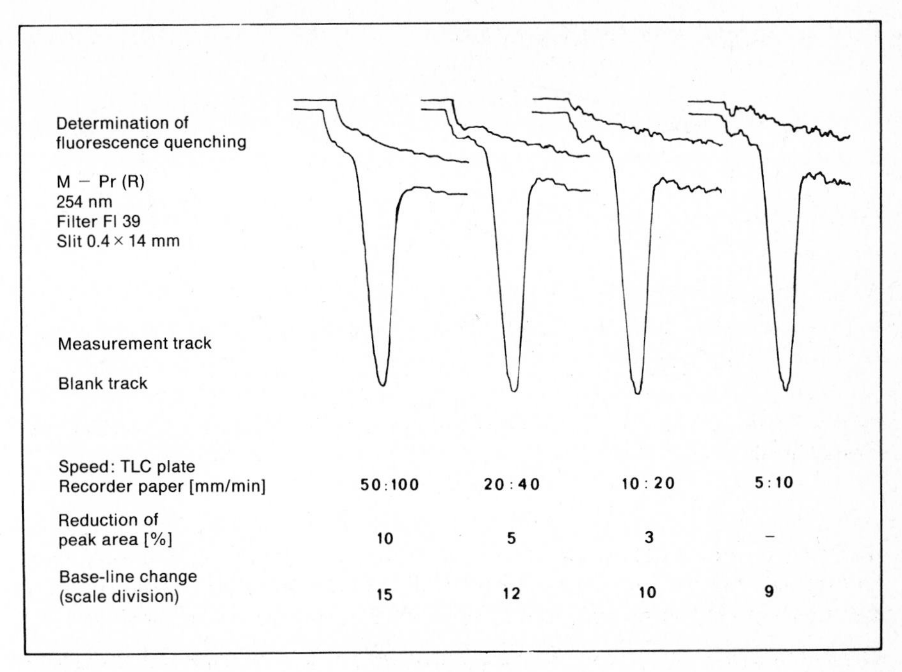

Fig. 26: Influence of scanning rate on signal size (peak area).

amount of substance applied. The limiting law is the KUBELKA-MUNK function [65, 66]:

$$\frac{k}{s} = \frac{(1 - R_\infty)^2}{2R_\infty}.$$

Where

k = absorption coefficient
s = coefficient of scattering of the stationary phase
R_∞ = absolute reflectance

According to the BEER-LAMBERT law

$$\frac{I}{I_0} = e^{-k\,d}$$

the coefficient of absorption k can be defined as

$$k = 2.303\ \varepsilon\ c$$

(ε = molar extinction coefficient; c = concentration of substance, in the present case the amount of substance applied).

Strictly speaking the KUBELKA-MUNK function is only applicable under the following conditions [67–73]:

- The irradiation of the sample must be diffuse.
- Monochromatic radiation must be employed for the analysis, so that the diffraction and refraction phenomena in the layer shall be as uniform as possible. This also means that the radiation reaching the detector in reflectance retains its "color value" and only changes in its intensity. This would not be the case for polychromatic light; since a certain proportion of the light is absorbed during the determination the composition of the light would change and would, amongst other things, alter the sensitivity of the photomultiplier to the remaining light.
- Mirror reflection (= regular reflection) must not occur.
- The layer thickness must be large in comparison with the wavelength employed so that no radiation can penetrate right through the layer and escape measurement.
- The particles must be randomly distributed in the layer to avoid interference effects.
- The particles making up the adsorbent must be very much smaller in size than the thickness of the chromatographic layer.

These general requirements also apply to adsorbents laden with substance. All these requirements are not fulfilled to the same extent in thin-layer chromatography. So the KUBELKA-MUNK function does not apply without qualification.

For this reason it is understandable that numerous empirical functions have been proposed as substitutes for use in practical analysis [5, 74].

2.3 The Detection of Fluorescent Substances

2.3.1 General

Molecules that absorb radiation are raised to an excited state. They can return to the ground state by

- converting the energy received into kinetic energy or converting it by means of impacts into thermal energy which is then slowly transmitted away (radiationless deactivation);
- emitting the absorbed energy instantaneously as radiation energy in the form of fluorescence [3, 75, 76].

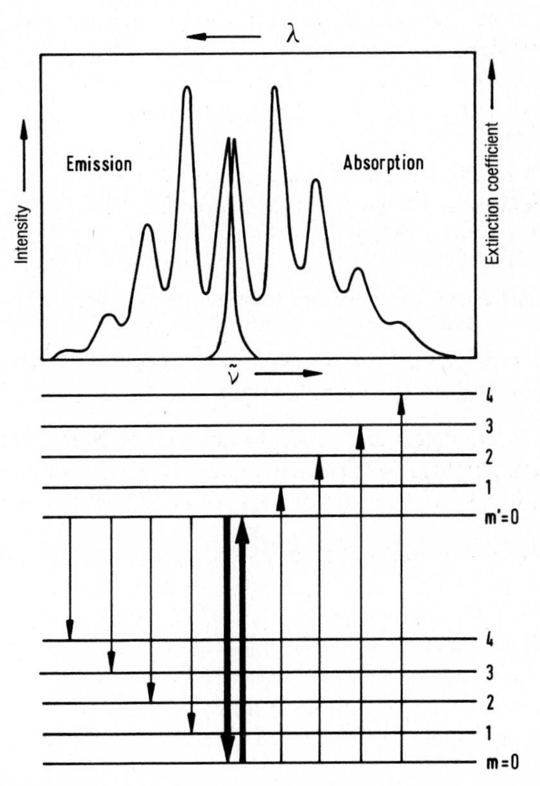

Fig. 27: Schematic representation of the relationship between absorption and fluorescence emission of the molecules – m and m' are the terms involved in the vibrational quantum numbers [4].

The radiation emitted is usually longer in wavelength (i.e. lower in energy) than the incident light (STOKE's law). It is only in the case of 0–0 transitions (shown in Figure 27 as thick arrows) that the wavelengths for fluorescence and activation are identical.

Since only relatively few substances are capable of emitting fluorescent radiation, they can be particularly selectively detected. This means that the selectivity of the chromatographic separation, which is always aimed at, is meaningfully extended by the selectivity of detection. Accompanying substances that absorb radiation but do not emit light do not interfere when the analysis is made by the selective determination of fluorescence!

The same UV lamps discussed in Section 2.2.3.1 are employed to excite fluorescence. Excitation is usually performed using long-wavelength radiation ($\lambda =$ 365 nm), shorter wavelengths are occasionally employed (e.g. $\lambda = 302$ nm, DNA analysis).

2.3.2 Visual Detection

If the *fluorescent radiation* lies above $\lambda = 400$ nm it is detectable to the naked eye. Light-bright zones are seen on a dark background. When this backgound does not appear black this is because the "black light" filter of the UV lamp is also more or less transparent to violet visible light. This covers the whole chromatogram and also distorts the visual impression of the color of the fluorescence. Table 1 lists the emission wavelengths of the various fluorescent colors, which can, just as R_f values be employed as an aid to substance identification (in fluorimetry the subjective color impression does not correspond to the complementary color!). If the evidence so obtained is insufficient then microchemical or physiological-biological detection can follow. It is also possible to record the absorption and fluorescence spectra.

2.3.3 Fluorimetric Determinations

The optical train employed for photometric determinations of fluorescence depends on the problem involved. A spectral resolution of the emitted fluorescence is not necessary for quantitative determinations. The optical train sketched in Figure 22B can, therefore, be employed. If the fluorescence spectrum is to be determined the fluorescent light has to be analyzed into its component parts before reaching the detector (Fig. 28). A mercury or xenon lamp is used for excitation in such cases.

Cut-off filters are employed to ensure that none of the excitation radiation can reach the detector. *Monochromatic filters* are used to select particular spectral

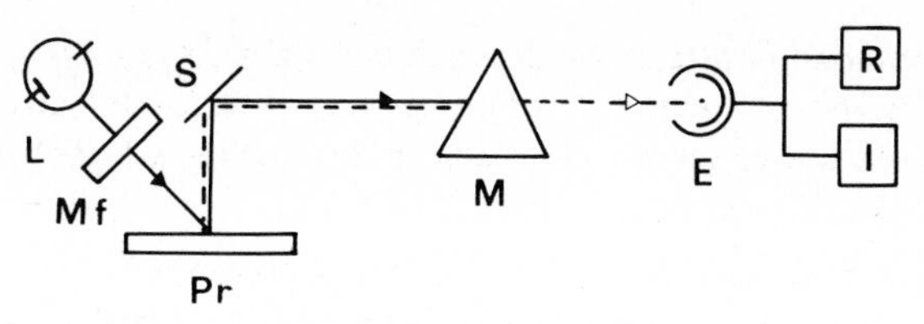

Fig. 28: Optical train for the recording of fluorescence spectra.

regions in a continuum (e.g. in the fluorescent radiation) or – employed on the excitation side – to pick out individual lines of a discontinuous spectrum. Their quality increases with the narrowness of their half-width and with their transparency at maximum.

Fluorescence scanning of chromatograms of polycyclic aromatic compounds is a vivid example of their employment. A careful choice of the wavelengths of exci-

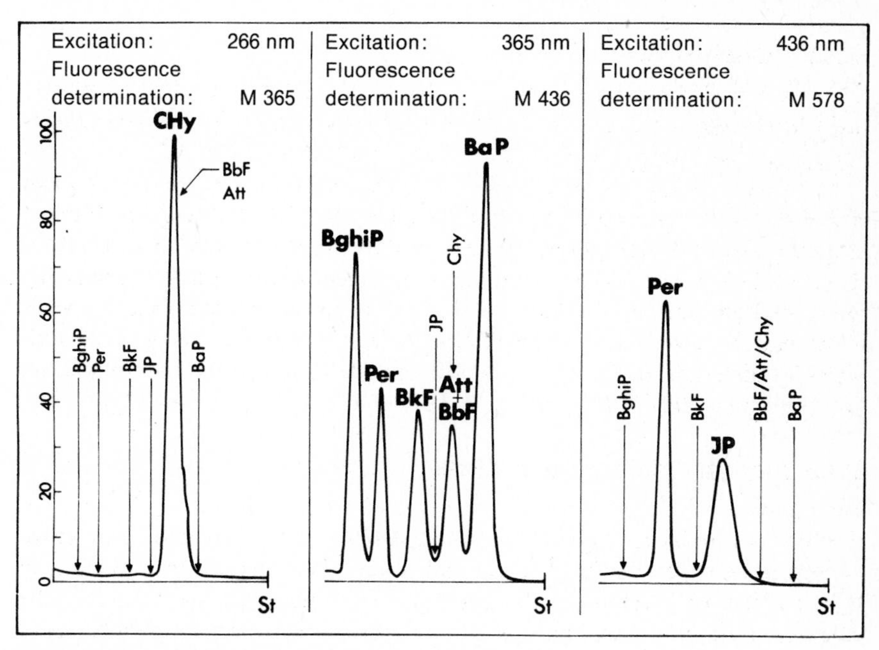

Fig. 29: Fluorescence scans of polycyclic aromatic hydrocarbons at various excitation wavelengths in combination with various secondary filters.
BghiP = benzo(g,h,i)perylene; CH_Y = chrysene; Per = perylene; BbF = benzo-(b)fluoranthene; BkF = benzo-(k)fluoranthene; BaP = benzo-(a)pyrene; IP = indeno-(1,2,3-cd)pyrene; Att = anthanthrene.

tation and recording (emission) allows selective detection of the individual components (Fig. 29).
The relationships between amount of substance applied and the heights or areas of the peaks in the chromatogram scan are employed for the quantitative determination of fluorescent substances. The following relationship is applicable when the amount of substance is small:

$$I_{fl} = k\, I_0\, \varepsilon\, a\, d$$

Where

I_{fl} fluorescence intensity
k proportionality factor
I_0 intensity of irradiating light
ε molar absorption coefficient
a amount of substance applied
d thickness of adsorbent layer

The intensity of the emitted fluorescence I_{fl} is, therefore, directly proportional to the amount of substance applied a. This relationship is much simpler than the KUBELKA-MUNK function and always leads to a linear calibration curve passing through the origin. If this is not true then interference is occurring [5].

When recording excitation and fluorescence spectra it must be ensured that monochromatic light falls on the detector. This can best be verified in instruments built up on the kit principle or in those equipped with two monochromators (spectrofluorimeters). The majority of scanners commercially available at the moment do not allow of such an optical train, which was realized in the KM3 chromatogram spectrometer (ZEISS). So such units are not able to generate direct absorption or fluorescence spectra for the characterization of fluorescent components.

2.4 Detection of Radioactively Labelled Substances

Radioactively labelled compounds have been employed in biology for the clarification of metabolic processes since the mid-1940s. It has, thus, been necessary to prepare such substances and to check on their purity.

All the usual detection methods are naturally suitable for the detection of such substances (Fig. 2). In addition, however, it is also possible to detect the β-radiation they produce. The selectivity of the detection is, thus, increased. This, however, does not provide any possibility of substance-specific detection, since all ^{14}C labelled compounds, for instance, give the same "reaction" at a detector.

The isotopes listed in Table 5 are the ones most frequently employed. They are all weak β-emitters with a relatively long half-life, so that they can unfortunately contaminate the organism in medical investigations.

Table 5. List of the most commonly employed β-radiation-emitting radioisotopes according to their half-lives [5].

Isotope	Energy [MeV]	Half-life
^{36}Cl	0.71	4.4×10^5 years
^{14}C	0.155	5.6×10^3 years
^{3}H	0.018	12.26 years
^{35}S	0.167	87.2 days
^{32}P	1.70	14.3 days
^{131}I	0.608	8.07 days

The isotopes can be detected by

- autoradiography,
- fluorography (scintillation autoradiography),
- spark chamber detection and
- scanning techniques.

Their sensitivities are given in Table 6. We will discuss these detection techniques in more detail in Volume 2.

Table 6. Summary of the ^{14}C detection limits of various radio TLC methods [77].

Method	Detection limit [nCi][a]
Liquid scintillation method	0.01
Autoradiography	0.05
Spark chamber method	0.05
Radioscanning	0.1 ... 0.2

[a] nCi, symbol for nano-curie; this unit corresponds to the amount of substance which produces as many disintegrations per second as 1 ng radium.

2.5 Nondestructive Detection Using Other Physical Methods

2.5.1 Spectral Phenomena

It is often of importance to detect compounds in the chromatogram without causing chemical change. In the case of substances containing conjugated double or triple bonds observation under UV light is to be recommended (Sec. 2.2.2). If the layer contains a "fluorescence indicator" that emits radiation on irradiation with short-wavelength UV light then the zone of absorbing substance appears as a dark spot on a yellowish-green or bluish emitting background.

Substances that are intrinsically fluorescent can often be excited with long-wavelength UV light. They absorb the radiation and then emit, usually in the visible region of the spectrum, so that they appear as bright luminous zones, which can frequently be differentiated by color. They, thus, set themselves apart from the multitude of substances that only exhibit absorption. This detection possibility is characterized by high specificity (Sec. 2.3).

Differences in solubility behavior or the influence on pH are also employed for nondestructive detection and they will now be discussed.

2.5.2 Wetting and Solubility Phenomena

Silica gel, kieselguhr and aluminium oxide are hydrophilic adsorbents. Spraying or dipping the chromatogram in water yields transparent layers, on which lipophilic substances are not wetted and they appear – if their concentration is sufficient – as dry "white zones". This effect can be recognized particularly well if the TLC plate is held against the light while the completely wetted layer is slowly allowed to dry. The zones can be marked and later employed for micropreparative investigations or for biological-physiological detection. If the compounds are radioactively labelled scintillation counting can also follow [78]. Examples of such detection are listed in Table 7.

Aqueous solutions of dyes can also be employed instead of water. In the case of *hydrophilic dyes* such as methylene blue or patent fast blue the transparent background of the TLC/HPTLC plate is stained blue. Pale spots occur where there are nonwetted zones. DÄUBLE [89] detected anion-active detergents in this way on silica gel layers as pale zones on a blue background with palatine fast blue GGN

Table 7. Nondestructive detection of lipophilic substances with water as detecting reagent.

Substances	References	Substances	References
Hydrocarbons	[79]	Cholestanone, α-cholestanol	[80]
Bile acids	[81, 82]	Triterpene derivatives	[83]
Sapogenins	[84]	Cyclohexanol	[85]
N-Aryl-N′,N′-dialkyl-urea herbicides	[86]	Sulfur-containing polysaccharides	[87]
Phosphoinositides	[88]		

and other dyestuffs. The detection limits for fatty acids lie at 1 to 2 μg per chromatogram zone [90]. The contrast is best recognizable immediately after dipping; it slowly disappears during drying.

On reversed phase layers, in contrast, detergents yield dark blue zones on a pale background. Here it is the lipophilic part of the detergent molecule that is aligned with the surface RP chain and the dye is attracted to the anionic part of the molecule. Steroid derivatives can also be detected with aqueous solutions of dyes [91].

Lipophilic dyes in aqueous alcoholic solutions can be employed in an analogous manner [92, 93]. They are enriched at the zones of lipophilic substances, so that these appear deeply colored on a pale background. This does not apply to fatty acids with less than 12 C atoms [94].

The same applies to *fluorescent substances.* These dissolve in the hydrophobic zones and lead to increased fluorescence when observed under long-wavelength UV light. MANGOLD and MALINS [95] were the first to exploit this principle. They sprayed their chromatograms with a solution of dichlorofluorescein and observed that lipophilic substances produced a yellow-green fluorescence on a purple-colored background. Their method was adopted by numerous groups of workers [96–106]. Mixtures with rhodamine B have also been employed [107, 108]. FROEHLING et al. [109] later employed an aqueous solution of Ultraphor WT to detect triglycerides. Glycoalkaloids can be detected fluorimetrically with alcoholic Blankophor-BA267 solution [110]. Such detection is even possible on paraffin-impregnated or RP layers [111].

Further examples illustrating the versatility of this nondestructive detection method are listed in Table 8.

Table 8. Employment of fluorescent substances for the nondestructive detection of lipophilic substances.

Reagent	Field of application and references
8-Anilinonaphthalenesulfonic acid ammonia salt (ANS reagent)	fatty acids [112, 113]; lecithin/sphingomyelin [114, 115]; cholesterol and its esters [116, 117]; steroids, detergents, hydrocarbons [118, 119]; prenol, prenylquinones [120]
Berberine	sterols [121 – 123]; saturated organic compounds [124]
Brilliant green	neutral esters of phosphoric acid [125], carbamate herbicides [34]
Eosin	condensation products of urea, formaldehyde and methanol [126], pesticide derivatives [127]; sweetening agents [128, 129]; anion-active and nonionogenic surface-active agents [130]
Flavonoids:	
Morin	steroids, pesticides [29, 132, 133]
Flavonol, fisetin, robinetin	pesticides [134 – 137]
Quercetin	vanadium in various oxidation states [138]
Rutin	uracil derivatives [139]
Fluorescein	paraffin derivatives, waxes, hydrocarbons [140, 141]; aliphatic acids [142]; hydroquinone and chlorinated derivatives [143]; isoprenoids, quinones [111, 144]; oxathizine fungicides [145]; barbiturates, phenothiazines [146]
Pinacryptol yellow	surface-active agents [147 – 151]; carbamate-based insecticides and herbicides [152]; organic anions [153]; sweeteners [129, 154]
Rhodamine B	vaseline [155]; diphenyl, polyphenols [156]; maleic and fumaric acids [162]; flavonoids [158]; alcohols as 3,5-dinitrobenzoates [159, 160]; gangliosides [161]; 1-hydroxychlorden [162]; carbamate pesticides [163]; parathion and its metabolites [164]; polyethylene and polypropylene glycols [165]; terpene derivatives [166]; menthol [167]
Rhodamine G	neutral steroids [168]
Rhodamine 6G	long-chain hydrocarbons [169]; squalene, α-amyrin [170]; methyl esters of fatty acids [171]; glycerides [91]; sterols [172, 173]; isoprenoids, quinones [111]; lipoproteins [174]; glycosphingolipids [175]; phenolic lipids [176]; phosphonolipids [177]; increasing the sensitivity after exposure to iodine vapor [178, 179]
6-*p*-Toluidino-2-naphthalene sulfonic acid (TNS reagent)	cholesterol [180]; phospho- and glycolipids [181]; neutral lipids [182]
Uranyl acetate	purines [36]

2.5.3 Acid/Base Properties

Acidic and basic substances can be detected using pH indicators. Indicators changing color in the acid region are primarily employed. They are applied to the chromatogram by dipping or spraying with 0.01 to 1% solutions. The pH is

Table 9. Use of indicator dyes to detect pH-active substances.

Indicator/pH transition range	Application and references
Bromocresol blue (3.8...5.4)	lichen acids [186]
Bromocresol green (3.8...5.4)	aliphatic carboxylic acids [103, 187–204]; triiodobenzoic acid [205], derivatives of barbituric acid [206]; amphetamine derivatives [207, 208]; phenazones, morazone [209]; alkaloids [91, 209]; nephopam [210]; phenyramidol metabolites [211]; diethylalkylacetamide derivatives [212]; zipeprol (Mirsol) [213]; thalidomide and hydrolysis products [214]; cyclohexylamine derivatives [215]; herbicide residues [216]
Bromocresol purple (5.2...6.8)	glutamic and ketoglutaric acids [217], halide and halate anions [91, 218, 219]; preservatives [220, 221]; products of pyrolysis of epoxy resins [222]; 5-aminodibenzocycloheptane derivatives [223]; phenylalkanolamines, ephedrine [224]
Bromophenol blue (3.0...4.6)	aliphatic carboxylic acids [225–228]; malonic and lactic acids [229]; palmitic and lactic acids [230]; malonic, glycolic, malic, citric, tartaric, ketoglutaric, galacturonic and oxalic acids [196]; dicarboxylic acids, succinic acid [231]; indoleacetic acid, trichloroacetic acid [232]; palmitic acid, palmityl- and stearyllactic acid [223]; benzoic, sorbic and salicylic acid [234]; metabolites of ascorbic acid [235]; chloropropionic acid [236]; oligogalacturonic acids [237]; amino acids, hydrocarbons, mono-, di- and triglycerides [238]; xylobiose, xylose, glucose and derivatives [239]; sugar alcohols [91]; toxaphene [240]
Bromothymol blue (6.0...7.6)	acid lipids, cholesterol glucuronides and gangliosides [241]; aryloxybutanolamine derivatives [242]; norfenfluramine derivatives [243]; ethylamphetamines [244]; involatile mineral oil hydrocarbons [245]; phospholipids [91]
Malachite green (0.0...2.0)	uracil derivatives, triazine herbicides [163]; polar lipids [246, 247]; phospholipids [248, 249]; fatty acids, fatty aldehydes, phospholipids and glycolipids [250]; microbiocidal isothiazolones [251]

adjusted to be near the indicator range by the addition of either boric or citric acid or borax, ammonium hydroxide or sodium hydroxide solution. Mixed indicators such as dimethyl yellow and pentamethylene red [183], or methyl red with bromocresol green [184] or with bromophenol blue [185] have also been employed in addition to the indicator dyes listed in Table 9.

2.5.4 Treatment with Iodine

Treatment of the solvent-free chromatogram with iodine vapor or by dipping in or spraying with iodine solution (0.5–1%) is a rapid and economical universal method of detecting lipophilic substances. Molecular iodine is enriched in the chromatogram zones and colors them brown.

In practice a few iodine crystals are usually placed on the bottom of a dry, closed trough chamber. After the chamber has become saturated with violet iodine vapor the solvent-free plates are placed in the chamber for 30 s to a few minutes. The iodine vapor condenses on the TLC layers and is enriched in the chromatogram zones. Iodine vapor is a universal detector, there are examples of its application for all types of substances, e.g. amino acids, indoles, alkaloids, steroids, psychoactive substances, lipids (a tabular compilation would be too voluminous to include in this section).

The chromatogram is observed and documented as soon as the spots are readily visible. The iodine can then be allowed to evaporate from the chromatogram (fume cupboard!). The chromatogram can then be subjected to further reactions or processes after this reversible reaction.

If it is desired to stabilize slightly yellow-colored iodine-containing chromatograms this is best done by treating them with dilute starch solution. This produces the well-known blue iodine inclusion compounds and these are stable over a long period.

Although, in most cases, iodine is a fairly inert halogen (in contrast to bromine) and does not normally react with the substances that have been chromatographed there are, nevertheless, examples where chemical changes have been observed. Oxidations can evidently take place (e.g. aromatic hydrocarbons and isoquinoline alkaloids [252, 253]) and additions and substitutions have also been observed. Pale zones then appear on a brown background.

Emetine and cephaeline, the two major alkaloids of ipecacuanha, begin to fluoresce after treatment with iodine vapor [254]. The molecular iodine, which acts as a quencher, must be removed by heating in the drying cupboard or on a hotplate (fume cuboard! 60–100 °C), before the yellow (emetine) and blue (cephaeline) fluorescence of the zones becomes visible. With appropriate standardization this reaction is suitable for in situ fluorimetric quantitation [255].

Other examples of irreversible derivatization on treatment with iodine have been described for phenolic steroids (estrone derivatives [256]), morphine [257] and 23 other pharmaceuticals [258]. These reactions are probably favored by the presence of silica gel as stationary phase and by the influence of light.

References

[1] Gauglitz, G.: *Praktische Spektroskopie,* Attempto Verlag, Tübingen 1983.
[2] Hezel, U. B., in: Zlatkis, A., Kaiser, R. E.: *HPTLC – High Performance Thin Layer Chromatography. J. Chromatogr. Library,* Vol. 9. Elsevier, Amsterdam 1977.
[3] Förster, Th.: *Fluoreszenz organischer Verbindungen*, Vandenhoek, Göttingen 1951.
[4] Zander, M.: *Fluorimetrie.* Springer, Heidelberg 1981.
[5] Jork, H., Wimmer, H.: *Quantitative Auswertung von Dünnschicht-Chromatogrammen* (TLC-Report), GIT-Verlag, Darmstadt 1982.
[6] Gänshirt, H. G., Poldermann, J.: *J. Chromatogr.* **1964,** *16,* 510-518.
[7] Tschesche, R., Biernoth, G. Wulff, G.: *J. Chromatogr.* **1963,** *12,* 342-346.
[8] Stahl, E.: *Chemiker Ztg.* **1958,** *82,* 323-329.
[9] Simpson, T. H., Wright, R. S.: *Anal. Biochem.* **1963,** *5,* 313-320.
[10] Machata, G.: *Mikrochim. Acta (Vienna)* **1960,** 79-86.
[11] Wieland, T., Heinke, B.: *Liebigs Ann. Chem.* **1958,** *615,* 184-202; *Experientia* **1962,** *18,* 430-432.
[12] Seher, A., Homberg, E.: *Fette, Seifen, Anstrichm.* **1971,** *73,* 557-560.
[13] Brown, J. L., Johnston, J. M.: *J. Lipid Res.* **1962,** *3,* 480-481.
[14] Kunkel, E.: *Mikrochim. Acta (Vienna)* **1977,** 227-240.
[15] Forney, F. W., Markovetz, A. J.: *Biochem. Biophys. Res. Commun.* **1969,** *37,* 31-38.
[16] Allebone, J. E., Hamilton, R. J., Bryce, T. A., Kelly, W.: *Experientia* **1971,** *27,* 13-14.
[17] Neissner, R.: *Fette, Seifen, Anstrichm.* **1972,** *74,* 198-202.
[18] Mangold, H. K.: *J. Am. Oil Chem. Soc.* **1961,** *38,* 708-727.
[19] Harvey, T., Matheson, T., Pratt, K.: *Anal. Chem.* **1984,** *56,* 1277-1281.
[20] Michalec, C., Sule, M., Mestan, J.: *Nature* **1962,** *193,* 63-64.
[21] Battaile, J., Dunning, R. L., Loomis, W. D.: *Biochem. Biophys. Acta* **1961,** *51,* 538-544.
[22] Copius-Peereboom, J. W., Beekes, H. W.: *J. Chromatogr.* **1964,** *14,* 417-423.
[23] Hunek, S.: *J. Chromatogr.* **1962,** *7,* 561-564.
[24] Avigan, J., Goodman, D. S., Steinberg, D.: *J. Lipid Res.* **1963,** *4,* 100-101.
[25] Scrignar, C. B.: *J. Chromatogr.* **1964,** *14,* 189-193.
[26] Nichaman, M. Z., Sweeley, C. C., Oldham, N. M., Olson, R. E.: *J. Lipid Res.* **1963,** *4,* 484-485.

[27] Schellenberg, P.: *Angew. Chem.* **1962,** *74,* 118-119.
[28] Cerny, V., Joska, J., Labler, L.: *Collect. Czech. Chem. Commun.* **1961,** *26,* 1658-1668.
[29] Kasal, A.: *Collect. Czech. Chem. Commun.* **1978,** *43,* 498-510.
[30] Copius-Peereboom, J. W.: *J. Chromatogr.* **1960,** *4,* 323-328.
[31] Halpaap, H.: *Chemiker Ztg.* **1965,** *89,* 835-849.
[32] Popov, A. D., Stefanov, K. L.: *J. Chromatogr.* **1968,** *37,* 533-535.
[33] Shealy, Y. F., O'Dell, C. A.: *J. Pharmac. Sci.* **1971,** *60,* 554-560.
[34] Abbott, D. C., Blake, K. W., Tarrant, K. R., Thomson, J.: *J. Chromatogr.* **1967,** *30,* 136-142.
[35] Kahan, I. L.: *J. Chromatogr.* **1967,** *26,* 290-291.
[36] Sarbu, C., Marutoiu, C.: *Chromatographia* **1985,** *20,* 683-684.
[37] Nakamura, H., Tamura, Z.: *J. Chromatogr.* **1974,** *96,* 195-210.
[38] Kirchner, J. G., Miller, J. M., Keller, G. J.: *Anal. Chem.* **1951,** *23,* 420-425.
[39] German Patent No. 2816574.4.
[40] Kortüm, G.: *Kolorimetrie, Photometrie und Spektrometrie.* 4th Ed., Springer, Berlin 1962.
[41] Reule, A.: *Zeiss-Mitteilungen* **1962,** *2,* 355-371.
[42] Kiefer, J. (Ed.): *Ultraviolette Strahlen*, De Gruyter, Berlin 1977.
[43] Bauer, A., Schulz, P.: *Ann. Phys.* **1956,** *18,* 227; *Z. Phys.* **1956,** *146,* 393.
[44] Rössler, F.: *Ann. Phys.* **1952,** *10,* 177.
[45] Elenbaas, W.: *Rev. Opt. Theor. Instrum.* **1948,** *27,* 603.
[46] Rick, W.: *Klin. Chemie und Mikroskopie.* Springer, Berlin 1972.
[47] Quarzlampen GmbH: *Company literature D 310531.*
[48] Kaase, H., Bischoff, K.: *Optik* (Stuttgart) **1977,** *48,* 451-458.
[49] Bischoff, K.: *Strahlungsnormale, ihre Darstellung und Anwendung,* Vortrag Technische Fachhochschule Eßlingen im Kurs „Strahl und optische Spektrometer", 1965.
[50] Hamamatsu: *Company literature S-C-1-2 T 78.4.130.*
[51] RCA: *Photomultiplier Manual.* Electronic Components, Harrison, USA, 1970.
[52] Schonkeren, J. M.: *Photomultipliers.* Philips Application Handbook, Eindhoven 1970.
[53] Schaetti, N., Baumgartner, W.: *Helv. Phys. Acta* **1952,** *25,* 605-611.
[54] Schaetti, N.: *Z. Angew. Math. Phys.* **1953,** *4,* 450.
[55] Cannon, C. G.: *Electronics for Spectroscopists*, London 1960.
[56] Sommer, A. H.: *Rev. Sci. Instr.* **1955,** *26,* 725-726.
[57] Spicer, W. E.: *Phys. Rev.* **1958,** *112,* 114-122.
[58] Frischmuth-Hoffmann, G., Görlich, P., Hora, H., Heimann, W., Marseille, H.: *Z. Naturforsch.* **1960,** *15a,* 648, 1014.
[59] Baumgartner, W.: *Chimia* **1957,** *11,* 88-91.
[60] Schwab, G.-M., Schneck, E.: *Z. Physikal. Chem. N. F.* **1958,** *18,* 206-222.
[61] Jork, H.: *Chromatogr. Electrophor. Symp. Int., 4th,* Brussels 1966, **1968,** p 227–239.
[62] Frei, R. W., Zeitlin, H.: *Anal. Chim. Acta* **1965,** *32,* 32-39.
[63] Robin, M., Trueblood, K. N.: *J. Am. Chem. Soc.* **1957,** *79,* 5138-5142.
[64] Ebel, S., Geitz, E., Hocke, H., Kaal, M.: *Kontakte (Darmstadt)* **1982,** *1,* 39-44.
[65] Kubelka, P., Munk, F.: *Z. Techn. Phys.* **1931,** *12,* 593-601.
[66] Kubelka, P., Munk, F.: *J. Opt. Soc. Amer.* **1948,** *38,* 448-453, 1067.
[67] Kortüm, G., Vogel, J.: *Z. Phys. Chem.* **1958,** *18,* 110-122.
[68] Kortüm, G.: *Trans. Faraday Soc.* **1962,** *58,* 1624-1631.
[69] Jork, H.: *Fresenius Z. Anal. Chem.* **1966,** *221,* 17-33.
[70] Ebel, S.: *Ullmanns Enzyklopädie der technischen Chemie.* 4th Ed. Vol. 5, p 205–215, Verlag Chemie, Weinheim 1980.
[71] Ebel, S., Geitz, E., Klarner, D.: *Kontakte (Darmstadt)* **1980,** *1,* 11-16.
[72] Pollak, V., Boulton, A. A.: *J. Chromatogr.* **1970,** *50,* 30-38.

[73] Jork, H.: *Qualitative und quantitative Auswertung von Dünnschicht-Chromatogrammen unter besonderer Berücksichtigung photoelektrischer Verfahren.* Professorial thesis, Universität des Saarlandes, Saarbrücken 1969.
[74] Ebel, S., Geitz, E.: *Kontakte (Darmstadt)* **1981**, *2*, 34-38.
[75] Winefordner, J. D., Schulman, S. G., O'Haver, T. C.: *Luminescence Spectrometry in Analytical Chemistry*, Wiley-Interscience, London 1972.
[76] Guilbout, G. G.: *Practical Fluorescence Theory, Methods and Techniques.* Marcel Dekker, New York 1973.
[77] Roberts, T. R.: *Radiochromatography.* Elsevier Scientific Publ. Co., Amsterdam 1978.
[78] Crosby, S. D., Dale, G. L.: *J. Chromatogr.* **1985**, *323*, 462-464.
[79] Tate, M. E., Bishop, C. T.: *Can. J. Chem.* **1962**, *40*, 1043-1048.
[80] Gritter, R. J., Albers, R. J.: *J. Chromatogr.* **1962**, *9*, 392.
[81] Gänshirt, H.: *Arch. Pharm.* **1963**, *296*, 73-79.
[82] Grutte, F. K., Gartner, H.: *J. Chromatogr.* **1969**, *41*, 132-135.
[83] Kikuchi, T., Yokoi, T., Shingu, T., Niwa, M.: *Chem. Pharm. Bull.* **1981**, *29*, 1819-1826; 2531-2539.
[84] Kartnig, T., Ri, C. Y.: *Planta Med.* **1973**, *23*, 269-271.
[85] Gritter, R. J., Albers, R. J.: *J. Org. Chem.* **1964**, *29*, 728-731.
[86] Soboleva, D. A., Makarova, S. V., Khlapova, E. P.: *J. Anal. Chem. (USSR)* **1977**, *32*, 1423-1425.
[87] Kochetkov, N. K., Usov, A. I., Miroshnikova, L. I.: *Zhurnal obshchej chimii* **1970**, *40*, 2473-2478.
[88] Crosby, S. D., Dale, G. L.: *J. Chromatogr.* **1985**, *323*, 462-464.
[89] Däuble, M.: *Tenside Deterg.* **1981**, *18*, 7-12.
[90] Kany, E., Jork, H.: *GDCh-training course Nr. 300 „Einführung in die Dünnschicht-Chromatographie"*, Saarbrücken 1986.
[91] Touchstone, J. C., Dobbins, M. F.: *Practice of Thin Layer Chromatography*, J. Wiley & Sons, New York 1978.
[92] Gregorowicz, Z., Sliwiok, J.: *Microchem. J.* **1970**, *15*, 545-547.
[93] Sliwiok, J., Macioszczyk, A.: *Microchim. J.* **1978**, *23*, 121-124.
[94] Sliwiok, J., Kocjan, B.: *Microchim. J.* **1972**, *17*, 273-276.
[95] Mangold, H. K., Malins, D. C.: *J. Am. Oil Chem. Soc.* **1960**, *37*, 383-385; 576-578.
[96] Kurucz, E., Lukacs, P., Jeranek, M., Prepostffy, M.: *Acta Alimentaria* **1975**, *4*, 139-150.
[97] Gerhardt, W., Harigopal, V. P., Süss, S.: *J. Assoc. Off. Anal. Chem.* **1974**, *51*, 479-481.
[98] Packter, N. M., Stumpf, P. K.: *Arch. Biochem. Biophys.* **1975**, *167*, 655-667.
[99] Gornall, D. A., Kuksis, A.: *Can. J. Biochem.* **1971**, *49*, 44-50.
[100] Urbach, G., Stark, W.: *J. Agric. Food Chem.* **1975**, *23*, 20-24.
[101] Streibl, M., Jarolim, V., Konecny, K., Ubik, U., Trka, A.: *Fette, Seifen, Anstrichm.* **1973**, *75*, 314-316.
[102] Polles, S. G., Vinson, S. B.: *J. Agric. Food Chem.* **1972**, *20*, 38-41.
[103] Regula, E.: *J. Chromatogr.* **1975**, *115*, 639-644.
[104] Wathana, S., Corbin, F. T.: *J. Agric. Food Chem.* **1972**, *20*, 23-26.
[105] König, H.: *Fresenius Z. Anal. Chem.* **1970**, *251*, 359-368.
[106] Wittgenstein, E., Sawicky, E.: *Mikrochim. Acta (Vienna)* **1970**, 765-783.
[107] Parodi, P. W.: *J. Assoc. Off. Anal. Chem.* **1976**, *53*, 530-534.
[108] Oosthuizen, M. M. J., Potgieter, D. J. J.: *J. Chromatogr.* **1973**, *85*, 171-173.
[109] Froehling, P. E., van den Bosch, G., Boekenoogen, H. A.: *Lipids* **1972**, *7*, 447-449.
[110] Jellema, R., Elema, E. T., Malingre, Th.: *J. Chromatogr.* **1981**, *210*, 121-129.
[111] Rokos, J. A.: *J. Chromatogr.* **1972**, *74*, 357-358.

[112] Ozawa, A., Jinbo, H., Takahashi, H.: *Bunseki Kagaku* **1985**, *34*, 707-711.

[113] Jork, H., Kany, E.: *GDCh-training course Nr. 300 „Einführung in die Dünnschicht-Chromatographie"*, Universität des Saarlandes, Saarbrücken 1984.

[114] Blass, G., Ho, C. S.: *J. Chromatogr.* **1981**, *208*, 170-173.

[115] Larsen, H. F., Frostmann, A. F.: *J. Chromatogr.* **1981**, *226*, 484-487.

[116] Vinson, J. A., Hooyman, J. E.: *J. Chromatogr.* **1977**, *135*, 226-228.

[117] Zeller, M.: Thesis, Fachhochschule Gießen, Fachbereich Technisches Gesundheitswesen, 1986.

[118] Gitler, C.: *Anal. Biochem.* **1972**, *50*, 324-325.

[119] Gitler, C., in: L. Manson, ed.: *Biomembranes.* Vol. 2, 41-47. Plenum, New York 1971.

[120] Lichtenthaler, H., Boerner, K.: *J. Chromatogr.* **1982**, *242*, 196-201.

[121] Rohmer, M., Ourisson, G., Benveniste, P., Bimpson, T.: *Phytochemistry* **1975**, *14*, 727-730.

[122] Huang, L. S., Grunwald, C.: *Phytochemistry* **1986**, *25*, 2779-2781.

[123] Misso, N. L. A., Goad, L. J.: *Phytochemistry* **1984**, *23*, 73-82.

[124] Mamlok, L.: *J. Chromatogr. Sci.* **1981**, *19*, 53.

[125] Lucier, G. W., Menzer, R. E.: *J. Agric. Food Chem.* **1971**, *19*, 1249-1255.

[126] Ludlam, P. R.: *Analyst* **1973**, *98*, 107-115.

[127] Kennedy, M. V., Stojanovic, B. J., Sauman, F. L.: *J. Agric. Food Chem.* **1972**, *20*, 341-343.

[128] Das, D. K., Mathew, T. V., Mitra, S. N.: *J. Chromatogr.* **1970**, *52*, 354-356.

[129] Nagasawa, K., Yoshidome, Y., Anryu, K.: *J. Chromatogr.* **1970**, *52*, 173-176.

[130] Farkas, L., Morgos, J., Sallai, P., Lantai, I., Rusznak, I.: *Kolorisztikai Ertesito* **1986**, *28*, 118-126.

[131] Schaetti, N., Baumgartner, W., Flury, Ch.: *Helv. Phys. Acta* **1953**, *26*, 380-383.

[132] Černý, V., Joska, J., Lábler, L.: *Coll. Czech. Chem. Commun.* **1961**, *26*, 1658-1668.

[133] Belliveau, P. E., Mallet, V. N., Frei, R. W.: *Abstr. Pittsburgh Conf. Anal. Chem. Appl. Spectros.* **1970**, (313), 151.

[134] Mallet, V., Frei, R. W.: *J. Chromatogr.* **1971**, *54*, 251-257.

[135] Mallet, V., Frei, R. W.: *J. Chromatogr.* **1971**, *56*, 69-77.

[136] Mallet, V., Frei, R. W.: *J. Chromatogr.* **1971**, *60*, 213-217.

[137] Schellenberg, P.: *Angew. Chem. Int. Ed. Engl.* **1962**, *1*, 114-115.

[138] Seiler, H.: *Helv. Chim. Acta* **1970**, *53*, 1423-1424.

[139] Kosinkiewicz, B., Lubczynska, J.: *J. Chromatogr.* **1972**, *74*, 366-368.

[140] Dietsche, W.: *Fette, Seifen, Anstrichm.* **1970**, *72*, 778-783.

[141] Weisheit, W., Eul, H.: *Seifen, Öle, Fette, Wachse* **1973**, *99*, 711-714.

[142] Canic, V. D., Perisic-Janjic, N. V.: *Fresenius Z. Anal. Chem.* **1974**, *270*, 16-19.

[143] Svec, P., Nondek, L., Zbirovsky, M.: *J. Chromatogr.* **1971**, *60*, 377-380.

[144] Whistance, G. R., Dillon, J. F., Threlfall, D. R.: *Biochem. J.* **1969**, *111*, 461-472.

[145] Mathre, D. E.: *J. Agric. Food Chem.* **1971**, *19*, 872-874.

[146] Kreysing, G., Frahm, M.: *Dtsch. Apoth. Ztg.* **1970**, *110*, 1133-1135.

[147] König, H.: *Fresenius Z. Anal. Chem.* **1970**, *251*, 167-171; **1971**, *254*, 337-345.

[148] Köhler, M., Chalupka, B.: *Fette, Seifen, Anstrichm.* **1982**, *84*, 208-211.

[149] Frahne, D., Schmidt, S., Kuhn, H.-G.: *Fette, Seifen, Anstrichm.* **1977**, *79*, 32-41; 122-130.

[150] Bey, K.: *Fette, Seifen, Anstrichm.* **1965**, *67*, 217-221.

[151] Matissek, R., Hieke, E., Baltes, W.: *Fresenius Z. Anal. Chem.* **1980**, *300*, 403-406.

[152] Nagasawa, K., Yoshidome, H., Kamata, F.: *J. Chromatogr.* **1970**, *52*, 453-459.

[153] Nagasawa, K., Ogamo, A., Anryu, K.: *J. Chromatogr.* **1972**, *67*, 113-119.

[154] Takeshita, R.: *J. Chromatogr.* **1972**, *66*, 283-293.

[155] Rincker, R., Sucker, H.: *Fette, Seifen, Anstrichm.* **1972**, *74*, 21-24.

[156] Thielemann, H.: *Z. Chem.* **1972**, *12*, 223; *Mikrochim. Acta (Vienna)* **1972**, 672-673; *Fresenius Z. Anal. Chem.* **1972**, *262*, 192; **1974**, *272*, 206; *Pharmazie* **1977**, *32*, 244.

[157] Thielemann, H.: *Mikrochim. Acta (Vienna)* **1973**, 521-522.
[158] Halbach, G., Görler, K.: *Planta Med.* **1971**, *19*, 293-298.
[159] Perisic-Janjic, N., Canic, V., Lomic, S., Baykin, D.: *Fresenius Z. Anal. Chem.* **1979**, *295*, 263-265.
[160] Canic, V. D., Perisic-Janjic, N. U., Babin, M. J.: *Fresenius Z. Anal. Chem.* **1973**, *264*, 415-416.
[161] Traylor, I. D., Hogan, E. L.: *J. Neurochem.* **1980**, *34*, 126-131.
[162] Bonderman, D. P., Slach, E.: *J. Agric. Food Chem.* **1972**, *20*, 328-331.
[163] Ebing, W.: *J. Chromatogr.* **1972**, *65*, 533-545.
[164] Tewari, S. N., Ram, L.: *Mikrochim. Acta (Vienna)* **1970**, 58-60.
[165] Salvage, T.: *Analyst* **1970**, *95*, 363-365.
[166] Groningsson, K., Schill, G.: *Acta Pharm. Suec.* **1969**, 447-468.
[167] Gleispach, H., Schandara, E.: *Fresenius Z. Anal. Chem.* **1970**, *252*, 140-143.
[168] McNamara, D. J., Proia, A., Miettinen, T. A.: *J. Lipid Res.* **1981**, *22*, 474-484.
[169] Nagy, S., Nordby, H. E.: *Lipids* **1972**, *7*, 722-727.
[170] Boskou, D., Katsikas, H.: *Acta Aliment.* **1979**, *8*, 317-320.
[171] Ellington, J. J., Schlotzhauer, P. F., Schepartz, A. I.: *J. Am. Oil Chem. Soc.* **1978**, *55*, 572-573.
[172] Garg, V., Nes, W.: *Phytochemistry* **1984**, *23*, 2925-2929.
[173] Burstein, S., Zamoscianyk, H., Kimball, H. L., Chaudhuri, N. K., Gut, M.: *Steroids* **1970**, *15*, 13-60.
[174] Gornall, D. A., Kuksis, A.: *Can. J. Biochem.* **1971**, *49*, 44-50.
[175] Hoffmann, L. M., Amsterdam, D., Brooks, S. A., Schneck, L.: *J. Neurochem.* **1971**, *29*, 551-559.
[176] Tyman, J. H. P.: *J. Chromatogr.* **1977**, *136*, 289-300.
[177] Moschids, M.: *J. Chromatogr.* **1984**, *294*, 519-524.
[178] Vroman, H. E., Baker, G. L.: *J. Chromatogr.* **1965**, *18*, 190-191.
[179] Milborrow, B. V.: *J. Chromatogr.* **1965**, *19*, 194-197.
[180] Jones, M., Keenan, R. W., Horowitz, P.: *J. Chromatogr.* **1982**, *237*, 522-524.
[181] Colarow, L., Pugin, B., Wulliemier, D.: *J. Planar Chromatogr.* **1988**, *1*, 20-23.
[182] Sherma, J., Bennett, S.: *J. Liq. Chromatogr.* **1983**, *6*, 1193-1211.
[183] Kucera, J., Pokorny, S., Coupek, J.: *J. Chromatogr.* **1974**, *88*, 281-287.
[184] Takeshita, R.: *J. Chromatogr.* **1972**, *66*, 283-293.
[185] Matin, A., Konings, W. N.: *Europ. Biochem.* **1973**, *34*, 58-67.
[186] Chawla, H. M., Gambhir, I., Kathuria, L.: *J. High Resolut. Chromatogr. Chromatogr. Commun.* **1979**, *2*, 673-674.
[187] Braun, D., Geenen, H.: *J. Chromatogr.* **1962**, *7*, 56-59.
[188] Miyazaki, S., Suhara, Y., Kobayashi, T.: *J. Chromatogr.* **1969**, *39*, 88-90.
[189] Hansen, S. A.: *J. Chromatogr.* **1976**, *124*, 123-126.
[190] Lukacova, M., Klanduch, J., Kovac, S.: *Holzforschung* **1977**, *31*, 13-18.
[191] Lupton, C. J.: *J. Chromatogr.* **1975**, *104*, 223-224.
[192] Dubler, R. E., Toscano jr., W. A., Hartline, R. A.: *Arch. Biochem. Biophys.* **1974**, *160*, 422-429.
[193] Kraiker, H. P., Burch, R. E.: *Z. Klin. Chem. Klin. Biochem.* **1973**, *11*, 393-397.
[194] Serova, L. I., Korchagin, V. B., Vagina, J. M., Koteva, N. I.: *Pharm. Chem. J. (USSR)* **1972**, *6*, 609-610.
[195] Bornmann, L., Busse, H., Hess, B.: *Z. Naturforsch.* **1973**, *28b*, 93-97.
[196] Chan, H. T., Chenchin, E., Vonnahme, P.: *J. Agric. Food Chem.* **1973**, *21*, 208-211.
[197] Chan, H. T., Chang, T. S., Stofford, A. E., Brekke, J. E.: *J. Agric. Food Chem.* **1971**, *19*, 263-265.

[198] Pfeifer, A.: *Seifen, Öle, Fette, Wachse* **1971**, *97*, 119-120.
[199] Stoll, U.: *J. Chromatogr.* **1970**, *52*, 145-151.
[200] Riley, R. T., Mix, M. C.: *J. Chromatogr.* **1980**, *189*, 286-288.
[201] Laub, E., Lichtenthal, H., Frieden, M.: *Dtsch. Lebensm. Rundsch.* **1980**, *76*, 14–16.
[202] Overo, K. F., Jorgensen, A., Hansen, V.: *Acta Pharmacol. Toxikol.* **1970**, *28*, 81-96.
[203] Sinsheimer, J. E., Breault, G. O.: *J. Pharm. Sci.* **1971**, *60*, 255-257.
[204] Kovalska, T.: *Chromatographia* **1985**, *20*, 434-438.
[205] McDowell, R. W., Landolt, R. R., Kessler, W. V., Shaw, S. M.: *J. Pharm. Sci.* **1971**, *60*, 695-699.
[206] Garrett, E. R., Bojarski, J. T., Yakatan, G. J.: *J. Pharm. Sci.* **1971**, *60*, 1145–1154.
[207] Cartoni, G. P., Lederer, U., Polidori, F.: *J. Chromatogr.* **1972**, *71*, 370-375.
[208] Gielsdorf, W., Klug, E.: *Dtsch. Apoth. Ztg.* **1981**, *121*, 1003-1005.
[209] Kung, E.: *Rechtsmedizin* **1972**, *71*, 27-36.
[210] Ebel, S., Schütz, H.: *Arch. Toxikol.* **1977**, *38*, 239-250.
[211] Goenechea, S., Eckhardt, G., Goebel, K. J.: *J. Clin. Chem. Clin. Biochem.* **1977**, *15*, 489-498.
[212] Klug, E., Toffel, P.: *Arzneim.-Forsch.* **1979**, *29*, 1651-1654.
[213] Giesldorfer, W., Toffel-Nadolny, R.: *J. Clin. Chem. Clin. Biochem.* **1981**, *19*, 25-30.
[214] Pischek, G., Kaiser, E., Koch, H.: *Mikrochim. Acta (Vienna)* **1970**, 530-535.
[215] Blumberg, A. G., Heaton, A. M.: *J. Chromatogr.* **1970**, *48*, 565-566.
[216] Smith, A. E., Fitzpatrick, A.: *J. Chromatogr.* **1971**, *57*, 303-308.
[217] Fortnagel, B.: *Biochim. Biophys. Acta* **1970**, *222*, 290-298.
[218] Thielemann, H.: *Mikrochim. Acta (Vienna)* **1970**, 645.
[219] Thielemann, H.: *Mikrochim. Acta (Vienna)* **1971**, 746-747.
[220] Tjan, G. H., Konter, T.: *J. Assoc. Off. Anal. Chem.* **1973**, *55*, 1223-1225.
[221] Yang, Z.: *Chinese Brew* **1983**, *2* (3), 32-34.
[222] Braun, D., Lee, D. W.: *Kunststoffe* **1972**, *62*, 517-574.
[223] Maulding, H. V., Brusco, D., Polesuk, J., Nazareno, J., Michaelis, A. F.: *J. Pharm. Sci.* **1972**, *61*, 1197-1201.
[224] Chafetz, L.: *J. Pharm. Sci.* **1971**, *60*, 291-294.
[225] Petrowitz, H. J., Pastuska, G.: *J. Chromatogr.* **1962**, *7*, 128-130.
[226] Chang, T. S., Chan jr., H. T.: *J. Chromatogr.* **1971**, *56*, 330-331.
[227] Gupta, S., Rathore, H., Ali, I., Ahmed, S.: *J. Liq. Chromatogr.* **1984**, *7*, 1321-1340.
[228] Selmeci, G., Aczel, A. Cseh, F.: *Budapest Chromatogr. Conf., June 2, 1983.*
[229] Selmeci, G., Hanusz, B.: *Elelmiszervizsgalati Közlemenyek* **1981**, *27*, 135-138.
[230] Sass, M., Vaczy, K.: *7. Kromatografias Vandorgyüles Elöadasai* **1979**, 127-131.
[231] Huxtable, R. J., Wakil, S. J.: *Biochim. Biophys. Acta* **1971**, *239*, 168-177.
[232] Rathore, H., Kumari, K., Agrawal, M.: *J. Liq. Chromatogr.* **1985**, *8*, 1299-1317.
[233] Orsi, F., Abraham-Szabo, A., Lasztity, R.: *Acta Aliment.* **1984**, *13*, 23-38.
[234] Zenen vidaud candebato, E., Garcia Roche, O.: *Elelmiszerviszgalati Közlemenyek* **1982**, *28*, 213-217.
[235] Huelin, F. E., Coggiola, I. M., Sidhu, G. S., Kennett, B. H.: *J. Sci. Food Agric.* **1971**, *22*, 540-542.
[236] Chalaya, Z. J., Gorbons, T. V.: *J. Anal. Chem. (USSR)* **1980**, *35*, 899-900.
[237] Liu, Y. K., Luh, B. S.: *J. Chromatogr.* **1978**, *151*, 39-49.
[238] Heinz, K. L., van der Velden, C.: *Fette, Seifen, Anstrichm.* **1971**, *73*, 449-454.
[239] Sinner, M., Parameswaran, N., Dietrichs, H. H., Liese, W.: *Holzforschung* **1972**, *26*, 218-228; **1973**, *27*, 36-42.
[240] Liebmann, R., Hempel, D., Heinisch, E.: *Arch. Pflanzenschutz* **1971**, *7*, 131-150.

[241] Hara, A., Taketomi, T.: *Lipids* **1982**, *17*, 515–518.
[242] Racz, I., Plachy, J., Mezei, J., Poor-Nemeth, M., Küttel, M.: *Acta Pharm. Hung.* **1985**, *55*, 17–24.
[243] Beckett, A. H., Shenoy, E. V., Brookes, L. G.: *J. Pharm. Pharmacol.* **1972**, *24*, 281–288.
[244] Beckett, A. H., Shenoy, E. V., Salmon, J. A.: *J. Pharm. Pharmacol.* **1972**, *24*, 194–202.
[245] Goebgen, H. G., Brockmann, J.: *Vom Wasser* **1977**, *48*, 167–178.
[246] Vaskovsky, V. E., Khotimchenko, S.: *J. High Resolut. Chromatogr. Chromatogr. Commun.* **1982**, *5*, 635–636.
[247] Vaskovsky, V. E., Latyshev, N. A.: *J. Chromatogr.* **1975**, *115*, 246–249.
[248] Latyshev, N. A., Vaskovsky, V. E.: *J. High Resolut. Chromatogr. Chromatogr. Commun.* **1980**, *3*, 478–479.
[249] Vaskovsky, V. E., Latyshev, N. A., Cherkassov, E. N.: *J. Chromatogr.* **1979**, *176*, 242–246.
[250] Teichman, R. J., Takei, G. H., Cummins, J. M.: *J. Chromatogr.* **1974**, *88*, 425–427.
[251] Matissek, R., Droß, A., Häussler, M.: *Fresenius Z. Anal. Chem.* **1984**, *319*, 520–523.
[252] Wilk, M., Hoppe, U., Taupp, W., Rochlitz, J.: *J. Chromatogr.* **1967**, *27*, 311–316.
[253] Wilk, M., Bez, W., Rochlitz, J.: *Tetrahedron* **1966**, *22*, 2599–2608.
[254] Stahl, E.: *Dünnschicht-Chromatographie, ein Laboratoriumshandbuch,* 2nd Ed., Springer, Berlin 1967.
[255] Jork, H., Kany, E.: *GDCh-Workshop Nr. 302 „Möglichkeiten der quantitativen Auswertung von Dünnschicht-Chromatogrammen“,* Saarbrücken 1986.
[256] Brown, W., Turner, A. B.: *J. Chromatogr.* **1967**, *26*, 518–519.
[257] Barrett, G. C. in: Giddings, J. C., Keller, R. A.: *Advances in Chromatography,* Vol. 11, 151. Marcel Dekker, New York 1974.
[258] Schmidt, F.: *Krankenhaus-Apoth.* **1973**, *23*, 10–11.
[259] Chen, T. I., Morris, M. D.: *Anal. Chem.* **1984**, *56*, 19–21; 1674–1677.
[260] Peck, K., Fotiou, F. K., Morris, M. D.: *Anal. Chem.* **1985**, *57*, 1359–1362.
[261] Tran, C. D.: *Appl. Spectrosc.* **1987**, *41*, 512–516.

3 Chemical Methods of Detection

Every analytical result forms the basis for a subsequent decision process. So the result should be subject to a high degree of precision and accuracy. This is also true of chromatographic methods. The physical detection methods described until now are frequently not sufficient on their own. If this is the case they have to be complemented by specific chemical reactions (derivatization).

These reactions can be carried out during sample preparation or directly on the layer at the start after application of the sample. Reactions have also been described in the capillaries employed for application.

Thus, MATHIS et al. [1, 2] investigated oxidation reactions with 4-nitroperbenzoic acid, sodium hypobromite, osmium tetroxide and ruthenium tetroxide. HAMANN et al. [3] employed phosphorus oxychloride in pyridine for dehydration. However, this method is accompanied by the disadvantages that the volume applied is increased because reagent has been added and that water is sometimes produced in the reaction and has to be removed before the chromatographic separation.

All cases involve prechromatographic derivatizations which introduce a chromophore leading to the formation of strongly absorbing or fluorescent derivatives

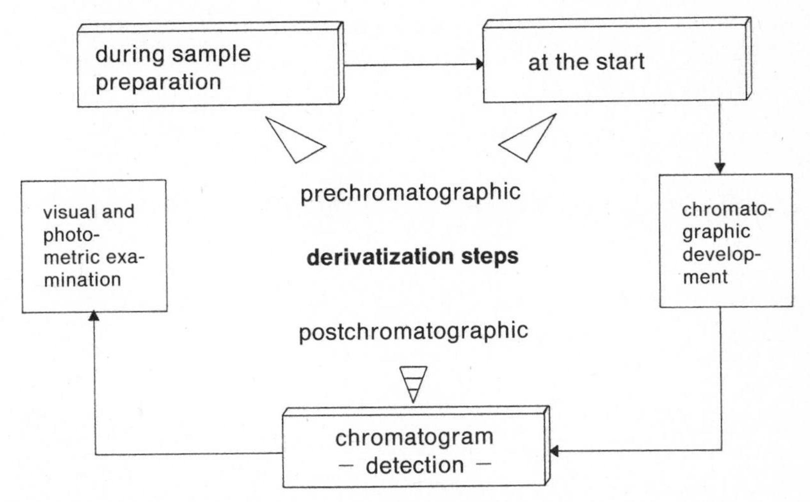

Fig. 30: Schematic representation of the position of the derivatization steps.

which increase the selectivity of the separation, increase the sensitivity of detection and improve the linearity [4]. Trace analyses often only become possible after chemical reaction of the substance to be detected. The aim of prechromatographic derivatization is, thus, rather different than that of postchromatographic derivatization, where the aim is first to detect the substance and then only secondarily to characterize it (Fig. 30).

3.1 In Situ Prechromatographic Derivatization

There has for some years been a considerable backlog in the development of practicable prechromatographic methods [5]. It is becoming more and more recognized that the future direction to be taken by trace analysts is to make improvements in the extraction, enrichment and clean-up of the sample and in the optimization of derivatization. It is only in this way that it is possible to employ the sensitive chromatographic techniques optimally for the solution of practically relevant problems.

About 100 000 new chemical compounds are synthesized every year [6]; these have to be recognized, identified and determined quantitatively. Ever more frequently this is only possible because of the employment of multiple chromatographic methods coupled with derivatization during or before the separation process.

For these reasons "Reaktions-Chromatographie" [7] ("Chromatographie fonctionelle sur couche mince" [1, 2]) is steadily gaining in importance. Here the reaction, which also then takes on the role of a clean-up step, is performed at the start or in the concentration zone of the TLC plate.

The requirements of such a reaction are [8]:

- single, stable reaction products,
- high yields in all concentration ranges,
- simplicity and rapidity in application,
- no interference by excess reagent with the chromatographic separation and analysis that follows.

Such in situ reactions are based on the work of Miller and Kirchner [9] and offer the following possibilities [10]:

- The reaction conditions can be selected so as to be able to separate substances with the same or similar chromatographic properties (critical substance pairs) by exploiting their differing chemical behavior, thus, making it easier to identify them. Specific chemical derivatization allows, for example, the esterification of primary and secondary alcohols which can then be separated from tertiary alcohols by a subsequent chromatographic development (Sec. 3.1.5). It is just as simply possible to separate aldehydes and ketones produced by oxidation at the start from unreactive tertiary alcohols and to detect them group-specifically.

- The stability of the compound sought (e.g. oxidation-sensitive substances) can be improved.
- The reactivity of substances (e.g. towards the stationary phase) can be reduced.
- The stability of the compound sought (e.g. oxidation-sensitive substances) can be improved.

This, on the one hand, reduces the detection limit so that less sample has to be applied and, thus, the amounts of interfering substances are reduced. On the other hand, the linearity of the calibration curves can also be increased and, hence, fewer standards need to be applied and scanned in routine quantitative investigations so that more tracks are made available for sample separations. However, the introduction of a large molecular group can lead to the "equalization" of the chromatographic properties.

In practice, reaction chromatography is usually performed by first applying spots or a band of the reagent to the start zone. The sample is usually then applied while the reagent zone is still moist. Care should be taken to ensure that the sample solvent does not chromatograph the reagent outwards. The reagent solution can be applied once more, if necessary, to ensure that it is present in excess. There is no problem doing this if it is employed as a band with the Linomat IV, (Fig. 31], for instance.

If heat is necessary to accelerate the reaction, the start zone should be covered by a glass strip before being placed on a hotplate or in a drying cupboard. After reaction is complete the TLC plate should be dried and development can begin.
There have also been repeated descriptions of coupling in the sense of a two-dimensional S – R – S (separation – reaction – separation) technique. In this case the chromatogram track from the first separation serves as the start zone for a second chromatographic development after turning the plate at 90°. The derivatization described above is performed between the two chromatographic separation steps.

Reactions can also occur during chromatographic development. These can either be undesired reactions or planned derivatizations. Thus, WEICKER and BROSSMER [11] have reported, for example, that hexoses, pentoses and disaccharides can be aminated when ammonia-containing mobile phases are employed on silica gel G layers. On the other hand, fluorescamine or ninhydrin have been added to the

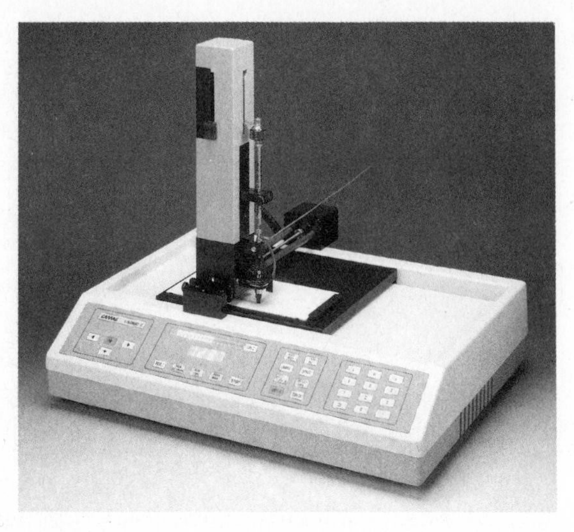

Fig. 31: Linomat IV (CAMAG).

mobile phase with the aim of converting peptides, amino acids or amines to fluorescent or colored derivatives. Unsaturated fatty acids or sterols have been brominated by adding bromine to the mobile phase or oxidized by the addition of peracetic acid. Other examples are to be found in Section 3.2.4.

When undertaking quantitative investigations it should be checked that the reaction on the TLC plate is complete – or at least stoichiometric and reproducible. In all cases it is also useful to apply reagent and sample solutions separately on neighboring tracks in order to be able to determine where the starting products appear in the chromatogram under the reaction conditions. In this way it is possible to decide whether additional by-products are produced.

3.1.1 Oxidation and Reduction

Oxidations and reductions are amongst the most frequent in situ prechromatographic reactions; they were exploited as early as 1953 by MILLER and KIRCHNER [9]. They characterized citral as an aldehyde by oxidizing it to geranic acid and reducing it to geraniol. Further examples are listed in Table 10.

Table 10: Selection of prechromatographic derivatizations involving oxidation and reduction reactions.

Substances	Method, reagent and end products	References
Oxidations		
Phenothiazines	Apply sample solution followed by 10–20% hydrogen peroxide solution, dry at 60 °C. Sulfoxides are produced.	[12]
Anthocyanins	Anthocyanins, which interfere with the chromatographic determination of other substances, are destroyed by "overspotting" with 3% ethanolic hydrogen peroxide solution.	[13]
α-Terpineol	Apply sample solution as spots followed by 4-nitroperbenzoic acid at the start, allow to react for several minutes, dry and develop.	[1]
Geraniol	Apply alcoholic solution, then 20% chromic acid in glacial acetic acid, allow to react and develop. Citral is formed.	[14]
Isopulegol, daucol, menthol, khusol etc.	The terpenoids are applied at the start together with 1,4-naphthoquinone potassium *tert*-butoxide, heated together to 120 °C for 24 h then developed.	[15]
Diosgenin, tigogenin, androst-5-en-17-on-3-β-ol	Moisten the dried sample zone with 20% aluminium isopropoxide in benzene, spray with acetone-benzene (4 + 1), heat to 55 °C in a benzene acetone (1 + 1) atmosphere (twin-trough chamber) for 2 h, then dry, spray with 10% aqueous silver nitrate solution and finally dry for 20 min at 80 °C. Diosgenone and tigogenone are produced, for example, (OPPENAUER reaction, anhydrous solvent!).	[16]
17-Hydroxy-corticosteroids	Apply sample then follow with 10% aqueous sodium periodate solution, allow to react, dry at 50 °C and develop. 17-Ketosteroids are produced.	[3]
Oleanolic acid, ursolic acid, betulic acid	Apply sample solution followed by 2% chromium(VI) oxide solution in acetic acid, spray with acetic acid and keep in an atmosphere of acetic acid for 30 to 50 min in a twin-trough chamber. Then heat to 50 °C for a few minutes, spray with methanol to destroy the excess of oxidizing agent, activate the TLC plate and develop. Oleanonic acid, 3-ketoursolic acid and 3-ketobetulic acid are produced.	[16]

Table 10: (Continued)

Substances	Method, reagent and end products	References
Alkaloids	a) 10% chromic acid in glacial acetic acid is applied on top of the sample spot. Development is performed after a brief reaction period.	[2]
	b) Dehydration by heating the applied sample solution on silica gel layers.	[18, 19]
Strychnine, brucine	Oxidation is performed with potassium dichromate solution. This oxidizes brucine to the *o*-quinone which can then be separated chromatographically.	[17]
Polyaromatic hydrocarbons (PAH)	a) Apply the sample solution, spray with trifluoroacetic acid solution, heat to 100 °C, cool and develop. Trifluoroacetic acid catalyzes oxidation by atmospheric oxygen.	[20]
	b)Apply sample solution and place in an iodine chamber for several hours, allow the iodine to evaporate. 3,4-Benzpyrene forms, for example, *bis*-3,4-benzpyrenyl.	[21]
Reductions		
Steroids	The applied steroids are reduced by means of a mixture of 10% ethanolic sodium borohydride solution and 0.1 N sodium hydroxide solution (1 + 1). The excess reagent is neutralized with acid after 30 min.	[3]
Strychnine	Sample applied as spots followed by 5% sodium borohydride solution, which is then dried and followed by development.	[22]
Oleanonic acid, tigogenone	Apply sample and then treat with 10% sodium borohydride solution in methanol – water (1 + 5). Spray TLC plate with methanol and store in a desiccator at 55 °C for 1.5 h over ethanol – methanol – dioxan (4 + 1 + 1); then dry the TLC plate (drying cupboard) and develop. Oleanolic acid and tigogenin are produced.	[16]
7-Ketocholesterol, sterol hydroperoxides	The applied sample is treated with 1% methanolic sodium borohydride. After allowing reaction to proceed for 5 min the TLC plate is dried and then developed.	[23, 24]
Alkaloids	Sodium borohydride solution is applied after the sample solution. The plate is dried and developed after a few minutes.	[2, 25]

Table 10: (Continued)

Substances	Method, reagent and end products	References
Disulfides	The applied sample solution is treated with 0.4% sodium borohydride solution in 95% ethanol. After 15 to 20 min reaction time the excess reagent is destroyed with acid.	[26]
Methyl glycyrrhetate, diosgenin	Apply 5% palladium or platinum chloride in 50% hydrochloric acid to the start, then spray with alkaline formaldehyde solution and dry in air, spray with 5% acetic acid solution and dry at 80 °C. Then apply the sample solution and lightly spray with ethyl acetate. Store the TLC plate for 50 to 72 h in a desiccator over ethyl acetate in a slight stream of hydrogen, then dry and develop. The product is, for example, methyl desoxyglycyrrhetate.	[16]
Fatty acids	Apply 1 drop colloidal palladium solution to the start zone and dry at 80 to 90 °C for 60 min. Then apply the sample solution, store the TLC plate for 60 min in a hydrogen-filled desiccator, then dry and develop.	[27, 28]
Maleic, fumaric, glutaconic, citraconic, mesaconic and itaconic acid	Apply colloidal palladium solution to the starting point (diameter 8 to 10 mm) and dry. Then apply sample solution and gas with hydrogen (desiccator) for 1 h. Maleic and fumaric acids yield succinic acid etc., which may also be separated chromatographically.	[29]
Amino acids	The configuration was determined by reacting with a carbobenzyloxy-L-amino acid azide and reductively removing the protective group with hydrogen/palladium chloride solution.	[30]
Nitro compounds	The sample was applied, followed by 15% zinc chloride solution and dilute hydrochloric acid. The reaction was allowed to proceed for a short time, the plate was then dried and the amino compounds so formed were chromatographed.	[31]
1-Nitropyrene	Extracts of diesel exhaust gases were applied to concentrating zone, platinum chloride solution was then applied followed by sodium borohydride. 1-Aminopyrene was formed.	[32]
Tetrazolium salts	Formazan dyes are produced on reaction of tetrazolium salts with ammonium sulfide.	[33]

Table 10: (Continued)

Substances	Method, reagent and end products	References
11-β-Hydroperoxy-lanostenyl acetate	The sample solution is applied and then treated with 5% iron(II) ammonium sulfate in water – methanol – ether (2 + 1 + 1). 11-Oxolanostenyl acetate is formed.	[34]

3.1.2 Hydrolysis

Hydrolytic reactions can also be performed at the start as well as oxidative and reductive ones. They can be carried out by "wet chemistry" or enzymatically. Examples are listed in Table 11.

Table 11: Prechromatographic derivatization involving hydrolytic and enzymatic cleavage reactions.

Substances	Method, reagent and end products	References
Acid hydrolysis		
Alkenylacyl- and diacylethanol amine phosphatides	Apply bands of sample solution, overspray with 12% hydrochloric acid, leave in an atmosphere of nitrogen for 2 min and then dry in a stream of nitrogen; then chromatograph. The vinyl ether linkages in the phosphatides are hydrolyzed.	[35]
Flavone, cumarin and triterpene glycosides, solamargine, solasonine	Spray the sample zone with 10% ethanolic hydrochloric acid. Then expose to the vapors of conc. hydrochloric acid – ethanol (1 + 1) in a twin-trough chamber, heat to 50 to 55 °C for 4 to 5 h, dry at 90 °C for 3 min, spray with 50% ethanolic ammonia solution and finally activate at 100 °C for 5 min. Solasodine and the corresponding aglycones and sugars are produced.	[16]
Cardenolide glycosides	Apply bands of sample solution containing ca. 25 µg glycoside; cover the layer, apart from the application zones, with a glass plate and place in a chamber over the vapors of 37% hydrochloric acid, allow to react and chromatograph after the removal of excess hydrochloric acid.	[36, 37]

Table 11: (Continued)

Substances	Method, reagent and end products	References
Epoxides	Apply two 5 μl portions of 10% phosphoric acid, allow to dry for 20 min then apply the sample solution and dry for 1 h. Epoxides including trisubstituted epoxides are completely ring-opened.	[38]
Sulfonamides, 1,2-diacetylhydrazine, procaine, benzocaine, etc.	Apply samples and dry, place TLC plate in a twin-trough chamber with fuming hydrochloric acid and heat to 100 °C. Then remove the acid in a stream of cold air and chromatograph.	[39]
Alkaline hydrolysis		
α-Amyrin benzoate, lupeol acetate, tigogenin acetate	Apply the sample solution to an aluminium oxide layer, apply 7% ethanolic potassium hydroxide solution on top and spray with methanol – water (2 + 1), then store for 4 h at 55 °C over ethanol – methanol – dioxan (4 + 4 + 1), finally dry at 100 °C and develop. The 3β-alcohol is formed in each case.	[16]
n-Hexadecyl esters	Apply sample solution, followed by methanolic sodium hydroxide solution, warm and then chromatograph.	[40]
Phenylurea and *N*-phenylcarbamate residues	Apply sample solution, overspot with 7% methanolic potassium hydroxide; cover the start zone with a glass plate and heat to 170 °C for 20 min. Primary arylamines are produced.	[41]
Digitalis glycosides	Apply the sample solution as a band, then cover the layer, apart from the application zone, with a glass plate and place it in an ammonia chamber for 24 to 48 h; remove excess ammonia and chromatograph. Acetyl groups are split off.	[37]
Enzymatic cleavage		
Cytidine-diphosphate glucose	Buffered phosphate diesterase is applied on top of the sample, covered with parafilm and warmed to 23 °C for 45 to 60 min. Cytidine-5′-monophosphate and glucose-1-phosphate are formed.	[42]
Cytidine-5′-monophosphate, glucose-1-phosphate	Prostate phosphate monoesterase is employed to hydrolyze to cytidine, glucose and orthophosphate.	[42]

Table 11: (Continued)

Substances	Method, reagent and end products	References
Phosphatidylcholine	Apply phospholipase C solution as a band, dry, apply sample solution to enzyme band, stop reaction with hydrochloric acid vapor. sn-1,2-Diglycerides are produced.	[43]
Digitalis glycosides	Apply sample solution as band and then luizyme solution over it; if necessary, moisten the application zone with water. Cover the layer, except for the application zone, with a glass plate and incubate at 39 °C for 2 to 5 h.	[37]

3.1.3 Halogenation

The treatment of unsaturated substances with halogen leads to addition to these molecules. This is true not only of bromine and chlorine vapor but also of the less reactive iodine. Substitution also occurs in the presence of light. Examples of such halogenations are listed in Table 12. Figure 32 illustrates the characterization of fluorescein in a bubble bath preparation. Bromination of the fluorescein in the start zone yields eosin.

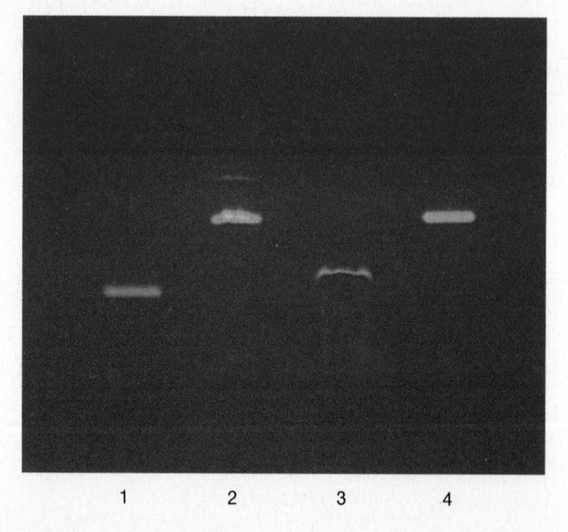

Fig. 32: Characterization of fluorescein in a foam bath by bromination. 1 = fluorescein, 2 = foam bath brominated, 3 = foam bath, 4 = eosin.

Table 12: Prechromatographic derivatization by halogenation.

Substances	Method, reagent and end products	References
Chlorination		
Cholesterol, glycyrrhetic acid acetate	Apply sample solution and moisten with anhydrous benzene, subject for 4 h to the vapors of thionyl chloride – benzene (1 + 1) in a desiccator, then dry and develop. A chlorinated cholesterol or the chloride of 3-β-acetoxyglycyrrhetic acid are formed.	[16]
Acetanilide, *p*-chloroacetanilide, 2,5-dichloroacetanilide	Apply sample solution and treat with chlorine vapor in the trough chamber for 20 s, then heat to 60 °C for 5 min in a ventilated drying cupboard. Various chlorination patterns are produced.	[44]
α-Bromoisovalerylurea, acetylbromodiethylacetylurea, caffeine, codeine phosphate	Generate atmosphere of chlorine with $KMnO_4$ and HCl. Expose TLC plate to chlorine for ca. 5 min. Remove excess chlorine completely by heating to 105 °C for 10 min. Spray with a mixture of 100 ml 0.5% benzidine solution in ethanol and 2 ml 10% KI solution. Caffeine and codeine phosphate only react on Al_2O_3 phases.	[209]
Bromination		
Cholestanol, cholesterol	Apply sample solution, treat with a 2 to 3-fold excess of 0.1% bromine in chloroform. Only cholesterol is derivatized.	[45]
Fluorescein	Apply sample solution, lead bromine vapor over it or apply 0.1% bromine in chloroform. Various intermediate bromination products are formed with eosin as the final product (Fig. 32).	[46]
Imperatorin	Apply sample solution, moisten with chloroform, place in the vapors of 10% bromine in chloroform and then dry and develop after an appropriate reaction time. Tribromoimperatorin is produced.	[16]
Sorbic acid	Treat with bromine solution or bromine vapor; di-, tri- and tetrabromocaproic acids are produced.	[47]
Capsaicinoids	Bromine vapor chamber: unsaturated capsaicinoids are completely brominated.	[48]
Phenylbutazone, prenazone	0.1% bromine in chloroform, 2 to 3-fold excess.	[49]

Table 12: (Continued)

Substances	Method, reagent and end products	References
Barbiturates, thiobarbiturates	Bromination, distinction between reacting and nonreacting barbiturates.	[50–52]
Iodination		
Pyridine, pyrrole, quinoline, isoquinoline and indole alkaloids	Apply sample solution and place the TLC plate in an iodine vapor chamber for 18 h, remove the excess iodine in a stream of warm air. Characterization on the basis of the iodination pattern.	[53]
Polycyclic aromatic hydrocarbons, naphthylamines	After application of the sample solution place the TLC plate in a darkened iodine vapor chamber (azulene a few minutes, PAH several hours). Then remove the excess iodine at 60 °C.	[20]
Dehydrated cholesterol	Apply sample solution; then place TLC plate in an iodine vapor chamber, blow off excess iodine. Di- and trimeric components are produced.	[54]
Phenolic steroids (estrone etc.)	Apply sample solution, then place TLC plate in an iodine vapor chamber. 2-Iodoestrone and 2,4-diiodoestrone are produced.	[55]

3.1.4 Nitration and Diazotization

Nitration and diazotization are often employed with the aim of producing colored, visually recognizable "derivatives" which are conspicuous amongst the majority of nonreacting compounds and can, thus, be specifically detected and investigated.

Aromatic nitro compounds are often strongly colored. They frequently produce characteristic, colored, quinoid derivatives on reaction with alkali or compounds with reactive methylene groups. Reduction to primary aryl amines followed by diazotization and coupling with phenols yields azo dyestuffs. Aryl amines can also react with aldehydes with formation of SCHIFF's bases to yield azomethines.

This wide range of reactions offers possibilities of carrying out substance-specific derivatizations. Some examples of applications are listed in Table 13.

Table 13: Prechromatographic derivatization by nitration and diazotization.

Substances	Method, reagent and end products	References
Nitration		
Polycyclic aromatic hydrocarbons (PAH)	Apply sample solution and dry. Place TLC plate for 20 min in a twin-trough chamber containing phosphorus pentoxide to which 2 to 3 ml conc. nitric acid have been added. PAH nitrated by nitrous fumes.	[20]
Phenols	Apply sample solution, dry, expose to nitrous fumes.	[56]
α-, β-Naphthol, 4-chloroaniline, chlorothymol, etc.	Apply sample solution and spray with 90% nitric acid, heat to 105 °C for 30 min, allow to cool and develop. Then reduce and diazotize.	[19, 57]
Marmesin, xanthotoxin	Apply sample solution and moisten with acetic acid and then expose to the vapors of conc. nitric acid and acetic acid (1 + 1) for 30 to 60 min at 55 °C in a desiccator. 6-Nitromarmesin and 4-nitroxanthotoxin are formed respectively.	[16]
Brucine	Apply a drop of conc. nitric acid to the spots of applied sample solution and allow to react for 15 min. Then activate the TLC plate at 120 °C for 15 min and develop after cooling.	[22]
Estrogens	Apply bands of sample solution, expose to ammonia vapor and dinitrogen tetroxide (from copper and conc. nitric acid), blow off excess and develop the nitroestrogens so formed. Detection by diazotization and coupling.	[58]
Diazotization and coupling		
o-, *m*-, *p*-anisidine, *o*-chloranil, 2,5-dimethoxyaniline	Apply bands of sample solution, spray with sodium nitrite in 1 mol/l hydrochloric acid solution and heat to 105 °C for 5 min. After cooling apply 5% α-naphthol solution and dry in a stream of warm air. Azo dyes are formed.	[60]
2,5-Dimethoxyaniline	Apply sample solution as spots; then apply diazonium chloride and α-naphthol solution and develop after 2 min.	[61]
Estriol	Dip the concentrating zone of a precoated HPTLC silica gel 60 plate in a saturated ethanolic solution of Fast Dark Blue R salt, allow the solvent to evaporate, apply the sample solution and dip once	[10, 59]

Table 13: (Continued)

Substances	Method, reagent and end products	References
	again into the reagent solution; dry the chromatogram and develop it.	
Estrone, estradiol, estriol	Dip silica gel foil 2 cm in saturated Fast Black Salt K solution and dry in a stream of warm air. Apply sample solution, dip again in reagent solution and dry. Dip the TCL plate 2 cm in 4% pyridine-cyclohexane solution, dry at 100 to 200 °C and develop the azo-dyestuffs that are formed.	[294]

3.1.5 Esterification and Etherification

Miller and Kirchner [9] and Mathis and Ourisson [1] have both already demonstrated that esterification at the start can be employed to distinguish primary, secondary and tertiary alcohols. Tertiary alcohols react much more slowly

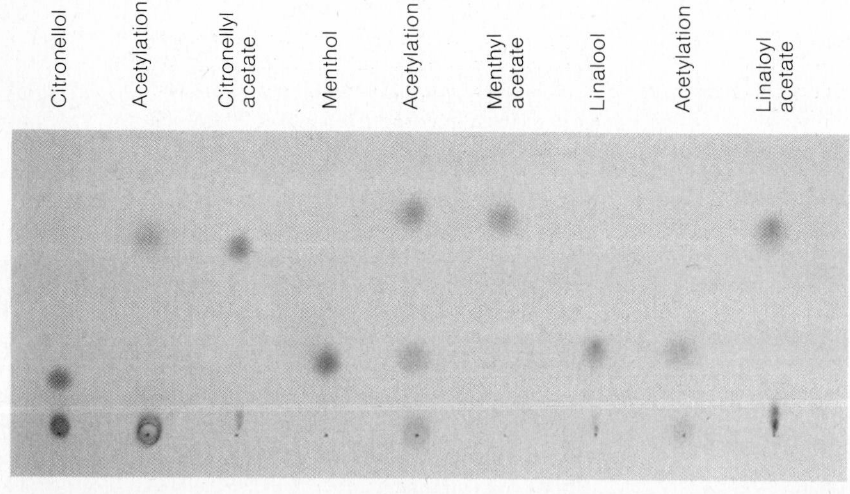

Fig 33: Differentiation of primary (citronellol), secondary (menthol) and tertiary alcohols (linalool) by in situ prechromatographic acetylation: citronellol reacts completely, menthol partially and linalool not at all.

or not at all and can, thus, be separated chromatographically from the esters that are formed (Fig. 33). Some examples are listed in Table 14. They reveal that acetic and trifluoroacetic anhydride have been employed almost exclusively for the esterification of alcohols, while acids have been esterified with diazomethane or sodium methylate.

Table 14: Prechromatographic derivatization by esterification and etherification.

Substances	Method, reagent and end products	References
Steroid sapogenins	Apply sample solution; then moisten with trifluoroacetic anhydride, dry and develop. Sapogenin trifluoroacetates are produced.	[62]
Aflatoxins	Apply extract and standard, then apply trifluoroacetic acid, allow to react at room temperature for 5 min then dry for 10 min at a max. of 40 °C and develop.	[63, 64]
Aflatoxins, ochratoxin A, sterigmatocystine, penicillic acid, patulin	Apply sample solution and dry; then apply trifluoroacetic anhydride; allow to react at room temperature for 45 min, develop. The derivatives of patulin and penicillic acid possess appreciably different hR_f values.	[65]
Ochratoxin A, citrinin, penicillic acid, sterigmatocystine, zearalenone	Apply extracts of cereals or fungal cultures; apply 50 µl pyridine – acetic anhydride (1 + 1) on top; remove the excess reagent in a stream of cold air and chromatograph. The reagents can also be applied via gas phase	[66]
Menthol, citronellol, linalool	Apply the alcohols followed by the acetylation mixture on the still damp spots. Repeated application of the reagent is necessary for complete reaction; heat to 100 °C for 15 min, then chromatograph.	[46]
Polyglycerol	Apply sample solution; then apply 60% acetic anhydride in anhydrous pyridine, heat to 95 to 100 °C for 15 min, allow the excess reagent to evaporate, develop.	[67]
Patulin	Apply sample solution, then apply acetic anhydride – pyridine (9 + 1), allow to react for 5 min and dry for 15 min in a stream of warm air.	[68, 69]
α-, β-Amyrin, 6-hydroxyflavones	Apply sample solution as band, followed by acetic anhydride – pyridine (6 + 1), warm to 40 °C for 1.5 h in a desiccator in an atmosphere of acetic	[16]

Table 14: (Continued)

Substances	Method, reagent and end products	References
	anhydride. Then heat to 100 °C for a few minutes. The corresponding acetates are produced.	
Alcohols	Apply sample solutions as bands followed by a suspension of sodium acetate in acetic anhydride – glacial acetic acid (3 + 1), then spray the plate lightly with acetic anhydride, allow to react for 3 h at 50 °C in a desiccator, evaporate off and develop. Here too the acetates are produced from the alcohols employed.	[16]
Sterols, triterpenoids (e.g. lupeol), primary and secondary alcohols	Apply sample solution followed by acetyl chloride and then remove the excess reagent in a stream of hot air.	[70]
Biogenic amines (e.g. serotonin)	Apply sample solution followed by benzoyl chloride solution (5% in toluene), dry and chromatograph.	[71]
Menthol, linalool, citronellol, geraniol, α-terpineol, cinnamic alcohol etc.	Apply 10% 3,5-dinitrobenzoyl chloride solution in p-xylene – tetrahydrofuran (15+2), followed by sample solution; allow to react; destroy excess reagent with 10% sodium hydroxide solution.	[72]
Carboxylic acids, organophosphoric acids	Apply sample solution followed by ethereal diazomethane solution, dry and develop.	[73, 74]
Alkyl and alkenyl acylglyceryl acetates	Apply sample solution, dry, blow on hydrochloric acid vapor; methylate with 2 mol/l sodium methylate solution.	[75]
Triglycerides	Methylation with 0.5 N potassium methylate solution in methanol (transesterification).	[76]
Peanut oil, glycerol phosphatides, cholesteryl esters etc.	Apply sample solution then spray with 2 mol/l sodium methylate solution, dry for 2 to 5 min and develop. Transesterification. Only ester linkages react and not acid amide linkages.	[77]
Oleanonic acid, 6-hydroxyflavone, xanthotoxol, glycyrrhetic acid acetate	Apply sample followed by 5% potassium carbonate solution in aqueous acetone, dry, apply 50% methyl iodide in acetone. Allow to react for 3 h at 50 °C in an atmosphere of methyl iodide – acetone (1 + 4), dry and develop. Oleanonic acid, for example, yields its methyl ester.	[16]
Phospholipids, free fatty acids	Apply sample solution, then 12% methanolic potassium hydroxide solution, keep moist with methanol for 5 min. Fatty acid methyl esters are produced, triglycerides do not react.	[7]

Table 14: (Continued)

Substances	Method, reagent and end products	References
Fatty acids, *n*-hydroxy acids, ursolic acid	Apply sample solution, followed by methanolic boron trifluoride solution, heat with a hot-air drier, allow to cool and develop.	[70]
Sorbic acid, benzoic acid	Apply sample solution in the form of a band, followed by 0.5% 4-bromophenacyl bromide in *N,N*-dimethylformamide. Heat to 80 °C for 45 min, dry and chromatograph.	[78]
Phenols	Apply sample solution and spray with saturated sodium methylate solution and then treat with 4% 2,4-dinitrofluorobenzene in acetone and heat to 190 °C for 40 min. Chromatograph the dinitrophenyl ethers so produced.	[79]

3.1.6 Hydrazone Formation

In order to characterize them and more readily separate them from interfering accompanying substances carbonyl compounds (aldehydes, ketones) can be converted to hydrazones at the start. The reagent mainly employed is 2,4-dinitrophenylhydrazine in acidic solution [70]. This yields osazones with aldoses and ketoses. Some examples are listed in Table 15.

Table 15: Prechromatographic derivatization by hydrazone formation.

Substances	Method, reagent and end products	References
Dipterocarpol, hecogenin, progesterone	Apply sample solution and moisten with 2,4-dinitrophenylhydrazine in acetic acid; then spray with acetic acid and store at room temperature or 55 °C for up to 1.5 h in a desiccator; then dry at 80° C and chromatograph.	[16]
Progesterone	Apply sample solution as a band followed by 0.1% ethanolic 2,4-dinitrophenylhydrazine (acidified with 0.1% conc. hydrochloric acid), allow to react at room temperature for 10 min and dry at 100 °C for 5 min.	[80]

Table 15: (Continued)

Substances	Method, reagent and end products	References
Steroid ketones	Apply the sample solution followed by GIRARD's reagent (0.1% trimethylacetyl hydrazide in 10% acetic acid) and allow to react for 15 h in an atmosphere of acetic acid. Then dry at 80 °C for 10 min and after cooling chromatograph the hydrazones that have been formed.	[81]
Aldehydes, ketones	Apply sample solution and moisten with 2 N 2,4-dinitrophenylhydrazine in acetic acid. After reacting, dry and chromatograph the 2,4-DNPH derivatives.	[14]
Carvone, menthone, acetophenone etc.	Apply an acidic solution of 2,4-dinitrophenylhydrazine, 4-nitrophenylhydrazine or 2,4-dinitrophenylsemicarbazide onto the previously applied sample solution. Aliphatic and aromatic hydrazones and carbazones can be differentiated by their colors.	[82]
Phenolic aldehydes	Apply sample solution and then acidic 2,4-dinitrophenylhydrazine solution; allow to react, dry and develop.	[15]
Chloro-, hydroxy-, and methoxybenzaldehyde derivatives	Derivatize with 2,4-dinitrophenylhydrazine solution in hydrochloric acid. Heat to 80 °C, cool and chromatograph.	[83]
p-Benzoquinone derivatives	Derivatization with 2,4-dinitrophenylhydrazine solution in hydrochloric acid. Heat to 80 °C, cool and chromatograph.	[84]
o-, *m*-, *p*-chloro-, 2,4- and 3,4-dichlorobenzoic acids	Apply *p*-bromophenacyl esters of the substances, followed by 0.5% 2,4-dinitrophenylhydrazine in 2 mol/l hydrochloric acid and allow to react for 10 to 15 min in the desiccator.	[85]

3.1.7 Dansylation

Prechromatographic dansylation has the advantage that chromatography separates excess reagent and also the fluorescent by-products (e.g. dansyl hydroxide) from the reaction products of the substances to be determined. In the case of postchromatographic dansylation the whole of the plate background fluoresces blue, so that in situ analysis is made more difficult.

Primary and secondary amines, amino acids and phenols react. In the case of long-wavelength UV light ($\lambda = 365$ nm) the DANS amides fluoresce yellow-green, while amines that have reacted at a phenolic OH group have an intense yellow to yellow-orange fluorescence. The detection limit for DANS amides is ca. 10^{-10} mol [86].

Acids can also be converted to fluorescent dansyl derivatives. The reaction of C_8 to C_{24} fatty acids with dansyl semipiperazide or semicadaveride provides an excellent example (Fig. 34) [87]. Odd-numbered and unsaturated fatty acids [88] and propionic, sorbic and benzoic acid [89] can be detected in the same manner.

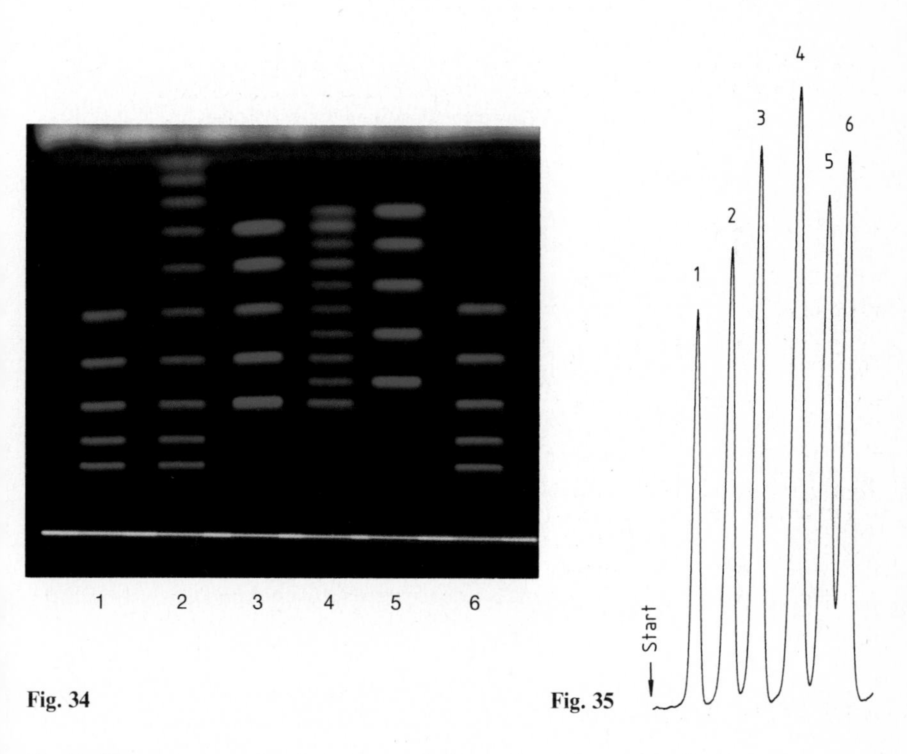

Fig. 34 **Fig. 35**

Fig. 34: Chromatogram of various even- and odd-numbered fatty acids after in situ derivatization with dansyl semicadaveride. The separations are with increasing R_f – track 1: C-24 to C-16; track 2: C-24 to C-6; track 3: C-20 to C-12; track 4: C-20 to C-11; track 5: C-19 to C-11; track 6: C-24 to C-16.

Fig. 35: Fluorescence scan of the dansyl semicadaveride derivatives of: 1 behenic acid, 2 erucic acid, 3 stearic acid, 4 oleic acid, 5 linoleic acid, 6 linolenic acid.

Table 16 lists some examples of dansylation.

Table 16: Prechromatographic derivatization by dansylation.

Substances	Method, reagent and end products	References
Carbamate and phenylurea herbicides	Treat the sample for 30 to 40 min with sodium hydroxide solution ($c = 1$ mol/l) at 80 °C. Apply sample solution and then 0.2% dansyl chloride over it. Cover with a glass plate and allow to react in the dark at room temperature for 60 min, then chromatograph.	[90]
Phenylurea herbicides	Treat sample with KOH, apply sample solution and overspot with dansyl chloride. Allow to react for 30 min in the dark at room temperature, then develop the chromatogram.	[91]
Metoxurone and degradation products	Phenylurea herbicides are first hydrolyzed to the corresponding aniline derivatives and then reacted at the start with 4 µl 0.25% dansyl chloride solution.	[92]
Urea herbicides, e.g. diurone, metoxurone, linurone	Apply sample solution to silica gel layer, cover with a glass plate, heat to 160 °C for 25 min and allow to cool. After this hydrolysis apply 0.2% dansyl chloride in acetone, cover with a glass plate and store in the dark for 1 h, then chromatograph.	[93]
Morphine, 6-monoacetylmorphine, morphine-6-nicotinate	Apply sample solution and apply to each spot 1 µl of dansyl chloride and two times 1 µl 8% sodium bicarbonate solution, allow to react for 7 min, dry at 70 °C and develop the chromatogram.	[94–96, 237]
β-Blockers	Apply the urine extract in the form of a band, overlay with 0.1% dansyl chloride in acetone followed by 8% aqueous sodium bicarbonate solution and heat to 120 °C for 15 min, allow to cool and develop the chromatogram.	[97]
Serum proteins, e.g. albumin	Allow to react at the start with 0.05% dansyl chloride solution in hexane, dry and expose for 10 h to the vapors of a triethylamine bicarbonate buffer ($c = 0.1$ mol/l; pH 8.5), then develop. Excess reagent and by-products run with the solvent front.	[98]
Even-numbered and odd-numbered fatty acids (C_6-C_{24})	Apply dansyl semipiperazide or dansyl semicadaveride solution as a 14 cm long band, followed by sample solution as short bands and then	[87, 88]

Table 16: (Continued)

Substances	Method, reagent and end products	References
	1% *N,N'*-dicyclohexylcarbodiimide solution as a 14 cm band. Dry and develop (Figs. 34, 35 and title picture).	
Preservatives (benzoic, sorbic, propionic acid)	Apply sample solution as spots or bands (3 to 4 mm) followed by dansyl semipiperazide and *N,N'*-dicyclohexylcarbodiimide solution, dry well and chromatograph.	[89]

3.1.8 Miscellaneous Prechromatographic Derivatizations

Other group-characteristic in situ prechromatographic reactions have been described in addition to the ones discussed above. They all serve to improve the characterization of the substances concerned and the selectivity of the subsequent chromatography. Table 17 provides an overview.

Table 17: Miscellaneous prechromatographic derivatizations.

Substances	Method, reagent and end products	References
Amino acids	Apply sample solution, dry, treat with 2,4-dinitrofluorobenzene solution. DNP-amino acids are produced, which are then separated chromatographically.	[99]
2-Anilino-5-thiazolinone derivatives of amino acids	Apply sample solution followed by heptafluorobutyric acid, heat to 140 °C for 10 min, rapidly cool to room temperature and develop the phenylthiohydantoins that are formed. No reaction occurs with, for example, threonine, serine, tryptophan or glutamic acid.	[100]
Amines	Apply sample solution and spray with 10% carbon disulfide in ethyl acetate. Place TLC plate in carbon disulfide vapor for 30 min, then spray with methanol – sulfuric acid (1 + 1). Heat to 100 °C for 10 min and chromatograph the isothiocyanates so formed.	[101]

Table 17: (Continued)

Substances	Method, reagent and end products	References
Amines	Apply sample solution then *p*-toluenesulfonic acid in pyridine, heat to 60 °C for 4 h, after cooling to room temperature develop the *p*-toluenesulfonates so formed.	[102]
Serotonin	Benzoylate the amino groups by overspotting at the start. This makes detection with GIBBS' reagent possible.	[103]
Sympathomimetics with free amino groups e.g. carbadrine, norfenefrine, noradrenaline, norephedrine	Apply sample solution (50 pg to 300 ng). Then apply fluorescamine (0.03% in acetone), dry and chromatograph.	[104]
Amines, Amino acids, peptides, e.g. tryptophan, tryptamine, peptides with terminal tryptophan groups	Apply indole derivatives dissolved in sodium borate buffer solution ($c = 0.2$ mol/l, pH 9.0) – ethanol (1 + 1). Dip TLC plate in fluorescamine solution to just above starting zone (15 s). Then dry at room temperature and develop. In case of indole amines followed by spraying with 40% perchloric acid.	[105]
Catecholamines	Apply sample solution followed by phosphate buffer ($c = 0.5$ mol/l, pH 8.0). Dip TLC plate into fluorescamine solution to just above starting zone, dry and develop. Blue fluorescence occurs after spraying with perchloric acid (70%).	[105]
Habituating drugs	Apply sample solution. Let it react with NBD-chloride or diphenylacetyl-1,3-indandion-1-hydrazone.	[236]
Desoxyribo-oligonucleotides, ribopolynucleotides	Layer: PEI cellulose. Complex formation of polyuridylic acid (6 mg/ml) with desoxyadenosine oligonucleotides.	[106]
Aliphatic and aromatic aldehydes	Apply sample solution, followed by 1% aniline in dichloromethane and chromatograph the SCHIFF's bases after 10 min.	[107]
Insecticides, e.g. eldrin, dieldrin, aldrin	Apply the sample solution, dry and then apply ethanolic zinc chloride solution, heat to 100 °C for 10 min, after cooling chromatograph the carbonyl compounds that have formed.	[108, 109]
Carveol, linalool, geraniol, α-terpineol, nerol etc.	Elimination of water with the aid of sulfuric acid and formation of the corresponding monoterpene hydrocarbons.	[9, 14]

Table 17: (Continued)

Substances	Method, reagent and end products	References
Alcohols	Apply sample solution, then nitrophenyl isocyanate solution (10% in benzene). Dry after reacting and develop.	[72]

Other possibilities are the reduction of nitro groups by applying the sample solutions to adsorbent layers containing zinc dust and then exposing to hydrochloric acid vapors [110]. 3,5-Dinitrobenzoates and 2,4-dinitrophenylhydrazones can also be reduced in the same way on tin-containing silica gel phases [111]. Cellulose layers are also suitable for such reactions [112]. SEILER and ROTHWEILER have described a method of "trans-salting" the alkali metal sulfates to alkali metal acetates [113].

3.2 Postchromatographic Detection

There is no difficulty in detecting colored substances or compounds with intrinsic fluorescence on TLC chromatograms. The same applies to components absorbing in UV light which have been separated on layers with incorporated "fluorescence indicators" and, hence, cause phosphorescence quenching in UV light so that the substances appear as dark zones on a bright emitting background.

Substances which do not exhibit such properties have to be transformed into detectable substances (derivatives) in order to evaluate the TLC separation. Such reactions can be performed as universal reactions or selectively on the basis of suitable functional groups. Substance-specific derivatization is practically impossible.

The aim of a postchromatographic derivatization is first

- the detection of the chromatographically separated substances in order to be able to evaluate the chromatogram better, visually.

But equally important are also

- increasing the selectivity, which is often associated with this and

- improving the detection sensitivity. In addition comes optimization of the subsequent in situ quantitation.

The separation is already complete when detection is undertaken. The solvent has been evaporated off, the substance is present finely distributed in the adsorbent. For a given amount of substance the smaller the chromatogram zone the greater is the concentration and, hence, the detection sensitivity. For this reason substances with low R_f values are more intensely colored than those present in the same quantity which migrate further.

Each reaction requires a minimum concentration for the detection to be possible at all. This concentration naturally varies from reagent to reagent (Fig. 36), so that every component cannot be detected to the same sensitivity with every reagent. This can lead to the appearance of a good separation that is, in fact, not good at all because the outer edge of a zone is not detected. So that the area occupied by the substance appears to be smaller than it actually is. The separation of the neighboring zones will, thus, seem better. It follows that the least sensitive reagents can counterfeit the "best separations".

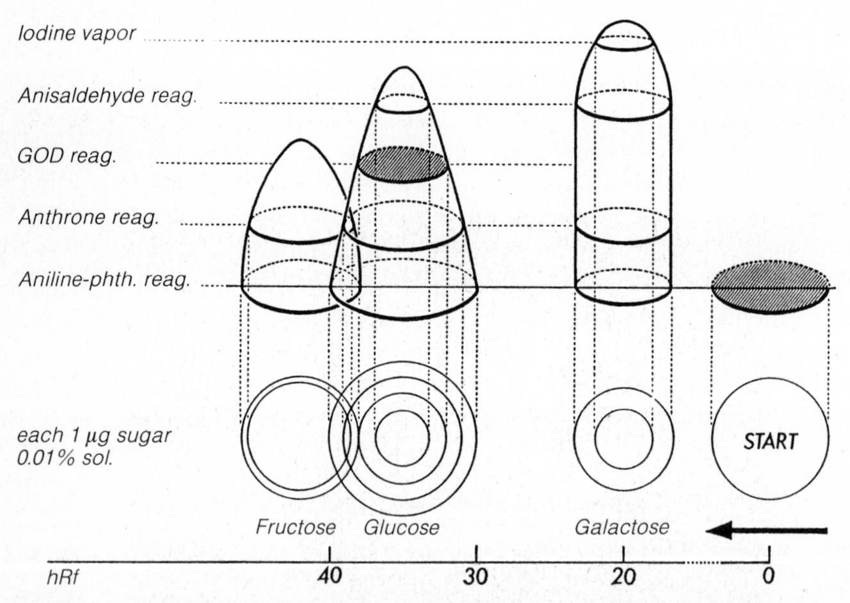

Fig. 36: Dependence of the area of the chromatogram zone at constant amount of applied substance (1 µg) on the reagent employed; top: relief representation; below: zone areas projected on one another. Iodine vapor reacts least sensitively here, aniline-phthalate most sensitively and the GOD reaction (glucose oxidase reaction) most specifically [216].

More sensitive detection methods and more objective recording methods (e.g. the employment of scanners) are constantly been striven for in order to overcome this illusion. It is for this reason too that fluorescent methods have been introduced to an increasing extent on account of their higher detection sensitivity. This allows an appreciable reduction in the amount of sample applied, so that possible interfering substances are also present in smaller quantities. This increases the quality of the chromatographic separation and the subsequent in situ analysis.

Various techniques are employed for applying the reagents to the TLC/HPTLC plate. The least satisfactory is spraying the reagent manually onto the chromatogram (Fig. 37). Dipping and evaporation methods are preferable with respect to precision and repeatability (Fig. 38). Methods have also been developed and described involving the addition of the reactants to the mobile or the stationary phase. These application techniques will be described below before discussing the influence of temperature on the reaction.

3.2.1 Spraying

Until a few years ago the most common method of rendering colorless substances on chromatograms visible was to spray them with reagent solutions [115]. An all-

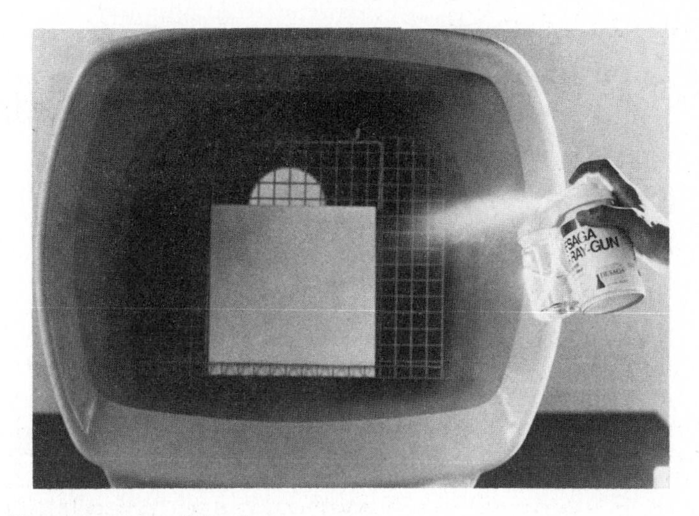

Fig. 37: Manual spraying of the chromatogram.

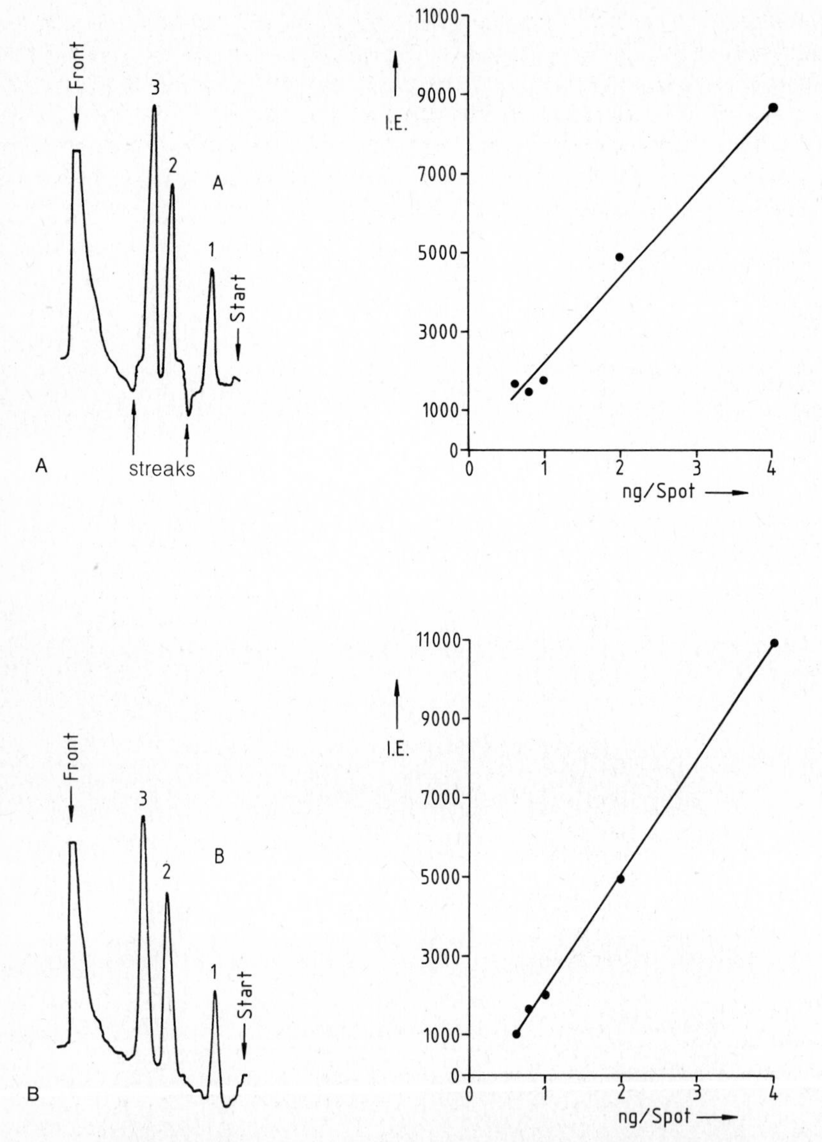

Fig. 38: Comparison of manual dipping (A) with mechanized dipping (B) on the basis of scans and calibration curves [114]. – 1 = *cis*-diethylstilbestrol, 2 = *trans*-diethylstilbestrol, 3 = ethinylestradiol. Scanning curve 2 ng of each substance per chromatogram zone. λ_{exc} = 313 nm, $\lambda_{fl} > 390$ nm. Dipping solution: water – sulfuric acid – methanol (85 + 15 + 1).

glass sprayer was normally employed (Fig. 39A), which was connected to a pressure supply (membrane pump) or cylinder of inert gas. It was necessary to use a jet so fine adjusted that it was possible to spray the reagent solution homogeneously. Spraying was carried out at a pressure of 0.6 to 0.8 bar from a distance of 20 to 30 cm in a suitable fume cupboard (Fig. 37). According to WALDI the spray should be applied in a meandering pattern with the return point of the spray outside the track of the chromatogram (Fig. 39B) [115].

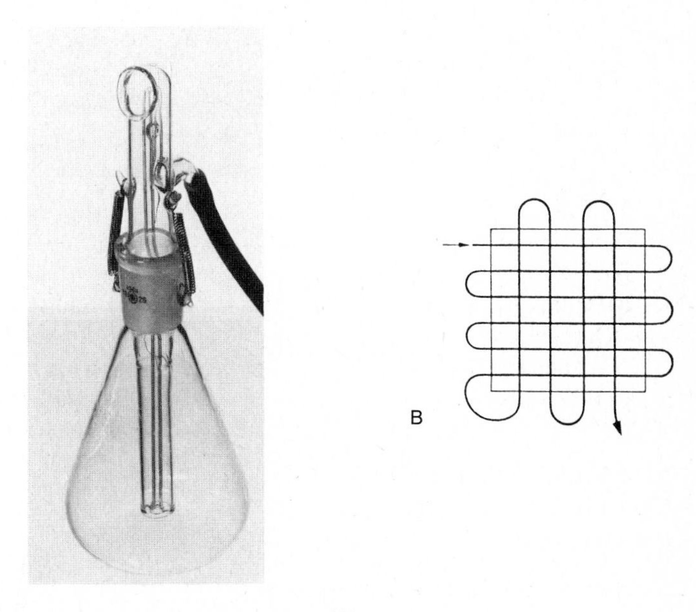

Fig. 39: All glass sprayer (A), spray pattern (B).

The same applies to the use of spray pistols (spray guns and aerosol cans), the frequency of whose use ought probably to be reduced on account of the propellant gas (chlorofluorohydrocarbons) employed. Manual depression of the button valve of the vertically held spray can "shoots" the propellant gas through a fine jet and drags the sucked-up reagent solution with it onto the vertically held chromatogram (water pump principle).

Because when operated manually the spray unit can never be moved so uniformly that the chromatogram is homogeneously covered with reagent and the amount of reagent applied to the TLC plate differs for each individual, SPITZ [116], KREUZIG [117–119] and others have developed and marketed automatic sprayers. In the first case the sprayer is moved over the fixed, stationary TLC plate. The second type of automatic sprayer works with a fixed spray jet and the TLC plate is moved by a motor.

Neither apparatus has, as yet, found general use in the laboratory, probably because too little attention was paid to the differing viscosities, surface tensions and polarities of the various solvents. Thus, in practice, these automatic sprayers repeatedly produced "sprinkled" zones because, in addition, to a fine mist individual larger drops also reached the zones and, thus, caused inhomogeneities of coverage and reaction and also because substances at the surface of the layer, where they are to be found after chromatography with readily volatile mobile phases [120–124, 128], are displaced in the direction of the supporting backing, so that in extreme cases detection from the rear is more sensitive than from the front.

The spraying of TLC/HPTLC plates should always be undertaken in a well-ventilated fume cupboard, so that the aerosols, some of which are damaging to health and aggressive, are not breathed in and the place of work is not contaminated. After the spraying is complete the spraying apparatus and fume cupboard should be cleaned with care so that undesired reactions do not occur with later reagents. Manganese heptoxide and perchloric acid-containing reagents, sodium azide and iodine azide solutions should never be sprayed as they can cause explosions in the exhaust ducts of fume cupboards. For the same reasons such reagents should only be made up in small quantities. In all these cases it is preferable to apply the reagent by dipping the chromatogram into it.

3.2.2 Dipping

It is becoming ever more usual to dip the solvent-free chromatograms into a suitable reagent solution [114]. The reasons for this are:

- The coating of the adsorbent layer with the reagent solution is more homogeneous than with even the most carefully carried out spraying process.
- The distribution of the reagent is no longer influenced by the manual dexterity of the operator, the performance of the spray apparatus, the viscosity of the reagent or the drop size of the spray mist.
- It is only since this method has come into use that the precise quantitative analysis of thin-layer chromatograms has become possible. For the increased

regularity of the wetting results in a baseline with less structure (Fig. 38B). This means that the detection limits are appreciably lower than they are in the case of sprayed chromatograms. The reproducibility of the result is also appreciably improved on account of the homogeneity of the reagent application (Tab. 18).

- The consumption of reagents is less particularly when series investigations are made (when the reagent is used repeatedly it is usual to cover the dipping chamber with a stainless steel lid).
- The contamination of the place of work with reagents that may be injurious to health or corrosive, is considerably less when dipping than when spraying.
- The complex spray facilities with integrated fume cupboard are unnecessary.

In spite of these acknowledged advantages, low volume dipping chambers and automated, time-controlled dipping apparatus (Fig. 40 and 41) have only been available commercially for a few years.

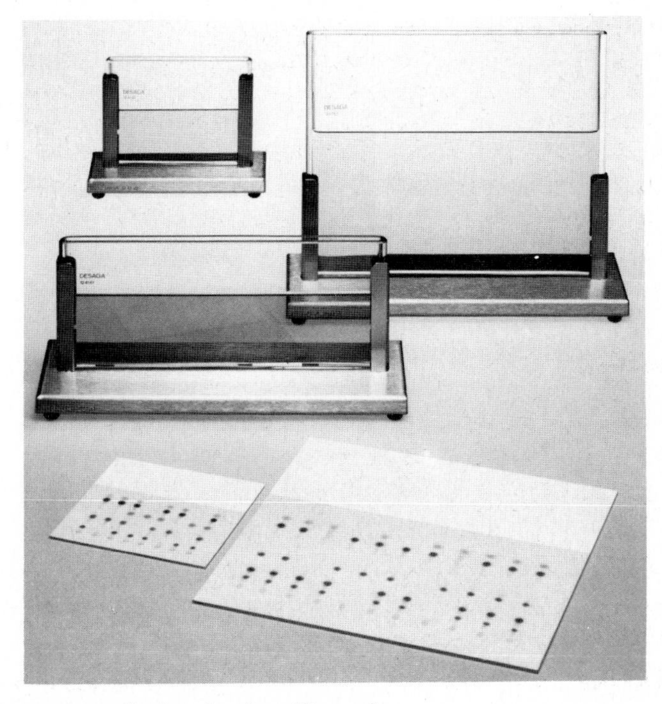

Fig. 40: Low volume dipping chambers (DESAGA).

Fig. 41: Automated dipping devices (DC Tauch-fix, BARON).

The advantage of such dipping apparatus is that the insertion and removal of the chromatogram is performed at a uniform speed and the time of immersion can be set as necessary. Interfering "ripple marks", such as are observed on manual dipping, do not occur. Care must be taken, however, to clean off the back of the

Table 18. Statistical comparison (F-test [125]) of the methods. Standard deviation s_{XO} of the calibration curves for diethylstilbestrol and ethinylestradiol [114].

Procedure	*trans*-diethylstilbestrol s_{XO} [ng per zone]	Ethinylestradiol s_{XO} [ng per zone]
Manual dipping	0.37	0.21
Mechanical dipping	0.096	0.044
Statistical difference (F-test)	significant	significant

TLC plate which is wetted with reagent solution when it leaves the dipping bath before laying it on the hotplate, laboratory bench or scanning stage.

The dipping solutions described in Part II of this book are usually less concentrated than the corresponding spray solutions. The solvents employed are specially chosen for their suitability to the special requirements of dipping solutions. Water, which on the one hand, can sit on the surface of RP plates and not penetrate them and, on the other hand, can cause disintegration of water-incompatible layers is usually replaced by alcohol or other lipophilic solvents.

In general care should be taken in the choice of solvent to ensure that neither the chromatographically separated substances nor their reaction products are soluble in the solvent of the dipping reagent.

It is probable that the solvents given in the individual reagent monographs are not suitable for all the substances with which the reagent will react. This point should be taken into account especially for quantitative work and the user should make appropriate modifications. In particular, there must be no loss of substance or reaction product by dissolution (formation of "comet tails" by the chromatographic zones).

When the plate is inspected the color intensity of the chromatogram zones must be more intense at the top surface of the layer than it is when viewed from the back of the TLC/HPTLC plate. If this is not the case the reagent must be made less polar to avoid a frontal development across the thickness of the layer.

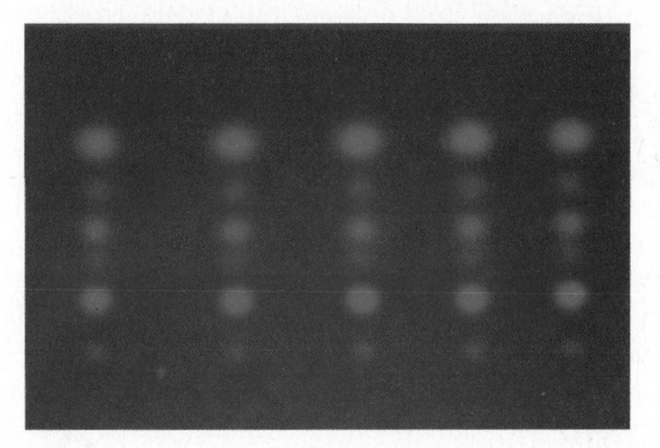

Fig. 42: Chromatogram of polycyclic aromatic hydrocarbons on caffeine-impregnated precoated silica gel 60 HPTLC plates with concentrating zone (MERCK). The following can be recognized in increasing R_f value. – 1. benzo(ghi)perylene, 2. indeno(1,2,3-cd)pyrene, 3. benzo(a)pyrene, 4. benzo(b)fluoranthene, 5. benzo(k)fluoranthene, 6. fluoranthene.

The times of immersion of the chromatogram in the reagent bath are usually short (< 5 s [126]) in order to avoid dissolving the substances out of the stationary phase. This is easily achieved if the Tauch-Fix (Fig. 41) is employed. The chromatogram is then laid horizontally and dried in a stream of air.

The dipping unit can also be employed to impregnate adsorbent layers. It is easy in this way to produce tungstate- [127] or silver nitrate-impregnated layers for separating oligosaccharides or unsaturated compounds.

FUNK et al. [128a] dipped silica gel plates in a 4% solution of caffeine in order to separate six polyaromatic hydrocarbons relevant in monitoring the quality of potable water (Fig. 42).

Such a dipping apparatus can also be employed with advantage for applying substances to preserve or intensify fluorescence after chromatography or derivatization is complete (cf. Section 3.2.7.3).

3.2.3 Exposure to Vapors

A layer can also be homogeneously treated with the reagent by exposure to its vapor. This is frequently carried out in a twin-trough chamber (Fig. 43A) or in a special conditioning chamber (Fig. 43B). In the case of the twin-trough chamber, for example, the reagent is placed in one of the troughs and the dried chromatogram plate in the other. Thin-layer chromatograms can be treated in this manner with the vapors of a large number of reagents.

Iodine vapor allows nonspecific, usually nondestructive detection of many substances (e.g. surface active agents [129], pharmaceuticals [130, 131], polyethylene glycols [132], see also Table 12). In addition, reactions have also been described with the vapors of bromine [133–135], cyanogen bromide [136], chlorine [137–141, 209], ammonia [142–147] (see also the reagent "Ammonia vapor"), diethylamine [148], ammonium hydrogen carbonate [149, 150], acids [145, 151–156] (see also reagent "Hydrochloric Acid Vapor"), *tert*-butyl hypochlorite [203], sulfuryl chloride [157, 158], sulfur dioxide [159, 160] oxides of nitrogen [161–169], hydrogen sulfide [170], formaldehyde [171–176], glyoxylic acid [177] and silicon or tin tetrachlorides [178].

SMITH [203] has described a special procedure for "distilling" reagents homogeneously onto a TLC plate. RIPPHAHN [179] later started with a TLC plate that had been dipped into ninhydrin solution, laid a 0.1 mm thick terephthalate

A

B

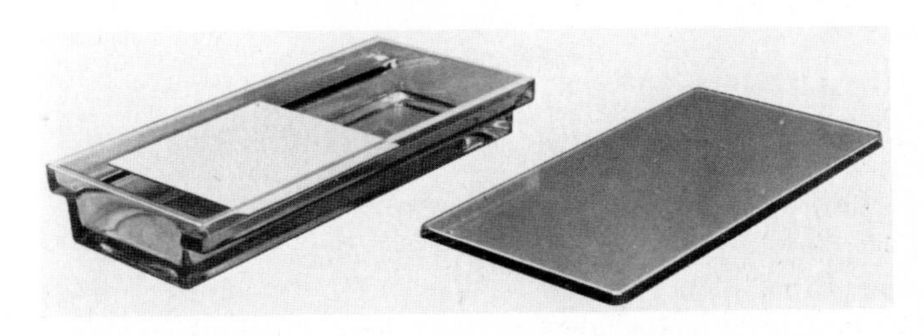

Fig. 43: Twin-trough chambers (A) and conditioning chamber (B) (CAMAG).

film frame round its edges and positioned the chromatogram bearing the amino acids to be detected so that the layer faced the "reagent plate". This "sandwich" was then placed, reagent side down, on a hotplate. The ninhydrin then "distilled" homogeneously onto the chromatogram and stained the amino acids (see reagent Ninhydrin). It is also possible to apply *tert*-butyl hypochlorite [203], sulfuric acid and ammonia vapor [180] or acetophenone [181] in an analogous manner.

MARTINEK [182] has described the reverse procedure for relatively volatile substances (e.g. essential oil components), where the compound to be detected is "distilled" onto the reagent plate and reacts with the reagent there.

PANDEY et al. [183] employed this idea of a sandwich configuration to transfer substances from one TLC plate to another for two- or multidimensional separations.

3.2.4 Reagent in Solvent

Another method of applying a reagent homogeneously to a TLC layer is to add it to the mobile phase. A necessary precondition is that the reagent is evenly spread over the layer (the reagent must "run" with the solvent front). Double developments have frequently been described, where the first mobile phase brought about the development and the second one an improvement in zone shape and a homogeneous application of reagent to the stationary phase.

The following are amongst the reagents that have been reported as being added to the mobile phase: acids for quinine alkaloids [184], ninhydrin for amino acids [185–187], fluorescamine for biogenic amines [188]. Fluorescein sodium [189], dichlorofluorescein [190], rhodamine 6G [191], ANS reagent [192] and bromine [193] have all been described as additives to mobile phases.

Dimethyl sulfoxide in the mobile phase acts as an "intrinsic detector" for certain phenols (e.g. dihydroxybenzenes) [194] on layers that have been treated with tin tungstate.

3.2.5 Stationary Phase as Reagent (Reagent in Adsorbent)

The exploitation of specific adsorbent properties can also lead to the same goal of homogeneous derivatization of separated substances.
Silica gel and aluminium oxide layers are highly active stationary phases with large surface areas which can, for example, – on heating – directly dehydrate, degrade and, in the presence of oxygen, oxidize substances in the layer. This effect is brought about by acidic silanol groups [93] or is based on the adsorption forces (proton acceptor or donor effects, dipole interactions etc.). The traces of iron in the adsorbent can also catalyze some reactions. In the case of testosterone and other Δ^4-3-ketosteroids stable and quantifiable fluorescent products are formed on layers of basic aluminium oxide [176, 195].

Derivatization can also be brought about by impregnating the adsorbent layer with a suitable reagent (frequently an inorganic substance) before application of the sample solutions. This impregnation is usually performed as the layer is prepared. Here care must be taken that the reagent does not dissolve in the mobile phase and migrate towards the solvent front in the subsequent development. Such transport towards the solvent front is especially to be expected in the case of organic reagents. It is for this reason that organic components are mainly not employed for later derivatization but for homogeneous detection by fluorimetric processes (cf. Sec. 2.2.2). Examples of inorganic reagents in the adsorbent are listed in Table 19.

Table 19: Derivatization by inorganic reagents incorporated in the layer.

Reagent	Substance detected, matrix	Reaction conditions, remarks	Reference
Ammonium sulfate	triglycerides, serum lipids	25–85 min at 150 °C; yields fluorescent derivatives	[197]
Ammonium sulfate	phosphatidyl glycerol derivatives	charring on heating	[198]
Ammonium sulfate	phosphatidyl glycerol, sphingomyelin	10 min at 280 °C densitometric in situ quantitation	[199]
Ammonium sulfate	detergents	SIL G 25 detergent plate (MACHEREY-NAGEL)	[200, 201]
Ammonium sulfate	lipids	charring on heating	[202]
Ammonium sulfate	triolein, oleic acid, androsten-3,17-dione, xanthanonic acid, cholesterol-propionate, N-methylphenylalanine, D-glucose	fluorescence after heating to 150–180 °C; exposure to *tert*-butyl hypochlorite	[203]
Ammonium monovanadate	organic nitrogen compounds	TLC plates prepared with 2% ammonium monovanadate solution	[204]
Aluminium oxide	deoxynivalenol in wheat	7 min at 120 °C; yields a fluorescent derivative under UV light ($\lambda = 365$ nm)	[193, 196]
Silver nitrate	1,2- and 1,4-dihydroxybenzene	oxidation to corresponding benzoquinones	[208]
Silver nitrate + phosphotungstic acid + cobalt nitrate	triglycerides, saturated and monounsaturated alcohols	heat to 230 to 250 °C	[205, 206]
Phosphomolybdic acid	essential oil components	stabilization of the silver nitrate-impregnated adsorbent layer	[207]
Zirconium(IV) oxychloride	estrone, 17-β-estradiol and estriol; plasma lipids	heat to 150 to 180 °C for 5 min. Fluorescent zones are produced – sometimes only after heating for longer period.	[178]

3.2.6 Sequences of Spraying or Dipping

It is desirable, in some cases, to apply different reagents to the developed chromatograms consecutively – e.g. with intermediate drying, heating or evaluation after each reagent application. It is possible in this way to detect and identify quite different substances and substance classes on the chromatogram, particularly in the case of complex mixtures of substances. Such series application of reagents increases the selectivity of the process and allows better differentiation of substances and groups of substances.

Series application of reagents has been particularly employed in toxicological analyses in cases of intoxication or drug abuse [277–279]. However, it can be a disadvantage in the use of reagents in series that the detection evidence produced by one reagent may be disturbed by the application of a later one. It is, therefore, necessary to carefully examine and document the chromatogram after the application of each reagent. Some typical reagent series are discussed in a later Volume.

3.2.7 Processing the Chromatogram

Fortunately in recent years there has been more consideration of the direct quantitative analysis rather than concentrating just on the question of substance-specific detection and increasing detection sensitivity. So that the reagents are no longer regarded merely as "visualizers" for chromatogram zones (qualitative evidence) but are used deliberately as tools to perform reproducible, stoichiometric reactions, which are a suitable basis for quantitative analyses. This means that it is necessary to pay greater attention to the processing of the chromatogram. This begins with the drying of the chromatogram after development and the homogeneous application of the reagent and continues with the reaction and stabilization of the products of reaction.

It is known that not all reactions proceed in the same manner on all adsorbent layers because the material in the layer may promote or retard the reaction. Thus, Gänshirt [209] was able to show that caffeine and codeine phosphate could be detected on aluminium oxide by chlorination and treatment with benzidine, but that there was no reaction with the same reagent on silica gel. Again the detection of amino acids and peptides by ninhydrin is more sensitive on pure cellulose than it is on layers containing fluorescence indicators [210]. The NBP reagent (*q.v.*) cannot be employed on Nano-Sil-C_{18}-100-UV_{254} plates because the whole of the plate background becomes colored.

The reasons for the above phenomena are to be found in differing configurations of hydrogen bonds, the effect of pH, differences in the structures of fluorescence indicators and binders and differences in surface area. For example, silica gel 60 possesses a surface area of 500 m^2/g [211] while that of Si 50 000 lies below 5 m^2/g [212].

In the case of substances whose structures are pH-dependent (e.g. phenols, carboxylic and sulfonic acids, amines etc.) it is possible to produce fluorescences or make them disappear by the deliberate manipulation of the pH [213] (Table 20). Shifts of the positions of the absorption and emission bands have also been reported. This is particularly to be observed in the case of modified silica gels, some of which are markedly acidic or basic in reaction (Table 25).

Table 20: Some substances, that change the color of their fluorescence at moderate pH range [214].

Compound	Color change		pH range
	from	to	
Fluorescein	weak yellow	yellow	4.0... 5.0
2′,7′-dichlorofluorescein	weak yellow	yellow	4.0... 6.0
Resorufin	colorless	orange	4.0... 6.0
Acridine	green	blue	4.5... 6.0
Quinine	blue	violet	5.9... 6.1
Thioflavine	colorless	green	6.5... 7.6
Umbelliferone	orange	blue	6.5... 8.0
4-Methyl umbelliferone	weak blue	blue	6.5... 8.0
2-Naphthol	weak blue	blue	7.0... 8.5
Morin	weak green	green	7.0... 8.5
1-Naphthol	colorless	blue green	7.0... 9.0
Cumarin	weak green	green	8.0... 9.5
Acridine orange	weak yellow-green	yellow	8.0...10.0

3.2.7.1 Drying the Chromatogram

After a chromatogram has been developed the TLC plate is removed from the developing chamber and the status quo is fixed by removing the mobile phase remaining in the layer as quickly as possible. This is properly performed in the fume cupboard so as not to contaminate the laboratory with solvent fumes. If possible the TLC plate should be laid horizontally because then as the mobile phase evaporates the separated substances will migrate evenly to the surface where they can be the more readily detected. A fan or hair dryer (hot or cold air stream) is often employed to increase the gradient of the solvent vapor over the surface of the layer. It must, however, be checked whether this evaporation affects the substances, for

- essential oil components may evaporate and produce mists in the direction of the air stream, some of which may redeposit on the active layers producing single-sided, fuzzy zone boundaries;
- oxygen-sensitive components may be destroyed by drying at elevated temperatures;
- particles of dust from the laboratory air can deposit on the chromatograms and may possibly affect the following analysis;
- chemical vapors are transported in the air stream onto the activated layers.

For these reasons many research groups prefer to dry the chromatograms in a vacuum desiccator with protection from light. Depending on the mobile phase employed phosphorus pentoxide, potassium hydroxide pellets or sulfuric acid can be placed on the base of the desiccator, to absorb traces of water, acid or base present in the mobile phase.

A further but also more time-consuming advance is to employ the AMD system (CAMAG). Here the mobile phase and mobile phase vapor is sucked out of the chamber and from the TLC plate after every development. This reduces to a minimum the contamination of the place of work with possibly injurious solvent vapors.

3.2.7.2 Effect of Heating after Application of Reagent

Almost all chemical reactions proceed more rapidly at elevated temperatures than in the cold and so it is recommended that the chromatogram treated with reagent be heated. Irradiation with UV light (high-pressure mercury lamp, Fig. 44) also promotes reaction. Heating to 100 to 120 °C for 5 to 10 min is often sufficient to ensure complete reaction. However, the pyrolysis of organic compounds requires temperatures of 200 to 250 °C, which would probably result in the whole background of the plate being darkened, because the binder in the layer also chars. The evaporation of volatile reagent during heating is avoided by laying on a covering plate [215].

Since it still is not a simple matter to heat a TLC plate really homogeneously there is a danger of reaction inhomogeneities on the plate. The usual types of apparatus employed for heat production and transfer are drying cupboards, hotplates and IR sources. The success obtained using microwaves has been modest up to now.

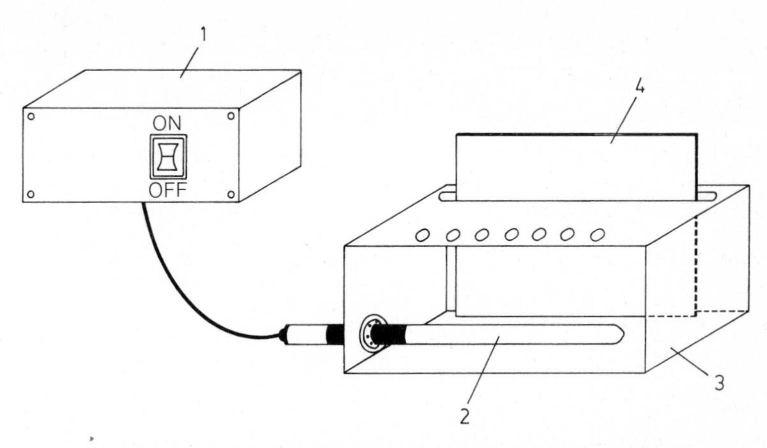

Fig. 44: Apparatus for irradiation with UV light. – 1 excitation apparatus (Heraeus, Hanau; OSRAM StE 501), 2 UV lamp (TNN 15-31001721), 3 housing, 4 TLC plate.

Drying Cupboards

Drying cupboards are often employed for heating chromatograms after they have been treated with reagents. The TLC plate should not be placed directly on the perforated shelf of the cupboard, since the heat transfer rate is appreciably greater where contact is made with metal than where the contact is with air and the pattern of the holes on the shelf would become visible on the chromatogram. There is, thus, the danger that the reaction of the substances would be dependent on the position on the chromatogram and, hence, the reproducibility of direct quantitative analysis would suffer.

It has been found to be better to set the TLC plate down vertically on the bottom of the drying cupboard and to lay the glass upper edge against the wall of the cupboard. In cupboards with air circulation the chromatogram should be "suspended" in the air and only supported at the corners in an analogous manner to the column in a gas chromatograph oven. It has been suggested that homogeneous heating be achieved by attaching the TLC plate to a turntable which is continuously rotated inside the oven. The results obtained with such a "carousel" are reported to be good, but the method has not come into general use at least until now.

Hotplates

Hotplates (Fig. 45) are coming into increasing use for heating chromatograms. They have the advantage that it is possible to follow the reaction visually and the

A

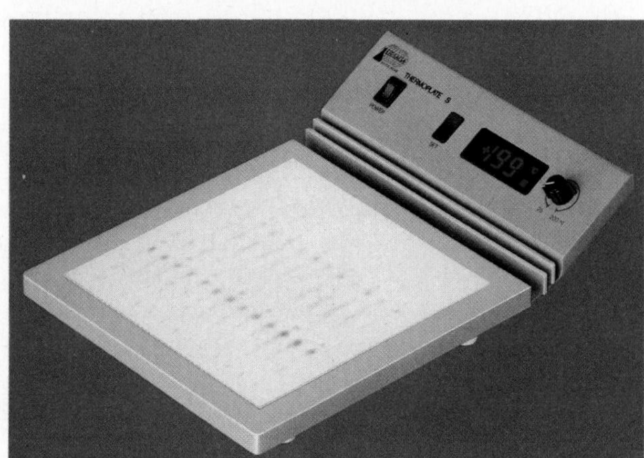

B

Fig. 45: Commercial hot plates. A) CAMAG, B) DESAGA.

reagent vapors can escape directly (fume cupboard!). The TLC plate is removed as soon as the optimal color development is produced.

Hotplates can normally be regulated over a temperature range of 30 to 190 °C. The temperature set should be maintained to an accuracy of within 2 °C, but this is not usually achieved in practice. Figure 46 shows the results of a representative range of measurements, where the effective temperature was determined as a function of the temperature setting. The temperatures were determined by means of 25 thermal sensors, which had previously been checked against each other. They were distributed over the hotplate according to the pattern shown in Figure 46B.

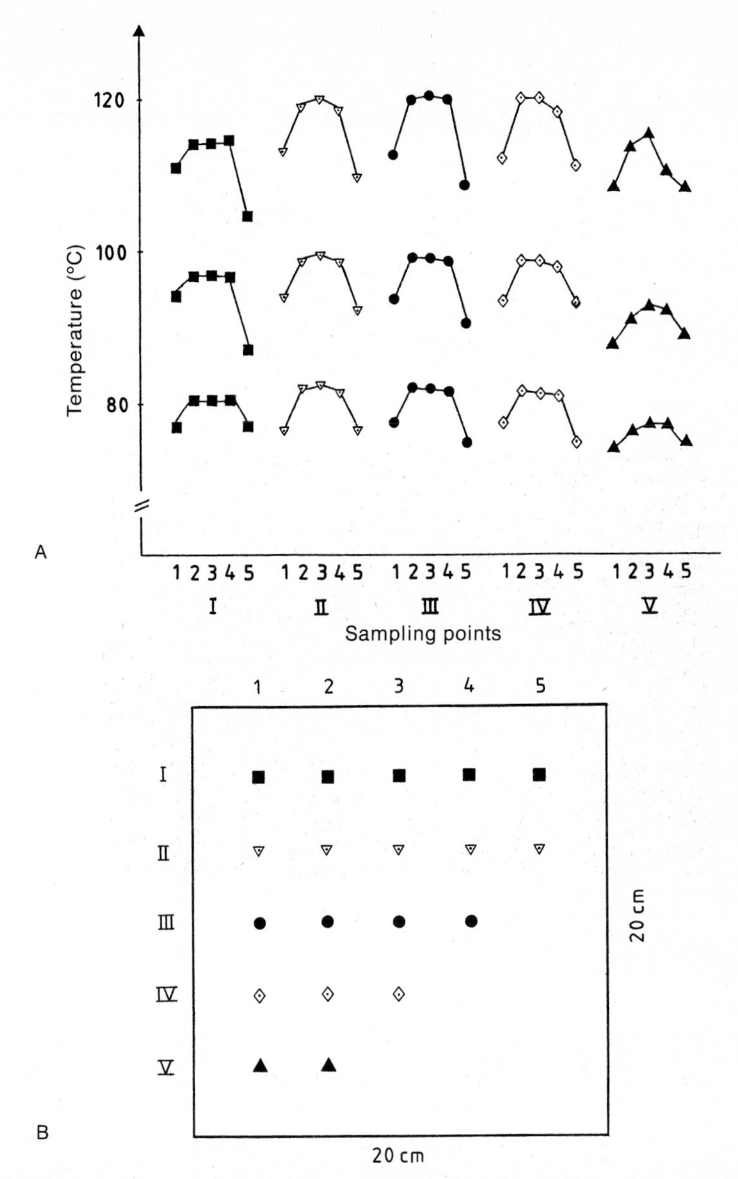

Fig. 46: Evaluation of the suitability of a hot plate for TLC by determination of the temperature distribution, A) results of 25 thermal elements at temperature settings of 80 °C, 100 °C and 120 °C, B) pattern of measuring points in five tracks (I–V) each with five measuring points.

At low temperatures the average temperatures calculated from the individual measurements corresponded to the temperature setting. They were appreciably lower at higher temperatures and it was found that the temperature setting corresponded to the highest temperature that could be reached in the individual measurements. It was also evident that the edge of the hotplate was colder than the middle, i.e. the effective measured temperature was not the same everywhere over the surface of the hotplate; a homogeneous temperature distribution is most likely to be found in the center of the plate.

In the derivatization of sugars with aniline-diphenylamine reagent for example, this leads to unsatisfactory irregular coloration. The standard deviation for the method deteriorates from 2 to 3% to 5 to 8%. For this reason color reactions should be avoided for direct quantitation if it is possible to scan in the UV range without derivatization.

IR Sources

Infrared sources are sometimes employed to heat thin-layer chromatograms. The chromatogram is laid on an insulating foil and irradiated from above at a distance of 10 to 20 cm by quartz spirals which are heated to 800 °C. Contact difficulties with the base naturally do not play any rôle here. But there is usually no regulation of the rate of heating so that such lamps can only be satisfactorily employed for pyrolysis investigations. They are insensitive to acid vapors and other aggressive reagents.

Microwave Apparatus

The methods of heating TLC/HPTLC plates described above depend on thermal conduction, convection or radiation. Microwave heating involves a special form

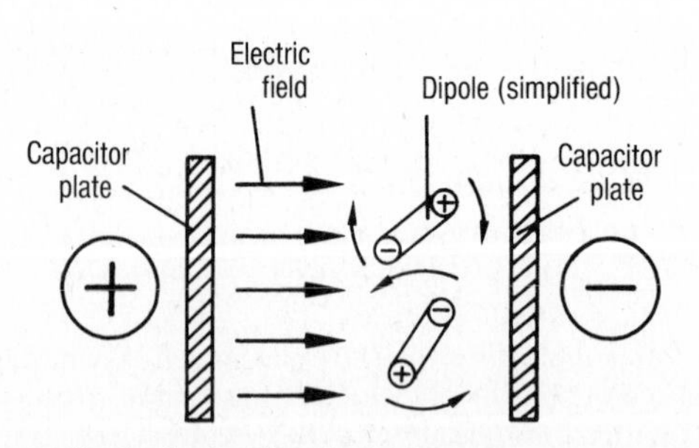

Fig. 47: Orientation of dipoles in the electric field [220].

of dielectric heating [217]. Here frictional heat is produced in an nonconducting or only slightly conducting body by a high frequency electromagnetic field. This frictional heating is a result of the fact that molecules with an intrinsic or induced (in the case of adsorbed molecules) dipole structure continually seek to align themselves to the alternating field (Fig. 47). Table 21 lists the dipole moments μ of some substances. The more symmetrically the charge is distributed in the molecule the smaller is the dipole moment.

Table 21: Dipole moments of some substances.

Substance	Dipole moment μ [D]*	Substance	Dipole moment μ [D]*
Chlorine gas, carbon dioxide	0	D-Ribose	5.1
Ammonia	1.48	D-Galactose	5.3
Sulfur dioxide	1.61	Glycine	13.3
Methanol	1.67	Alanine	17.5
Water	1.84	Myoglobin (whale)	155
Acetone	3.7	Oxyhemoglobin, carboxyhemoglobin	400
D-Glucose	4.7		

* 1 D (Debye) = $3.33 \cdot 10^{-3}$ cm

The frequency of microwave radiation lies between that of IR radiation and high frequency radio waves and the boundaries between these regions are not fixed [221]. The microwaves are generated in a transmitter (magnetron) which possesses a "stalk" which penetrates like a radio antenna into a hollow energy guide (Fig. 48). This leads the electromagnetic waves into the reaction chamber (power about

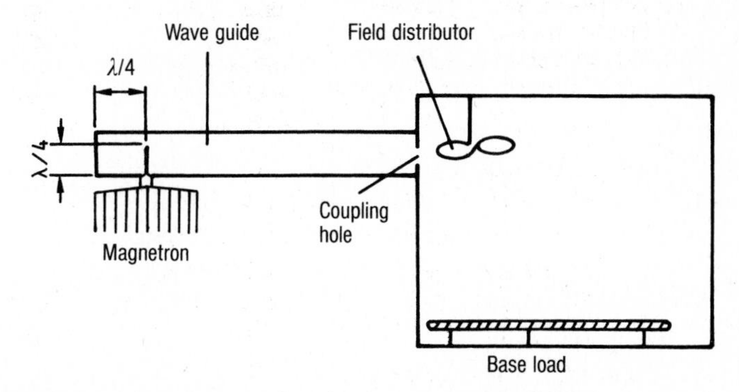

Fig. 48: Schematic representation of a microwave apparatus [222].

600 W). At the entrance an aperture focuses the microwaves on a rotating metal propeller (wave mixer), which distributes the radiation throughout the reaction chamber and prevents the setting up of standing waves which would result in uneven heating of the TLC plates [217, 223].

Modern apparatus are equipped with a rotating table to accommodate the TLC plate above the base of the reaction chamber. This means that microwaves can penetrate the TLC plate from below through the glass plate or plastic film. Aluminium foil backings are not suitable! They reflect the radiation and high potentials are built up between the aluminium foil and the wall of the reaction chamber; these result in electrical discharges.

So long as water is present in the adsorbent layer, the temperature does not rise above 100 °C. However, microwave heaters will even perform pyrolyses when this has evaporated.

It follows from the depth of penetration of the microwaves, which is calculated from the formula $d = \frac{c}{f \cdot \sqrt{\varepsilon}}$ (f = frequency, ε = dielectric constant of the adsorbent, c = velocity of propagation of electromagnetic waves in vacuo) [225], that practically the same temperature is reached at every depth in the TLC layer during microwave heating. This is an advantage compared with hotplates. In addition, the reaction is completed more rapidly (after 1 – 2 min) than in the drying cupboard. In spite of these advantages microwave heaters have found scarcely any application in TLC analysis.

3.2.7.3 Stabilization of Developed Zones

Treatment of the chromatogram with a reagent results in the production of colored or fluorescent chromatogram zones, which are used to evaluate the success of the separation and for quantitative analysis. For this purpose it is necessary that the color or fluorescence intensities remain stable for about 30 minutes.

There are no general recommendations applicable to the stabilization of the color of *colored chromatogram zones* apart from that of storing the chromatograms in an atmosphere of nitrogen and protected from light until they are evaluated. There are naturally other color stabilization methods which are applicable. Well known is the addition of cadmium or copper salts in the case of the ninhydrin reagent (*q.v.*) and spraying with sodium nitrite solution after the van URK reaction for lysergic acid derivatives [226]. The blue color of tryptamine after reaction with 2,6-dibromoquinone-4-chloroimide (*q.v.*) can also be stabilized, this time by exposing to ammonia vapor or spraying with ammonia solution [103].

In the case of *fluorescent chromatogram zones* there is also, in addition to storage of the chromatogram in the absence of light and oxygen mentioned above (Fig. 49 and 50), another method of stabilization, namely treatment of the chromatogram with viscous lipophilic or hydrophilic agents. These evidently reduce the ease with which parts of the molecules rotate and keep out the laboratory air. Singlet oxygen, which is the primary agent in the oxidative degradation of substances, is not

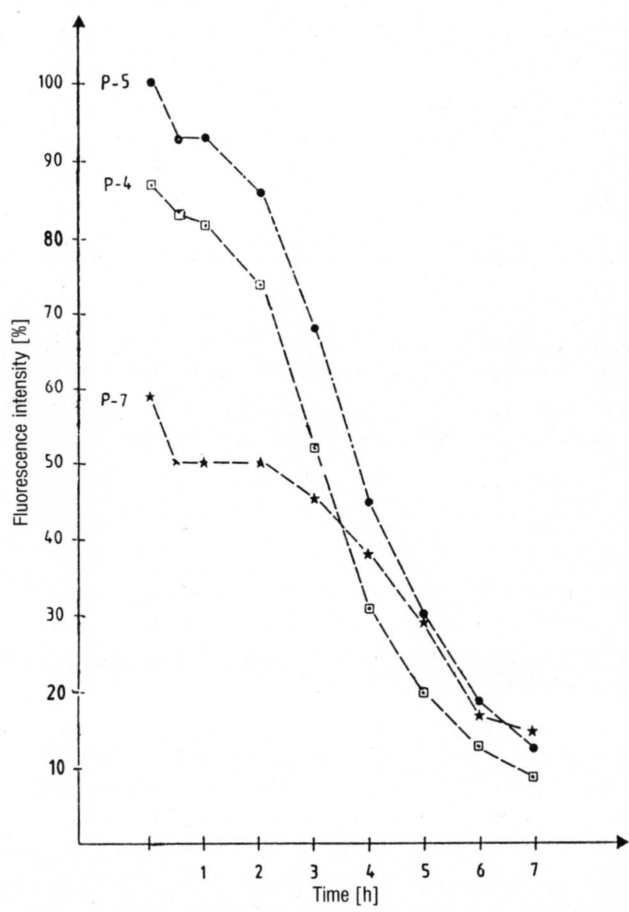

Fig. 49: Reduction in the fluorescence of coproporphyrin (P-4), pentaporphyrin (P-5) and heptaporphyrin (P-7) as a function of time (storage of chromatogram in the air and in daylight).

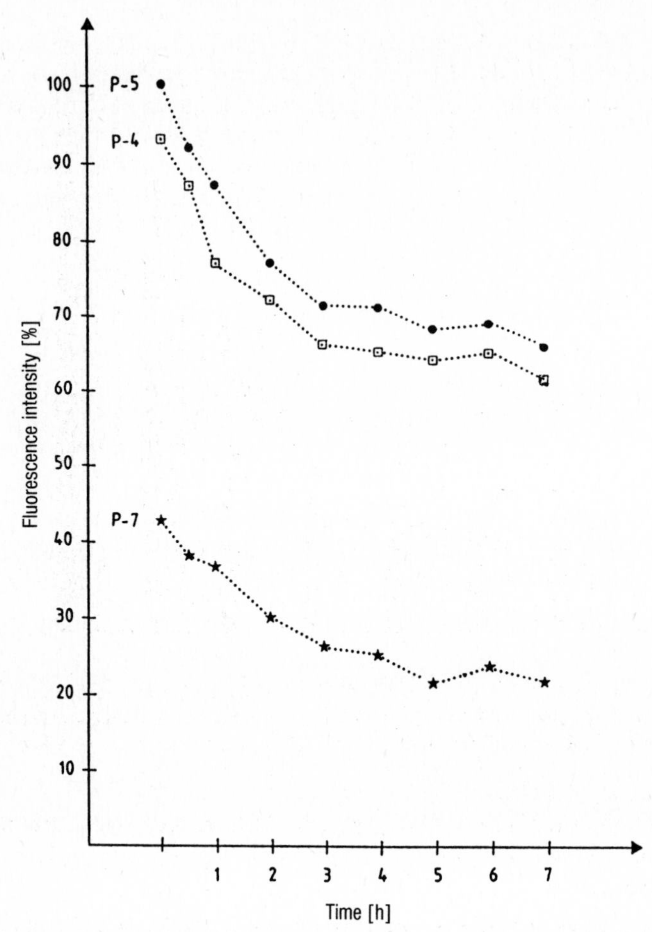

Fig. 50: Reduction in the fluorescence of coproporphyrin (P-4), pentaporphyrin (P-5) and heptaporphyrin (P-7) as a function of time (storage of chromatogram in the dark).

excluded [227], but it is converted to less aggressive triplet oxygen during transport through the lipophilic phase.

In 1967 spraying with a solution of paraffin wax allowed the recording of the fluorescence spectrum of anthracene directly on the TLC plate without any difficulties [228]. HELLMANN too was able to stabilize emissions by the addition of 2% paraffin to the solvent [229]. Low concentrations evidently serve primarily to *stabilize the fluorescence* – this "stabilization concentration" extends up to ca.

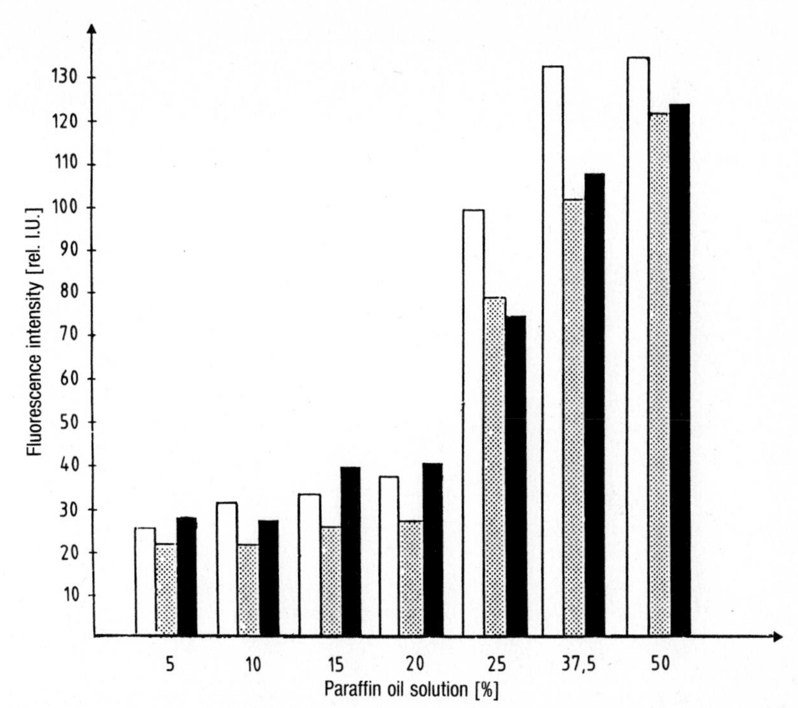

Fig. 51: Fluorescence intensities of porphyrins as a function of the concentration of the paraffin oil dipping solution: ■ mesoporphyrin, ▩ coproporphyrin, □ pentaporphyrin.

20% in the case of porphyrins (Fig. 51). If the chromatograms are dipped in more concentrated solutions the fluorescence yield suddenly jumps. It is not clear whether micelle formation plays a rôle in this intensification.

Figure 49 shows that porphyrins are decomposed in the layer within a few hours if no special measures are taken, but that they can be stabilized for more than 24 hours if the layers are dipped in 50% paraffin solution and stored in the dark. This was true of all six porphyrins investigated (Fig. 52). Quantitation should not be undertaken less than an hour after dipping the chromatograms, because it takes so long for the fluorescence emission to stabilize [230].

Similar fluorescence-stabilization has been reported for polyethylene glycol 4000 by Wintersteiger [291].

Funk et al. [231] have demonstrated that the sensitivity of the analysis can be extended down into the femtogram range for the determination of selenium in water and biological matrices (Fig. 53).

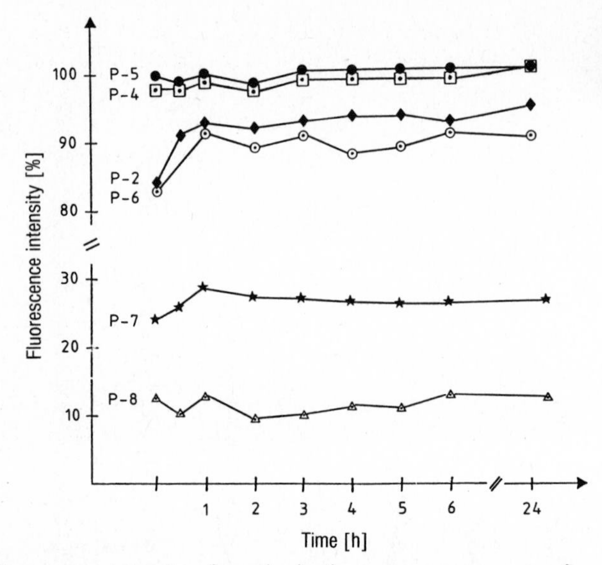

Fig. 52: Fluorescence intensity of porphyrin chromatogram zones as a function of time after dipping in 50% liquid paraffin solution and storage in darkness. –
P-2: mesoporphyrin, P-4: coproporphyrin, P-5: pentaporphyrin, P-6: hexaporphyrin, P-7: heptaporphyrin, P-8: uroporphyrin.

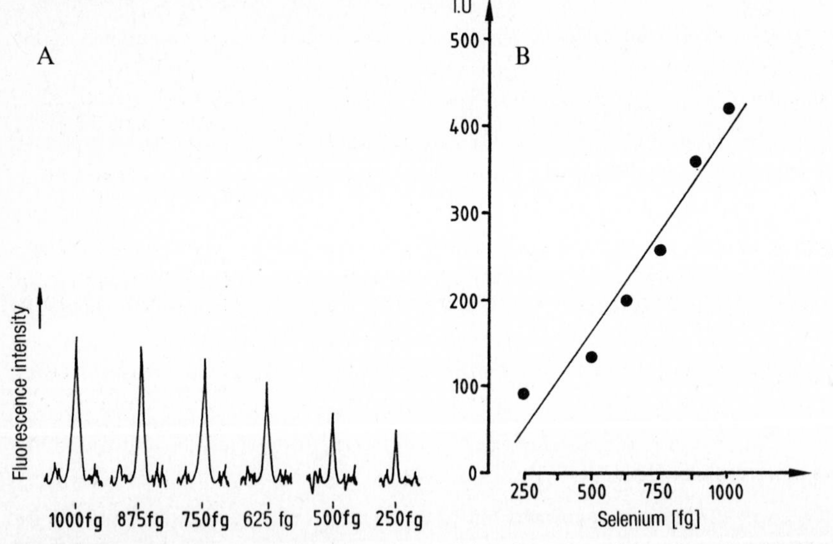

Fig. 53: Fluorescence scan of femtogram quantities of 2,1,3-naphthoselenodiazole (A) and associated calibration curve (B).

Further examples of fluorescence stabilization and intensity augmentation as a result of treatment of the chromatogram with viscous, lipophilic liquids are listed in Table 22. The alteration of the pH [293] or the addition of organic acids or bases [292] have also been found to be effective. WINTERSTEIGER [291] has also described the effect that the TLC layer itself (binder) can influence the fluorescence intensity.

Table 22: Lipophilic fluorescence intensifiers and their areas of application.

Fluorescence intensifier	Substances	Sensitivity increase/ stabilization	Remarks/references
Isooctane	porphyrin methyl esters	reactivation of faded fluorescence	benzene and petroleum ether can be employed in the same way [233, 289]
Dodecane	polycyclic aromatic hydrocarbons	2 to 2.5-fold	50% in *n*-hexane, appreciable time-dependant zone enlargement [234]
Palmitic and stearic acids	aflatoxins B_1 and B_2	15 to 35%	unseparated fatty acids intensified the fluorescence when investigating corn extracts [224]
Kerosine	porphyrins	appreciably	[235]
Fomblin Y-Vac	polycyclic aromatic hydrocarbons	up to 100-fold	35% in 1,1,2-trichloro-trifluoroethane [234]
Fomblin H-Vac	1-aminopyrene	stabilization and enhancement	50% in 1,1,2-trichloro-trifluoroethane [238]
Liquid paraffin	benzo(a)pyrene	35-fold	spray solution, 67% in *n*-hexane; the fluorescence is stable for more than 10 h [245]
Liquid paraffin	aflatoxins	3 to 4-fold	dipping solution, 33% in *n*-hexane, investigation of fungal nutrients [239]
Liquid paraffin	aflatoxin B_1	2.5-fold	spray solution, 67% in *n*-hexane [245]
Liquid paraffin	aflatoxins, sterigmatocystine	10 to 100-fold	foodstuff investigations [65]

Table 22: (Continued)

Fluorescence intensifier	Substances	Sensitivity increase/ stabilization	Remarks/references
Liquid paraffin	selenium as 2,1,3-naphthoselenodiazole	25-fold	spray solution, 67% in *n*-hexane (Fig. 53 [240])
Liquid paraffin	carbamate and urea herbicides	stabilization and enhancement	spray solution, 20% in toluene, water investigation [241]
Liquid paraffin	digitalis glycosides	stabilization and enhancement	30% in chloroform [242]
Liquid paraffin	digitoxin, digoxin, methyldigoxin	enhancement	the chromatogram was coated with a film of paraffin liquid [243]
Liquid paraffin	luteoskyrin, rugulosin	stabilization	spray solution, 50% in toluene – ethyl acetate – formic acid (16 + 4 + 1) [244]
Liquid paraffin	Δ^4-3-ketosteroids	10-fold	dipping solution, 33% in *n*-hexane [10]
Liquid paraffin	Δ^4-3-ketosteroids (testosterone isonicotinic hydrazone, testosterone dansyl hydrazone)	> 10-fold	dipping solution, 33% in *n*-hexane [232]
Liquid paraffin	cholesterol, coprostanone, coprostanol etc.	2 to 8-fold	dipping solution, 33% in *n*-hexane [246]
Liquid paraffin	estriol as dansyl derivative	10-fold	dipping solution, 67% in *n*-hexane [247]
Liquid paraffin	cortisol as dansyl hydrazone	10-fold	dipping solution, 67% in *n*-hexane, serum investigations [248, 249]
Liquid paraffin	dansylamides	10-fold, stabilization > 10 h	spray solution, 67% in *n*-hexane [245]
Liquid paraffin	amiloride	80% increase	dipping solution, 33% in cyclohexane, human plasma investigation [250]
Liquid paraffin	fluphenazine	enhancement	dipping solution, 5% in toluene, plasma investigation [251]

Table 22: (Continued)

Fluorescence intensifier	Substances	Sensitivity increase/ stabilization	Remarks/references
Liquid paraffin	piroxicam	enhancement	dipping solution, 10% in *n*-pentane, urine, tissue and plasma investigations [252]
Liquid paraffin	gentamycins, netilmicin	50 to 65% enhancement	dipping solution, 15% in *n*-hexane [253]
Liquid paraffin	gentamycins	stabilization	dipping solution, 33% in *n*-hexane [254]
Liquid paraffin	morphine as dansyl derivative	stabilization	dipping solution, 20% in *n*-hexane [291]
Liquid paraffin – triethanolamine	carbamazepine	30-fold	chloroform – liquid paraffin – triethanolamine (60 + 10 + 10) [255, 256]
Liquid paraffin – triethanolamine	cis/trans-diethylstilbestrol, ethynylestradiol	3-fold	chloroform – liquid paraffin – triethanolamine (60 + 10 + 10) [257]
Liquid paraffin – triethanolamine	vitamin B_1	2-fold	chloroform – liquid paraffin – triethanolamine (60 + 10 + 10) [258]
Silicone DC 200	sterigmatocystine	10-fold	18% in diethyl ether, cheese investigation [259]

Hydrophilic liquids can also cause stabilization and amplification of fluorescence. Thus, DUNPHY et al. employed water or ethanol vapor to intensify the emissions of their chromatograms after treatment with 2′,7′dichlorofluorescein [260]. Some groups of workers have pointed out that the layer material itself can affect the yield of fluorescent energy [261 – 263]. Thus, polyamide and cellulose layers were employed in addition to silica gel ones [245]. The fluorescence yield was generally increased by a factor of 5 to 10 [264], but the increase can reach 100-fold [234, 265].

Some examples of fluorescence amplification with the aid of hydrophilic liquids are listed in Table 23.

Table 23: Hydrophilic fluorescence intensifiers and their fields of application.

Fluorescence intensifier	Substances	Sensitivity increase/ stabilization	Remarks/references
Ethylene glycol	furosemide and its metabolites	no information	10% citric acid in ethylene glycol – water (1 + 1); plasma investigations [266]
2-Ethoxyethanol	gramine	stabilization	6% in mobile phase [267]
Glycerol	ethoxyquin, dansyl amides	20-fold	spray solution, 33% in methanol [292] or 50% ethanol [245]
Polyethylene glycol 400	cysteine adducts of α-,β-unsaturated aldehydes as dansyl hydrazones	stabilization and enhancement	dipping solution, 25% in chloroform [268]
Polyethylene glycol 400	compounds with alcoholic OH groups	20 to 25-fold	dipping solution, 10% in methanol [269]
Polyethylene glycol	primary amines, indole derivatives, sympathomimetics	stabilization and enhancement	dipping solution, 20% in methanol [270]
Polyethylene glycol 4000	alcohols, amines	stabilization and enhancement	spray solution, 50% in methanol; best results on silica gel [271]
Polyethylene glycol 4000	flavonoide glycosides	no information	dipping solution, 5% in ethanol, phytochemical investigations [273]
Polyethylene glycol 4000	silymarin	enhancement	5% in ethanol; optimum for fluorescence after 24 h [274]
Polyethylene glycol 4000	primary, secondary, tertiary alcohols as anthracene-urethane derivatives	stabilization and enhancement	saturated dipping solution in methanol [275]
Polyethylene glycol 4000	compounds with alcoholic OH groups	20 to 25-fold	dipping solution, 10% in methanol [269]
Polyethylene glycol 4000	cetanol after reaction with 8-bromomethyl-benzo-d-pyrido(1,2-a)pyrimidin-6-one	stabilization > 15 d	dipping solution, 10% in chloroform [291]

Table 23: (Continued)

Fluorescence intensifier	Substances	Sensitivity increase/ stabilization	Remarks/references
Triethylamine	phenylurea and N-phenylcarbamate pesticides	stabilization and enhancement	spray solution, 10% in dichloromethane; 15 min delay before analysis [41]
Triethylamine	aminoglycoside antibiotics	stabilization	spray solution, 10% in dichloromethane; investigation of solutions for injection [276, 278]
Triethylamine	amino acids as fluorescamine derivatives	stabilization	spray solution, 10% in dichloromethane; employed before and after reaction with fluorescamine [280]
Monoethanolamine	dansyl amides	10-fold	spray solution [245, 272]
Triethanolamine	dansyl amino acids, dansyl amides	enhancement	spray solution, 20% in 2-propanol; fluorimetric evaluation after 16 h storage in vacuo [281]
Triethanolamine	gramine	enhancement	20% in water; sodium hydroxide solution ($c =$ 1 mol/l) or 20% monoethanolamine solution can also be employed [267].
Triethanolamine	ephedrine, effortil and estriol as dansyl derivatives	stabilization > 100 min	20% in 2-propanol [282]
Triethanolamine	spermine, spermidine	no information	spray solution, 20% in 2-propanol; fluorimetric evaluation after 16 h storage in vacuo [283]
Triethanolamine	N-nitrosamines	no information	spray solution, 10% in dichloromethane [284]
Triethanolamine	thiourea	enhancement	spray solution, 20% in 2-propanol [285]
Triethanolamine	carbamate and urea herbicides as dansyl derivatives	stabilization	20% in 2-propanol, soil investigations [286, 287]
Triethanolamine	matacil, zectran	stabilization	spray solution, 20% in 2-propanol [288]

Table 23: (Continued)

Fluorescence intensifier	Substances	Sensitivity increase/ stabilization	Remarks/references
Triethanolamine	propham, chloropropham, swep, linurone, maloron	stabilization and enhancement	spray solution, 20% in 2-propanol, water investigations [241]
Triton X-100	dansyl amides and amino acids	30 to 110-fold	spray solution, 20% in chloroform [265, 290]
Triton X-100	ethoxyquin (antioxidant in spices)	> 200-fold, stabilization > 15 h	spray solution, 33% in benzene; the fluorescence of aflatoxin B_1 is reduced by 10 to 15% [292].
Triton X-100	polycyclic aromatic hydrocarbons	10-fold	1% solution in *n*-hexane; optimal emission after 60 min; 10% zone enlargement [234]
Triton X-100	testosterone dansyl hydrazone	> 25-fold	dipping solution, 20% in chloroform [232]
Triton X-100	selenium as 2,1,3-naphthoselenodiazole	90-fold	dipping solution, 20% in chloroform (Fig. 53 [231])
Sodium dodecylsulfate, cetyltrimethylammonium chloride, sodium cholate, β-cyclodextrin	dansylated amino acids and polycyclic aromatic hydrocarbons	> 45-fold	1% in water; the greatest enhancement of fluorescence is that of sodium cholate on pyrene [263]
Dioctyl sulfosuccinate	codeine, morphine, monoacetyl-morphine, heroin	stabilization	dipping solution, 20% in ethanol [94]

Although there is ample experimental evidence confirming the stabilization and amplification of fluorescence by means of viscous lipophilic and hydrophilic liquids there is as yet no convincing physicochemical explanation of the phenomenon. Wetting phenomena and pH changes could also play a rôle alongside solubilization phenomena and micelle formation in the liquid film. As a nonionogenic detergent Triton X-100 possesses wetting properties, so that here a surface-active effect can be brought into the discussion. The amines that have been employed certainly displace the pH. Further investigations are required to clarify the phenomenon.

3.3 Biological-Physiological Methods of Detection

Just like the physical and microchemical methods of detection, the indirect, biological-physiological detection procedures are very selective when applied to thin-layer chromatography. Here it is not chemical functional groups or particular physical properties that are selectively detected but effects on highly sensitive "biodetectors". The following detection techniques have been employed:

- *Manual transfer* of the chromatographically separated substance to the "detector". These include, for example, the detection of antibiotically active substances, plant and animal hormones, mycotoxins, insecticides, spice and bitter principles and alkaloids. The frequency distribution of their employment is shown in Figure 54 [295].
- *Bioautographic determinations*, where test organisms, tissue homogenates or cell organelles are applied in agar or gelatine solution as detectors directly on the surface of the developed chromatogram. The detection of antibiotics, fungicides, saponins, vitamins etc. have been described using this method.
- *Reprint methods* where the developed dried chromatogam is laid on the prepared agar layer "detector" with the exclusion of air bubbles. In this and in the

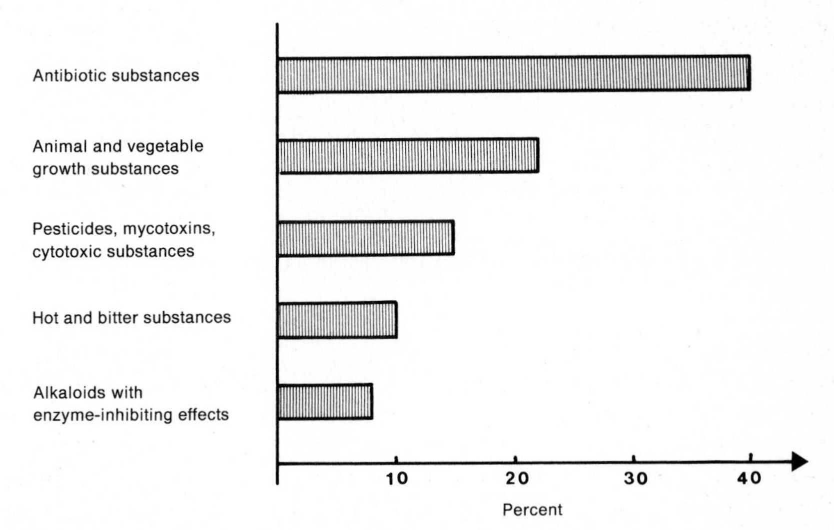

Fig. 54: Fields of application and frequency distribution of biological-physiological detection methods.

previous method the "active agents" diffuse out of the TLC layer into the 1 to 4 mm thick agar detector layer and there promote or inhibit the growth of the indicator organism during the period of incubation.

- *Enzymatic determinations of the detection limit* where the chromatograms are first sprayed with an enzyme solution. Then after appropriate incubation the enzymatically altered components are detected by reaction with a suitable reagent.

These methods are naturally subject to the degree of variation usual in biology and a degree of manual dexterity is often required. Their employment in TLC is, nevertheless, justified because

- these methods are highly specific (Fig. 36),
- inactive accompanying substances do not interfere,
- the detection limits of the compounds are sometimes so low that it is scarcely possible to obtain better results by the classical methods of physical chemistry without resorting to enrichment.

These topics will be treated in depth in Volume 3.

References

[1] Mathis, C., Ourisson, G.: *J. Chromatogr.* **1963,** *12*, 94–96.
[2] Kaess, A., Mathis, C.: *Chromatogr. Electrophor. Symp. Int. 4th, 1966*, Ann Arbor Science Publishers, Michigan 1968.
[3] Hamann, B. L., Martin, M. M.: *Steroids* **1967,** *10*, 169–183.
[4] Funk, W.: *Mitt. Gebiete Lebensm. Hyg.* **1982,** *73*, 139–154.
[5] Dünges, W.: *Prächromatographische Mikromethoden.* Hüthig-Verlag, Heidelberg-Basel-New York 1979.
[6] Gubor, L. A., Elyashberg, M. E.: *Fresenius Z. Anal. Chem.* **1977,** *32*, 2025–2043.
[7] Kaufmann, H. P., Radwan, S. S., Ahmad, A. K. S.: *Fette, Seifen, Anstrichmittel* **1966,** *68*, 261–268.
[8] Dünges, W.: *GIT Fachz. Lab. Supplement „Chromatographie"* **1982,** 17–26.
[9] Miller, J. M., Kirchner, J. G.: *Anal. Chem.* **1953,** *25*, 1107–1109.
[10] Funk, W.: *Fresenius Z. Anal. Chem.* **1984,** *318*, 206–219.
[11] Weicker, H., Brossmer, R.: *Klin. Wochenschrift* **1961,** *39*, 1265–1266.

[12] Kofoed, J., Korczak-Fabierkiewicz, C., Lucas, G. H. W.: *Nature* **1966,** *211*, 147-150; *J. Chromatogr.* **1966,** *23*, 410-416.
[13] Stijve, T.: *Dtsch. Lebensm. Rundsch.* **1980,** *76*, 234-237.
[14] Mathis, C.: *Ann. Pharm. Fr.* **1965,** *23*, 331-334.
[15] Froment, P., Robert, A.: *Chromatographia* **1971,** *4*, 173.
[16] Elgamal, M. H. A., Fayez, M. B. E.: *Fresenius Z. Anal. Chem.* **1967,** *226*, 408-417.
[17] Rusiecki, W., Henneberg, M.: *Ann. Pharm. Fr.* **1963,** *21*, 843-846.
[18] Frijns, J. M. G. J.: *Pharm. Weekbl.* **1971,** *106*, 605-623; *CA 75, 121441 b* (1971).
[19] Polesuk, J., Ma, T. S.: *Mikrochim. Acta (Vienna)* **1970,** 677-682.
[20] Wilk, M., Hoppe, U., Taupp, W., Rochlitz, J.: *J. Chromatogr.* **1967,** *27*, 311-316.
[21] Wilk, M., Bez, W., Rochlitz, J.: *Tetrahedron* **1966,** *22*, 2599-2608.
[22] Kaess, A., Mathis, C.: *Ann. Pharm. Fr.* **1966,** *24*, 753-762.
[23] Smith, L. L., Price, J. C.: *J. Chromatogr.* **1967,** *26*, 509-511.
[24] Van Lier, J. E., Smith, L. L.: *J. Chromatogr.* **1969,** *41*, 37-42.
[25] Polesuk, J., Ma, T. S.: *J. Chromatogr.* **1971,** *57*, 315-318.
[26] Glaser, C. B., Maeda, H., Meienhofer, J.: *J. Chromatogr.* **1970,** *50*, 151-154.
[27] Kaufmann, H. P., Makus, Z., Khoe, T. H.: *Fette, Seifen, Anstrichm.* **1962,** *64*, 1-5.
[28] Kaufmann, H. P., Khoe, T. H.: *Fette, Seifen, Anstrichm.* **1962,** *64*, 81-85.
[29] Knappe, E., Peteri, D.: *Fresenius Z. Anal. Chem.* **1962,** *190*, 380-386.
[30] Wieland, T., Ottenheym, H.: *Pept. Proc. Eur. Pept. Symp., 8th, 1966,* 195. North-Holland, Amsterdam 1967.
[31] Graf, E., Hoppe, W.: *Dtsch. Apoth. Ztg.* **1962,** *102*, 393-397.
[32] Brown, K., Poole, C.: *J. High Resolut. Chromatogr. Chromatogr. Commun.* **1984,** *7*, 520-524.
[33] Tyrer, J. H., Eadie, M. J., Hooper, W. D.: *J. Chromatogr.* **1969,** *39*, 312-317.
[34] Scotney, J., Truter, E. V.: *J. Chem. Soc. C* **1968,** 1911-1913.
[35] Viswanathan, C. V., Basilo, M., Hoevet, S. P., Lundberg, W. O.: *J. Chromatogr.* **1968,** *34*, 241-245.
[36] Junior, P., Krüger, D., Winkler, C.: *Dtsch. Apoth. Ztg.* **1985,** *125*, 1945-1949.
[37] Krüger, D., Wichtl, M.: *Dtsch. Apoth. Ztg.* **1985,** *125*, 55-57.
[38] Bierl, B. A., Beroza, M., Aldridge, M. H.: *Anal. Chem.* **1971,** *43*, 636-641.
[39] Baggiolini, M., Dewald, B.: *J. Chromatogr.* **1967,** *30*, 256-259.
[40] Purdy, S. J., Truter, E. V.: *Proc. R. Soc. London, B.* **1963,** *158*, 536-543.
[41] Young-Duck Ha, Bergner, K.-G.: *Dtsch. Lebensm. Rundsch.* **1981,** *77*, 102-106.
[42] Randerath, K., Randerath, E.: *Angew. Chem. Int. Ed. Engl.* **1964,** *3*, 442.
[43] Dutta, J., Das, A. K., Ghosh, R.: *J. Chromatogr.* **1981,** *210*, 544-549.
[44] Polesuk, J., Ma, T. S.: *Mikrochim. Acta (Vienna)* **1971,** 662-666.
[45] Cargill, D. I.: *Analyst (London)* **1962,** *87*, 865-869.
[46] Jork, H., Kany, E.: *GDCH-Workshop Nr. 301 „Dünnschicht-Chromatographie für Fortgeschrittene“,* Saarbrücken 1986.
[47] Lück, E., Courtial, W.: *Dtsch. Lebensm. Rundsch.* **1965,** *61*, 78-79.
[48] Jork, H.: *GDCh-Workshop Nr. 301 „Dünnschicht-Chromatographie für Fortgeschrittene“,* Universität des Saarlandes, Saarbrücken 1985.
[49] Schütz, C., Schütz, H.: *Arzneim. Forsch.* **1973,** *23*, 428-431.
[50] De Zeeuw, R. A., Wijsbeek, J.: *J. Chromatogr.* **1970,** *48*, 222-224.
[51] Schütz, C., Schütz, H.: *Arch. Toxikol.* **1972,** *28*, 286-295.
[52] Schütz, C., Schütz, H.: *Dtsch. Apoth. Ztg.* **1973,** *113*, 1559-1562.
[53] Wilk, M., Brill, U.: *Arch. Pharm.* **1968,** *301*, 282-287.
[54] Wilk, M., Taupp, W.: *Z. Naturforsch.* **1969,** *24 B*, 16-23.

[55] Brown, W., Turner, A. B.: *J. Chromatogr.* **1967**, *26*, 518–519.
[56] Klesment, I. R.: *Gazov. Kromatogr.* **1966**, *No. 4*, 102; *CA 66, 26752a* (1967).
[57] Polesuk, J., Ma, T. S.: *Mikrochim. Acta (Vienna)* **1969**, 352–357.
[58] Lisboa, B. P., Diczfalusy, E.: *Acta Endocrinol.* **1962**, *40*, 60–81.
[59] Lewitzky, E.: Thesis, Fachhochschule Gießen, Fachbereich Technisches Gesundheitswesen 1979.
[60] Marcus, B. J., Fono, A., Ma, T. S.: *Mikrochim. Acta (Vienna)* **1967**, 960–966.
[61] Fono, A., Sapse, A.-M., Ma, T. S.: *Mikrochim. Acta (Vienna)* **1965**, 1098–1104.
[62] Bennett, R. D., Heftmann, E.: *J. Chromatogr.* **1962**, *9*, 353–358.
[63] *Bundesgesundheitsbl. 18,* **1975**, 231–233.
[64] Przybylski, W.: *J. Assoc. Off. Anal. Chem.* **1975**, *58*, 163–164.
[65] Gertz, C., Böschemeyer, L.: *Z. Lebensm. Unters. Forsch.* **1980**, *171*, 335–340.
[66] Golinski, F., Grabarkiewicz-Szczesna, J.: *J. Assoc. Off. Anal. Chem.* **1984**, *67*, 1108–1110.
[67] Dallas, M. S. J.: *J. Chromatogr.* **1970**, *48*, 225–230.
[68] Koch, C. E., Thurm, V., Paul, P.: *Nahrung* **1979**, *23*, 125–130.
[69] Leuenberger, U., Gauch, R., Baumgartner, E.: *J. Chromatogr.* **1978**, *161*, 303–309.
[70] Holloway, P. J., Challen, S. B.: *J. Chromatogr.* **1966**, *25*, 336–346.
[71] Studer, A., Traitler, H.: *Proc. Int. Symp. Instrum. High Perform. Thin-Layer Chromatogr., 2nd, Interlaken,* 1982.
[72] Minyard, J. P., Tumlinson, J. H., Thompson, A. C., Hedin, P. A.: *J. Chromatogr.* **1967**, *29*, 88–93.
[73] Riess, J.: *J. Chromatogr.* **1965**, *19*, 527–530.
[74] Maruyama, Y.: *Igaku To Seibutsugaku* **1966**, *73*, 20; *CA 69, 92757c* (1968).
[75] Viswanathan, C. V., Phillips, F., Lundberg, W. O.: *J. Chromatogr.* **1968**, *38*, 267–273.
[76] Saha, S., Dutta, J.: *Lipids* **1973**, *8*, 653–655.
[77] Oette, K., Doss, M.: *J. Chromatogr.* **1968**, *32*, 439–450.
[78] Robbiani, R., Büchi, W.: *Proc. Euro Food Chem. III,* Vol. 2, 216–223; Antwerpen, March 1985.
[79] Cohen, I. C., Norcup, J., Ruzicka, J. H. A., Wheals, B. B.: *J. Chromatogr.* **1969**, *44*, 251–255.
[80] Marcus, B. J., Ma, T. S.: *Mikrochim. Acta (Vienna)* **1968**, 436–441.
[81] Lisboa, B. P.: *J. Chromatogr.* **1966**, *24*, 475–477.
[82] Tumlinson, J. H., Minyard, J. P., Hedin, P. A., Thompson, A. C.: *J. Chromatogr.* **1967**, *29*, 80–87.
[83] Wilczynska, I.: *Chem. Anal. (Warsaw)* **1972**, *17*, 21–30; *CA 77, 42872f* (1972).
[84] Wilczynska, I.: *Chem. Anal. (Warsaw)* **1971**, *16*, 69–76; *CA 75, 5401q* (1971).
[85] Stedman, E. D.: *Analyst (London)* **1969**, *94*, 594–598.
[86] Schorn, P. J.: in: Stahl, E.: *Dünnschicht-Chromatographie, ein Laboratoriumshandbuch.* 2nd Ed., Springer, Berlin–Heidelberg–New York 1967.
[87] Junker-Buchheit, A., Jork, H.: *Fresenius Z. Anal. Chem.* **1988**, *331*, 387–393.
[88] Junker-Buchheit, A., Jork, H.: *Spectrum* (Darmstadt) **1988**, *2/88*, 22–25.
[89] Hänsel, W., Strömmer, R.: *GIT Fachz. Lab. Supplement 3 „Chromatographie“* **1987**, 21–26.
[90] Frei, R. W., Lawrence, J. F., le Gay, D. S.: *Analyst (London)* **1973**, *98*, 9–18.
[91] Lawrence, J. F., Laver, G. W.: *J. Assoc. Off. Anal. Chem.* **1974**, *57*, 1022–1025.
[92] Lantos, J., Brinkman, U. A. T., Frei, R. W.: *J. Chromatogr.* **1984**, *292*, 117–127.
[93] Scholten, A. H. M. T., Buuren, C. van, Lawrence, J. F., Brinkman, U. A. T., Frei, R. W.: *J. Liq. Chromatogr.* **1979**, *2*, 607–617.
[94] Patzsch, K., Funk, W., Schütz, H.: *GIT Fachz. Lab. Supplement 3 „Chromatographie“* **1988**, 83–91.

[95] Rippstein, S.: *Forensische Probleme des Drogenmißbrauchs.* Verlag Dr. D. Helm, Heppenheim 1985, S. 163-166.
[96] Tagliaro, F., Dorizzi, R., Plescia, M., Pradella, M., Ferrari, S., Lo Cascio, V.: *Fresenius Z. Anal. Chem.* **1984,** *317,* 678-679.
[97] Bernhard, W., Fuhrer, A. D., Jeger, A. N., Rippstein, S. R.: *Fresenius Z. Anal. Chem.* **1988,** *330,* 458–459.
[98] Nakajima, T., Endou, H., Sakai, F., Tamura, Z.: *Chem. Pharm. Bull. (Tokyo)* **1970,** *18,* 1935.
[99] Pataki, G., Borko, J., Curtius, H. C., Tancredi, F.: *Chromatographia* **1968,** *1,* 406-417.
[100] Inglis, A. S., Nicholls, P. W., Strike, P. McK.: *J. Chromatogr.* **1975,** *107,* 73-80.
[101] Lin, R.-L., Narasimhachari, N.: *Anal. Biochem.* **1974,** *57,* 46-58.
[102] Parihar, D. B., Sharma, S. P., Tewari, K. C.: *J. Chromatogr.* **1966,** *24,* 443-447.
[103] Studer, A., Traitler, H.: *J. High Resolut. Chromatogr. Chromatogr. Commun.* **1982,** *5,* 581-582.
[104] Wintersteiger, R., Gübitz, G., Hartinger, A.: *Chromatographia* **1980,** *13,* 291-294.
[105] Nakamura, H., Pisano, J. J.: *J. Chromatogr.* **1976,** *121,* 33-40; **1978,** *152,* 153-165; **1978,** *154,* 51-59.
[106] Randerath, K., Weimann, G.: *Biochem. Biophys. Acta* **1963,** *76,* 129-131.
[107] Young, J. C.: *J. Chromatogr.* **1977,** *124,* 17-28; **1977,** *130,* 392-395.
[108] Kurhekar, M. P., D'Souza, F. C., Meghal, S. K.: *J. Chromatogr.* **1978,** *147,* 432-434.
[109] Kurhekar, M. P., D'Souza, F. C., Pundlik, M. D., Meghal, S. K.: *J. Chromatogr.* **1981,** *209,* 101-102.
[110] Yasuda, S. K.: *J. Chromatogr.* **1964,** *13,* 78-82.
[111] Thawley, A. R.: *J. Chromatogr.* **1968,** *38,* 399-400.
[112] Barton, G. M.: *J. Chromatogr.* **1968,** *34,* 562.
[113] Seiler, H., Rothweiler, W.: *Helv. Chim. Acta* **1961,** *44,* 941-942.
[114] Funk, W., Heiligenthal, M.: *GIT Fachz. Lab. Supplement 5 „Chromatographie"* **1984,** 49-51.
[115] Krebs, K. G., Heusser, D., Wimmer, H., in: E. Stahl: *Dünnschicht-Chromatographie, ein Laboratoriumshandbuch.* 2nd Ed., Springer, Berlin-Heidelberg-New York 1967.
[116] Spitz, H. D.: *J. Chromatogr.* **1972,** *72,* 403-404.
[117] Kreuzig, F.: *Fresenius Z. Anal. Chem.* **1976,** *282,* 457-458.
[118] Kreuzig, F.: *Chromatographia* **1980,** *13,* 238-240.
[119] Kreuzig, F.: *J. Chromatogr.* **1977,** *142,* 441-447.
[120] Jork, H.: *Qualitative und quantitative Auswertung von Dünnschicht-Chromatogrammen unter besonderer Berücksichtigung photoelektrischer Verfahren.* Professorial thesis, Universität des Saarlandes, Saarbrücken 1969.
[121] Hezel, U.: *Angew. Chemie* **1973,** *85,* 334-342.
[122] Rimmer, J. G.: *Chromatographia* **1968,** *1,* 219-220.
[123] Getz, M. E., in: J. C. Touchstone, D. Rogers: *Thin-Layer Chromatography: Quantitative Environmental and Clinical Application.* J. Wiley & Sons, New York-Chichester-Brisbane-Toronto 1980.
[124] Hulpke, H. R., Stegh, R., in: W. Bertsch, S. Hara, R. E. Kaiser, A. Zlatkis: *Instrumental HPTLC.* Hüthig-Verlag, Heidelberg-Basel-New York 1980.
[125] Doerffel, K.: *Statistik in der analytischen Chemie.* 4th Ed., VCH Verlagsgesellschaft, Weinheim 1987.
[126] Kwan Young Lee, Nurok, D., Zlatkis, A.: *J. Chromatogr.* **1979,** *174,* 187-193.
[127] Mezetti, T., Lato, M., Rufini, S., Ciuffini, G.: *J. Chromatogr.* **1971,** *63,* 329-342.
[128] Bagger Hansen, A., Schytt Larsen, S.: *Dan. Tidsskr. Farm.* **1972,** *46,* 105-113.
[128a] Funk, W., Glück, V., Schuch, B.: *J. Planar Chromatogr.* **1989,** *2,* 28–32.

[129] Armstrong, D. W., Stine, G. Y.: *J. Liq. Chromatogr.* **1983**, *6*, 23-33.
[130] Schekerdjiev, N., Dshoneydi, M., Koleva, M., Budewski, O.: *Pharmazie* **1973**, *28*, 199-201.
[131] Stahl, E., in: K. Paech, M. V. Tracy: Moderne Methoden der Pflanzenanalyse, Vol. 5. Springer, Berlin-Göttingen-Heidelberg 1962.
[132] Favretto, L., Pertoldi Marletta, G., Favretto Gabrielli, L.: *J. Chromatogr.* **1970**, *46*, 255-260.
[133] Egli, R. A.: *Fresenius Z. Anal. Chem.* **1972**, *259*, 277-282.
[134] Grusz-Harday, E.: *Pharmazie* **1971**, *26*, 562-563.
[135] Lucier, G. W., Menzer, R. E.: *J. Agric. Food Chem.* **1971**, *19*, 1249-1255.
[136] Banci, F., Grande, P. del, Monai, A.: *Arzneim. Forsch.* **1970**, *20*, 1030-1037.
[137] Brtnik, F., Barth, T., Jost, K.: *Coll. Czech. Chem. Commun.* **1981**, *46*, 1983-1989.
[138] Pauncz, J. K.: *J. High Resolut. Chromatogr. Chromatogr. Commun.* **1981**, *4*, 287-291.
[139] Procházka, Z., Jošt, K.: *Coll. Czech. Chem. Commun.* **1980**, *45*, 1305-1314; *45*, 1982-1990; **1981**, *46*, 947-956.
[140] Lebl, M., Machová, A., Hrbas, P., Barth, T., Jošt, K.: *Coll. Czech. Chem. Commun.* **1980**, *45*, 2714-2723.
[141] Procházka, Z., Lebl, M., Barth, T., Hlaváček, J., Trka, A., Buděšínsky, M., Jost, K.: *Coll. Czech. Chem. Commun.* **1984**, *49*, 642-652.
[142] Reimerdes, E. H., Engel, G., Behnert, J.: *J. Chromatogr.* **1975**, *110*, 361-368.
[143] Methees, D.: *J. Agric. Food Chem.* **1983**, *31*, 453-454.
[144] Gimeno, A.: *J. Assoc. Off. Anal. Chem.* **1984**, *67*, 194-196.
[145] Stahr, H. M., Hyde, W., Pfeiffer, R., Domoto, M.: *Proc. Int. Symp. Instrum. High Perform. Thin-Layer Chromatogr. (HPTLC), 3rd, Würzburg,* 1985, 447-468.
[146] Stahr, H. M., Hyde, W., Pfeiffer, R.: *Vet. Hum. Toxicol.* **1981**, *23*, 433-436.
[147] Booth, J., Keysell, G. R., Sims, P.: *Biochem. Pharmacol.* **1974**, *23*, 735-744.
[148] Juvvik, P., Sundry, B.: *J. Chromatogr.* **1973**, *76*, 487-492.
[149] Funk, W.: *Fresenius Z. Anal. Chem.* **1984**, *318*, 206-219.
[150] Segura, R., Gotto, A. M. jr.: *J. Chromatogr.* **1974**, *99*, 643-657.
[151] Matrka, M., Rambousek, V., Divis, J., Zverina, V., Marhold, J.: *Coll. Czech. Chem. Commun.* **1971**, *36*, 2725-2728.
[152] Kany, E., Jork, H.: *GDCh-Workshop Nr. 300 „Einführung in die Dünnschicht-Chromatographie“,* Universität des Saarlandes, Saarbrücken 1988.
[153] Adams, J. B.: *J. Sci. Food Agric.* **1973**, *24*, 747-762.
[154] Chalam, R. V., Stahr, H. M.: *J. Assoc. Off. Anal. Chem.* **1979**, *62*, 570-572.
[155] Bottler, R.: *Kontakte (Darmstadt)* **1978**, (2), 36-39.
[156] Winsauer, K., Buchberger, W.: *Chromatographia* **1981**, *14*, 623-625.
[157] Chobanov, D., Tarandjiska, R., Chobanova, R.: *J. Assoc. Off. Anal. Chem.* **1976**, *53*, 48-51.
[157a] Schlemmer, W.: *J. Chromatogr.* **1971**, *63*, 121-129.
[158] Biernoth, G.: *Fette, Seifen, Anstrichm.* **1968**, *70*, 402-404.
[159] Geissler, H. E., Mutschler, E.: *Z. Klin. Chem. Klin. Biochem.* **1974**, *12*, 151-153.
[160] Kosinkiewicz, B., Lubczynska, J.: *J. Chromatogr.* **1972**, *74*, 366-368.
[161] Huck, H., Dworzak, E.: *J. Chromatogr.* **1972**, *74*, 303-310.
[162] Nagy, A., Treiber, L.: *J. Pharm. Pharmacol.* **1973**, *25*, 599-603.
[163] Jauch, R., Bozler, G., Hammer, R., Koss, F. W.: *Arzneim. Forsch.* **1978**, *28*, 904-911.
[164] Bottler, R., Knuhr, T.: *Fresenius Z. Anal. Chem.* **1980**, *302*, 286-289.
[165] Jain, R., Agarwal, D. D., Goyal, R. N.: *J. Liq. Chromatogr.* **1980**, *3*, 557-560; *Fresenius Z. Anal. Chem.* **1981**, *307*, 207-208.
[166] Jain, R., Agarwal, D. D.: *J. Liq. Chromatogr.* **1981**, *4*, 2229-2232; **1982**, *5*, 1171-1175.
[167] Takacs, M., Kertesz, P.: *Fresenius Z. Anal. Chem.* **1971**, *254*, 367-368.
[168] Brantner, A., Vamos, J.: *Proc. Int. Symp. Chromatogr. 6th,* Bruxelles 1970, 401-407.

[169] Delfel, N. E., Tallent, W. H.: *J. Assoc. Off. Anal. Chem.* **1969**, *52*, 182-187.
[170] Lederer, M., Rinalduzzi, B.: *J. Chromatogr.* **1972**, *68*, 237-244.
[171] Aures, D., Fleming, R. Håkanson, R.: *J. Chromatogr.* **1968**, *33*, 480-493; *Fresenius Z. Anal. Chem.* **1968**, *243*, 564-567.
[172] Dell, H. D., Fiedler, J., Wäsche, B.: *Arzneim. Forsch.* **1977**, *27*, 1312-1316.
[173] Schwartz, D. P., McDonough, F. E.: *J. Assoc. Off. Anal. Chem.* **1984**, *67*, 563-565.
[173a] Edvinsson, L., Håkanson, R., Sundler, F.: *Anal. Biochem.* **1972**, *46*, 473-481.
[174] Bell, C. E., Sommerville, A. R.: *Biochem. J.* **1966**, *98*, 1c-3c.
[175] Björklund, A., Falck, B., Håkanson, R.: *J. Chromatogr.* **1969**, *40*, 186-188; **1970**, *47*, 530-536; *Anal. Biochem.* **1970**, *35*, 264-276.
[176] Håkanson, R., Lombard des Gouttes, M.-N., Owman, C.: *Life Sci.* **1967**, *6*, 2577-2585.
[177] Axelsson, S., Björklund, A., Lindvall, O.: *J. Chromatogr.* **1975**, *105*, 211-214.
[178] Segura, R., Navarro, X.: *J. Chromatogr.* **1981**, *217*, 329-340.
[179] Ripphahn, J.: *Advances in Quantitative Analysis by HPTLC, Danube Symp. Chromatogr.*, 1st, Szeged (Hungary) 1976.
[180] Rücker, G., Neugebauer, M., El Din, M. S.: *Planta Med.* **1981**, *43*, 299-301.
[181] Norpoth, K., Addicks, H. W., Wittig, M.: *Arzneim. Forsch.* **1973**, *23*, 1529-1535.
[182] Martinek, A.: *J. Chromatogr.* **1971**, *56*, 338-341.
[183] Pandey, R. C., Misra, R., Rinehart, K. L.: *J. Chromatogr.* **1979**, *169*, 129-139.
[184] Jork, H., Kany, E.: *GDCh-training course Nr. 302 „Möglichkeiten der quantitativen Direktauswertung von Dünnschicht-Chromatogrammen"*, Universität des Saarlandes, Saarbrücken 1985.
[185] Kynast, G., Dudenhausen, J. W.: *Z. Klin. Chem. Klin. Biochem.* **1972**, *10*, 573-576.
[186] Omori, T., in: W. Bertsch, S. Hara, R. E. Kaiser, A. Zlatkis: *Instrumental HPTLC.* Hüthig-Verlag, Heidelberg-Basel-New York 1980, 275-279.
[187] MERCK, E.: Company literature *Information on Thin Layer Chromatography XIII Aminoacids in Plasma*, 1979.
[188] Abe, F., Samejima, K.: *Anal. Biochem.* **1975**, *67*, 298-308.
[189] Gentile, I. A., Passera, E.: *J. Chromatogr.* **1982**, *236*, 254-257.
[190] Rasmussen, H.: *J. Chromatogr.* **1967**, *26*, 512-514.
[191] Rokos, J. A. S.: *J. Chromatogr.* **1972**, *74*, 357-358.
[192] Blass, G., Ho, C. S.: *J. Chromatogr.* **1981**, *208*, 170-173.
[193] Copius-Peereboom, J. W., Beekes, H. W.: *J. Chromatogr.* **1962**, *9*, 316-320.
[194] Nabi, S. A., Farooqui, W. U., Rahman, N.: *Chromatographia* **1985**, *20*, 109-111.
[195] Egg, D., Huck, H.: *J. Chromatogr.* **1971**, *63*, 349-355.
[196] Trucksess, M., Nesheim, S., Eppley, R.: *J. Assoc. Off. Anal. Chem.* **1984**, *67*, 40-43.
[197] Mlekusch, W., Truppe, W., Paletta, B.: *J. Chromatogr.* **1972**, *72*, 495-497; *Clin. Chim. Acta* **1973**, *49*, 73-77.
[198] Tsao, F. H. C., Zachman, R. D.: *Clin. Chim. Acta* **1982**, *118*, 109-120.
[199] Mitnick, M. A., De Marco, B., Gibbons, J. M.: *Clin. Chem.* **1980**, *26*, 277-281.
[200] Mutter, M.: *Tenside* **1968**, *5*, 138-140.
[201] König, H.: *Fresenius Z. Anal. Chem.* **1970**, *251*, 167-171; **1970**, *251*, 359-368; **1971**, *254*, 337-345.
[202] Touchstone, J. C., Murawec, T., Kasparow, M., Wortman, W.: *J. Chromatogr. Sci.* **1972**, *10*, 490-493.
[203] Smith, B. G.: *J. Chromatogr.* **1973**, *82*, 95-100.
[204] Nürnberg, E.: *Arch. Pharmaz.* **1959**, *292*, 610-620.
[205] Andreev, L. V.: *J. High Resolut. Chromatogr. Chromatogr. Commun.* **1983**, *6*, 575-576.
[206] Andreev, L. V.: *J. Anal. Chem. (USSR)* **1983**, *38*, 871-873.

[207] Glass, A.: *J. Chromatogr.* **1973,** *79*, 349.
[208] Thielemann, H.: *Pharmazie* **1977,** *32*, 244.
[209] Gänshirt, H.: *Arch. Pharmaz.* **1963,** *296*, 73–79.
[210] MACHEREY-NAGEL: Catalogue *„Fertigprodukte für die DC" UD/dl/07/0/*4.85.
[211] MERCK, E.: Company brochure *"Standardized Silica Gels for Chromatography"*, 1973.
[212] Unger, K. K.: *Porous Silica.* Elsevier Sci. Publ., Amsterdam–Oxford–New York 1979.
[213] Jork, H., Wimmer, H.: *Quantitative Auswertung von Dünnschicht-Chromatogrammen.* GIT-Verlag, Darmstadt 1986.
[214] Udenfried, S.: *Fluorescence Assay in Biology and Medicine.* Academic Press, London–New York 1962.
[215] Porgesova, L., Porges, E.: *J. Chromatogr.* **1964,** *14*, 286–289.
[216] Stahl, E.: *Fresenius Z. Anal. Chem.* **1968,** *236*, 294–310.
[217] Grünewald, T., Rudolf, M.: *ZFL* **1981,** *32*, 85–88.
[218] Bomar, M. T., Grünewald, T.: *Lebensm. Wiss. Technol.* **1972,** *5*, 166–171.
[219] Lambert, J. P.: *J. Food Prot.* **1980,** *43*, 625–628.
[220] Pichert, H.: *Hauswirtsch. Wiss.* **1977,** *25*, 83–89.
[221] Wallhäuser, K. H.: *Sterilisation, Desinfektion, Konservierung.* 2nd Ed., G. Thieme, Stuttgart 1978.
[222] Klatte-Siedler, K.: *Hauswirtsch. Wiss.* **1981,** *29*, 110–115.
[223] Dehne, L., Bögl, W., Großklaus, D.: *Fleischwirtschaft* **1983,** *63*, 231–237.
[224] Zennie, T. M.: *J. Liq. Chromatogr.* **1984,** *7*, 1383–1391.
[225] Pichert, H.: *Haushaltstechnik.* (Uni-pocket books) E. Ulmer, Stuttgart 1978.
[226] Genest, K.: *J. Chromatogr.* **1965,** *19*, 531–539.
[227] Funk, W.: Private communication, Fachhochschule Gießen, Fachbereich Technisches Gesundheitswesen, 1988.
[228] Jork, H.: Dia-Serie *„Quantitative Direktauswertung von Dünnschicht-Chromatogrammen"*, C. ZEISS, Oberkochen 1967.
[229] Hellmann, H.: *Fresenius Z. Anal. Chem.* **1983,** *314*, 125–128.
[230] Junker-Buchheit, A., Jork, H.: *J. Planar Chromatogr.* **1988,** *1*, 214–219.
[231] Funk, W., Dammann, V., Couturier, T., Schiller, J., Völker, L.: *J. High Resolut. Chromatogr. Chromatogr. Commun.* **1986,** *9*, 224–235.
[232] Funk, W., Schanze, M., Wenske, U.: *GIT Fachz. Lab. Supplement 3 „Chromatographie"* **1983,** 8–16.
[233] Götz, W., Sachs, A., Wimmer, H.: *Dünnschicht-Chromatographie.* G. Fischer Verlag, Stuttgart–New York 1978.
[234] Shaun, S. J. H., Butler, H. T., Poole, C. F.: *J. Chromatogr.* **1983,** *281*, 330–339.
[235] Fuhrhop, J.-H., Smith, K. M.: *Laboratory Methods in Porphyrin and Metalporphyrin Research.* Elsevier Publ., Amsterdam 1975, 243 ff.
[236] Gübitz, G., Wintersteiger, R.: *Anal. Toxikol.* **1980,** *4*, 141–143.
[237] Wintersteiger, R.: *Analyst (London)* **1982,** *107*, 459–461.
[238] Brown, K., Poole, C. F.: *LC Mag.* **1984,** *2*, 526–530.
[239] Chalela, G., Schwantes, H. O., Funk, W.: *Fresenius Z. Anal. Chem.* **1984,** *319*, 527–532.
[240] Funk, W., Vogt, H., Dammann, V., Weyh, C.: *Vom Wasser* **1980,** *55*, 217–225.
[241] Frei, R. W., Lawrence, J. F., LeGay, D. S.: *Analyst (London)* **1973,** *98*, 9–18.
[242] Reh, E., Jork, H.: *Fresenius Z. Anal. Chem.* **1984,** *318*, 264–266.
[243] Faber, D. B., de Kok, A., Brinkman, U. A. T.: *J. Chromatogr.* **1977,** *143*, 95–103.
[244] Takeda, Y., Isohata, E., Amano, R., Uchiyama, M.: *J. Assoc. Off. Anal. Chem.* **1979,** *62*, 573–578.
[245] Uchiyama, S., Uchiyama, M.: *J. Chromatogr.* **1978,** *153*, 135–142.

[246] Schade, M.: Thesis, Fachhochschule Gießen, Fachbereich Technisches Gesundheitswesen, 1986.
[247] Arndt, F.: Thesis, Medizinisches Zentrum für Klinische Chemie und Fachhochschule Gießen, 1983.
[248] Kerler, R.: Thesis, Fachhochschule Gießen, Fachbereich Technisches Gesundheitswesen, 1981.
[249] Funk, W., Kerler, R., Boll, E., Dammann, V.: *J. Chromatogr.* **1981,** *217,* 349-355.
[250] Reuter, K., Knauf, H., Mutschler, E.: *J. Chromatogr.* **1982,** *233,* 432-436.
[251] Davis, C. M., Fenimore, D. C.: *J. Chromatogr.* **1983,** *272,* 157-165.
[252] Riedel, K.-D., Laufen, H.: *J. Chromatogr.* **1983,** *276,* 243-248.
[253] Kunz, F. R., Jork, H.: *Fresenius Z. Anal. Chem.* **1988,** *329,* 773-777.
[254] Kiefer, U.: Thesis, Fachhochschule Gießen, Fachbereich Technisches Gesundheitswesen, 1984.
[255] Funk, W., Canstein, M. von, Couturier, T.: *Proc. Int. Symp. Instrum. High Perform. Thin-Layer Chromatogr., 3rd, Würzburg* 1985, 281-311.
[256] Canstein, M. von: Thesis, Fachhochschule Gießen, Fachbereich Technisches Gesundheitswesen, 1984.
[257] Sommer, D.: Thesis, Fachhochschule Gießen, Fachbereich Technisches Gesundheitswesen, 1984.
[258] Derr, P.: Thesis, Fachhochschule Gießen, Fachbereich Technisches Gesundheitswesen, 1985.
[259] Francis, O. J., Ware, G. M., Carman, A. S., Kuan, S. S.: *J. Assoc. Off. Anal. Chem.* **1985,** *68,* 643-645.
[260] Dunphy, P. J., Whittle, K. J., Pennock, J. F.: *Chem. Ind. (London)* **1965,** 1217-1218.
[261] Mangold, H. K., in: Stahl, E.: *Dünnschicht-Chromatographie, ein Laboratoriums-Handbuch.* 2nd Ed., Springer, Berlin-Heidelberg-New York, 1967.
[262] Malins, D. C., Mangold, H. K.: *J. Am. Oil Chem. Soc.* **1960,** *37,* 576-581.
[263] Alak, A., Heilweil, E., Hinze, W. L., Oh, H., Armstrong, D. W.: *J. Liq. Chromatogr.* **1984,** *7,* 1273-1288.
[264] Wintersteiger, R.: *GIT Fachz. Lab. Supplement 3 „Chromatographie",* **1988,** 5-11.
[265] Uchiyama, S., Uchiyama, M.: *J. Liq. Chromatogr.* **1980,** *3,* 681-691.
[266] Schäfer, M., Geissler, H. E., Mutschler, E.: *J. Chromatogr.* **1977,** *143,* 636-639.
[267] Majak, W., McDiarmid, R. E., Bose, R. J.: *Phytochemistry* **1978,** *17,* 301-303.
[268] Tillian, H., Gübitz, G., Korsatko, W., Wintersteiger, R.: *Arzneim. Forsch.* **1984,** *35,* 552-554.
[269] Wintersteiger, R., Wenninger-Weinzierl, G.: *Fresenius Z. Anal. Chem.* **1981,** *309,* 201-208.
[270] Gübitz, G.: *Chromatographia* **1979,** *12,* 779-781.
[271] Wintersteiger, R., Gamse, G., Pacha, W.: *Fresenius Z. Anal. Chem.* **1982,** *312,* 455-461.
[272] Seiler, N.: *J. Chromatogr.* **1971,** *63,* 97-112.
[273] Bauer, R., Berganza, L., Seligmann, O., Wagner, H.: *Phytochemistry* **1985,** *24,* 1587-1591.
[274] Wagner, H., Diesel, P., Seitz, M.: *Arzneim. Forsch.* **1974,** *24,* 466-471.
[275] Wintersteiger, R.: *J. Liq. Chromatogr.* **1982,** *5,* 897-916.
[276] Breitinger, M., Paulus, H., Wiegrebe, W.: *Dtsch. Apoth. Ztg.* **1980,** *120,* 1699-1702.
[277] Breiter, J.: *Kontakte (Darmstadt)* **1974,** *3,* 17-24.
[278] Breiter, J., Helger, R., Interschick, E., Wüst, H.: *J. Clin. Chem. Clin. Biochem.* **1978,** *16,* 127-134.
[279] Stead, A. H., Gill, R., Wright, T., Gibbs, J. P., Moffat, A. C.: *Analyst* **1982,** *107,* 1106-1168.
[280] Felix, A. M., Jimenez, M. H.: *J. Chromatogr.* **1974,** *89,* 361-364.
[281] Seiler, N., Wiechmann, M.: *Fresenius Z. Anal. Chem.* **1966,** *220,* 109-127.

[282] Dertinger, G., Scholz, H.: *Dtsch. Apoth. Ztg.* **1973**, *113*, 1735–1738.
[283] Seiler, N., Wiechmann, M.: *Hoppe Seyler's Z. Physiol. Chem.* **1967**, *348*, 1285–1290.
[284] Young, J. C.: *J. Chromatogr.* **1976**, *124*, 17–28.
[285] Klaus, R.: *Chromatographia* **1985**, *20*, 235–238.
[286] Frei, R. W., Lawrence, J. F.: *J. Chromatogr.* **1971**, *61*, 174–179.
[287] Lawrence, J. F., Frei, R. W.: *J. Chromatogr.* **1972**, *66*, 93–99.
[288] Frei, R. W., Lawrence, J. F.: *J. Assoc. Off. Anal. Chem.* **1972**, *55*, 1259–1264.
[289] Doss, M.: *Z. klin. Chem. u. klin. Biochem.* **1970**, *8*, 197–207.
[290] Uchiyama, S., Kondo, T., Uchiyama, M.: *Bunseki Kagaku* **1977**, *26*, 762–768.
[291] Wintersteiger, R.: *GIT Fachz. Lab. Supplement 3 „Chromatographie"* **1988**, 5–11.
[292] Uchiyama, S., Uchiyama, M.: *J. Chromatogr.* **1983**, *262*, 340–345.
[293] Nakamura, H., Pisano, J. J.: *J. Chromatogr.* **1978**, *154*, 51–59; **1978**, *152*, 153–165; **1976**, *121*, 79–81.
[294] Weiß, P. A. M.: *Endokrinologie* **1972**, *59*, 273–278.
[295] Jork, H.: *GIT Fachz. Lab. Supplement 3 „Chromatographie"* **1986**, 79–87.

4 Documentation and Hints for Chromatography Experts

Every chromatographic investigation begins with the preparation of the sample and the chromatographic system. This is followed by the crux of the separation process (development of the chromatogram) which is in turn followed by the visualization of the separated substances and the preservation of the chromatogram and finally by the analysis of the results.

Each of these steps must be so documented that it can always be repeated simply from the protocol. The most important steps will be discussed below, paying special attention to the processing of the chromatogram after development.

4.1 Preparations for Chromatography

A description of the preparatory steps before chromatography is performed from an integral part of the complete documentation. A record is necessary of the

- sampling process,
- sample storage (cool room, protection from light, inert gas atmosphere etc.),
- comminution of the sample (sieve size etc.) and
- sample preparation (extraction, distillation, sublimation etc.).

In addition, information must be provided concerning the enrichment and clean up of the sample. If possible the sample solution prepared should be adjusted to a particular concentration, so that the application of the chosen volume gives a preliminary idea of the amounts in the chromatogram produced.

Particular pieces of apparatus and chemicals and solvents of exactly defined quality were employed in the examples that follow and it is necessary to define them precisely. Manufacturers' names have been given when necessary as defining a quality criterion. The purity of solvents and chemicals is a particularly important point.

4.1.1 Solvent Quality

The higher the demands made on the analysis the higher must be the quality of the solvents employed. Since the substances are present in dissolved form during most stages of the analysis it must, for example, be known what additives have been employed to stabilize sensitive solvents (Tab. 24).

Only particular solvents are suitable for certain purposes. The choice depending, for instance, on their residual water content or their acid-base nature if R_f values are to be reproduced [1, 2]. Halogen-containing solvents may not be employed for the determination of chlorinated pesticides. Similar considerations apply to PAH analyses. "Pro analysi" grades are no longer adequate for these purposes. It is true that it would be possible to manufacture universally pure solvents that were adequate for all analytical purposes, but they would then be too expensive for the final user [3, 4].

Table 24: List of some solvents and their stabilizers

Solvent	Quality	Stabilizer	Company
Chloroform	HPLC solvent	ethanol, 0.5 to 1%	BAKER
	hydrocarbon stabilized	amylene, 0.01 to 0.02%	BAKER
	GR	ethanol, 0.6 to 1.0%	MERCK
	LiChrosolv	amylene	MERCK
	for analysis	ethanol, ca. 1%	RIEDEL-de HAËN
	Chromasolv	amylene	RIEDEL-de HAËN
Dichloro-methane	HPLC solvent	cyclohexane, 100 to 350 ppm	BAKER
	GR	amylene, ca. 20 ppm	MERCK
	LiChrosolv	amylene, ca. 20 ppm	MERCK
	for analysis	amylene, ca. 25 ppm	RIEDEL-de HAËN
	Chromasolv	amylene, ca. 25 ppm	RIEDEL-de HAËN
Diethyl ether	GR	BHT, 7 ppm	MERCK
	dried	no information	MERCK
	for analysis	BHT, 5 ppm	RIEDEL-de HAËN
	Chromasolv	ethanol, ca. 2%	RIEDEL-de HAËN
Dioxane	GR	BHT, 25 ppm	MERCK
	LiChrosolv	BHT, 1.5 ppm	MERCK
	for analysis	BHT, 25 ppm	RIEDEL-de HAËN
	Chromasolv	BHT, 1.5 ppm	RIEDEL-de HAËN
Diisopropyl ether	GR	BHT, ca. 5 ppm	MERCK
	for analysis	BHT, 5 ppm	RIEDEL-de HAËN
Tetrahydrofuran	HPLC solvent	BHT, 0.02 to 0.03%	BAKER
	GR	BHT	MERCK
	LiChrosolv	no information	MERCK
	for analysis	BHT, 250 ppm	RIEDEL-de HAËN
	Chromasolv	no information	RIEDEL-de HAËN

Large quantities of solvents are employed for sample preparation, in particular, and these are then concentrated down to a few milliliters. So particularly high quality materials that are as free as possible from residual water and especially free from nonvolatile or not readily volatile impurities ought to be employed here; such impurities are enriched on concentration and can lead to gross contamination. The same considerations also apply to preparative chromatography. Special solvents of particular purity are now available.

4.1.2 Choice of Stationary Phase

If sufficient knowledge is available concerning the previous history and chemical nature of the sample, then it is possible to choose the type of stationary and mobile phase according to the triangular scheme [5, 40] (Fig. 55). Silica gel and aluminium oxide are active adsorbents. They should be characterized more precisely when documenting the experiment, most simply by noting exactly the product designation and source, since every manufacturing company takes care to guarantee the properties and qualities of its products.

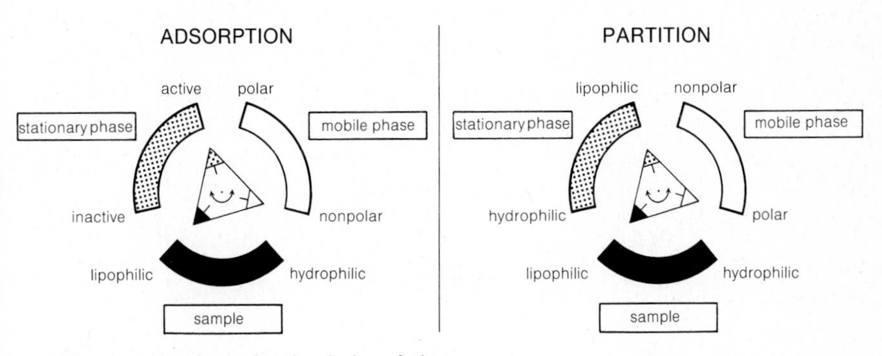

Fig. 55: Triangle scheme for the choice of phases.

Thus, for instance, 10% aqueous suspensions of the adsorbents produced by the various manufacturers are different (Table 25). It is not a matter of indifference whether Silica gel 60 or Silica gel 60 W (water-resistant) layers are employed, for the two differ appreciably both in pH and in running time and, hence, in selectivity [6].

Table 25: Summary of the pH values of some layer materials of precoated plates, determined as 10% aqueous suspensions (duplicate determination: two different TLC/HPTLC plates from the same batch).

Layer type	Company and Catalogue No.	pH	Layer type	Company and Catalogue No.	pH
Aluminium oxide 60, type E	MERCK 5713	9.7	Silica gel G 1500	SCHLEICHER & SCHÜLL	6.2
Aluminium oxide 150, type T	MERCK 5727	8.9	Silica gel G 1570	SCHLEICHER & SCHÜLL	7.5
Cellulose	MERCK 5786	6.6	Kieselguhr	MERCK 5738	7.6
Cellulose	SCHLEICHER & SCHÜLL 3793	6.8	Si 50 000	MERCK 15132	7.9
Silica gel	MACHEREY & NAGEL 809023	6.4	Silica gel 60 silanized	MERCK 5747	7.6
Silica gel	MACHEREY & NAGEL 811022	6.3	Silica gel RP-8	MERCK 13725	4.7
Silica gel 60	MERCK 5713	7.8	Silica gel RP-18	MERCK 13724	5.0*
Silica gel 60	MERCK 5721	7.6	Silica gel RP-18 W	MERCK 13124	4.4
Silica gel 60	MERCK 5628	7.7	CN phase	MERCK 16464	5.1
Silica gel 60 purest	MERCK 15552	4.7	Diol phase	MERCK 12668	3.6
Silica gel 60 with conc. zone	MERCK 11798	6.8	NH_2 phase	MERCK 15647	9.9
Silica gel 60	RIEDEL-de HAËN 37643	7.6	Polyamide 11	MERCK 5557	6.9
Silica gel	RIEDEL-de HAËN 37601	6.4			

* Suspension well shaken, afterwards the RP-18 material floated on the surface.

Aluminium oxide is available in grades with neutral, acidic and basic reactions, which can also vary in the specific surface area and pore size. This makes the separations achieved vary and care must be taken to document precisely.

Polyamide is available commercially as polyamide-6 (based on ε-aminocaprolactam, MACHEREY-NAGEL) and as polyamide-11 (based on 11-aminoundecanoic acid, MERCK). The lipophilic properties of these are different, thus altering their chromatographic selectivity.

Cellulose layers are produced from native, fibrous or microcrystalline cellulose (Avicel^{R}). The separation behaviors of these naturally vary, because particle size (fiber length), surface, degree of polycondensation and, hence, swelling behavior are all different.

The most important thing to pay attention to in the case of *RP phases* is the chain length. It is often forgotten, however, that RP phases are available with differing degrees of surface modification and which also differ in their hydrophobicity and wettability and separation behavior (R_f values, development times). These details should, therefore, also be documented.

Differences in the materials employed for the layers can also become evident when chemical reactions are performed on them. Thus, MACHEREY-NAGEL report that the detection of amino acids and peptides by reaction with ninhydrin is less sensitive on layers containing luminescent or phosphorescent indicators compared to adsorbents which do not contain any indicator [7].

It is for this reason that the details given in the monographs are, on the one hand, obtained by reviewing the literature while, on the other hand, the "Procedure Tested" section reports the results we ourselves obtained, which are not necessarily in complete agreement with the literature reports.

The fact that the binder used in the layer can affect the reagent is shown in the monograph on 4-(4-Nitrobenzyl)pyridine reagent. It is not possible to employ this reagent on Nano-SIL C 18-100 UV_{254} plates (MACHEREY-NAGEL) because the whole surface of the layer is colored bluish-violet. The corresponding water-wettable layers produced by the same manufacturer do not present any difficulties.

These few remarks should suffice to demonstrate the importance of the precise knowledge of the various layer materials and the precise documentation of their use. Such differences should also be taken into account when choosing the stationary phase so that the impression is not later produced that phase A is better or worse than phase B.

4.1.3 Prewashing the Layer

All commercially available precoated plates are manufactured with great care. But they are active layers which, on account of the numbers and structures of their pores, possess a very large internal surface area, on which water vapor and other volatile substances can condense, particularly once the packaging has been opened. In order to prevent this as far as possible the precoated plates are packed with the glass or foil side upwards.
It is possible that plasticizers or monomeric components of the packing materials (adhesives etc.) are also absorbed during storage. In order to stop such "impurities" or oligomeric components of the binder interfering with the development, the plates are often prewashed before the actual chromatography, particularly during quantitative work. This is done by developing the chromatogram one or more times with either a mixture of methanol and chloroform (1 + 1) or with the mobile phase that is to be employed later. When doing this the mobile phase should be allowed to climb appreciably above the solvent front of the subsequent chromatographic run. Acidic or basic mobile phases can cause difficulties because they are not completely removed during the subsequent activation (30 min at 110 °C) in the drying cupboard and, thus, "impregnate" the stationary phase.

It is always inadvisable to activate in a stream of hot or cold air (hair drier), because laboratory air is then blown over the layer. Such details also belong in the documentation of working instructions.

4.1.4 Choice of Chamber System

There is no other facet where thin-layer chromatography reveals its paper-chromatographic ancestry more clearly than in the question of development chambers (Fig. 56). Scaled-down paper-chromatographic chambers are still used for development to this day. From the beginning these possessed a vapor space, to allow an equilibration of the whole system for partition-chromatographic separations. The organic mobile phase was placed in the upper trough after the internal space of the chamber and, hence, the paper had been saturated, via the vapor phase, with the hydrophilic lower phase on the base of the chamber.

In the case of thin-layer chromatography there is frequently no wait to establish complete equilibrium in the chamber before starting the development. The chamber is usually lined with a U-shaped piece of filter paper and tipped to each side after adding the mobile phase so that the filter paper is soaked with mobile phase and adheres to the wall of the chamber. As time goes on the mobile phase evaporates from the paper and would eventually saturate the inside of the chamber.

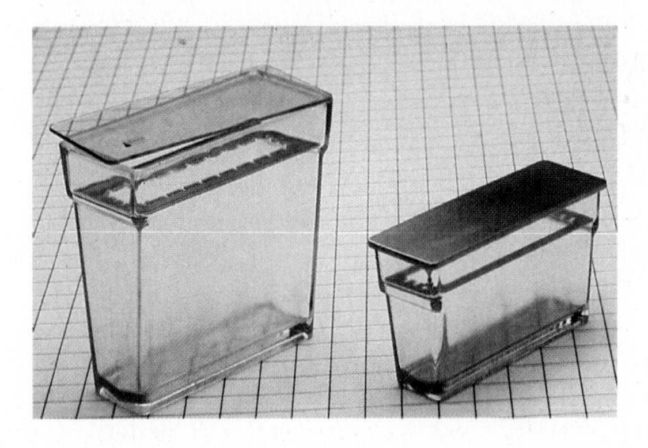

Fig. 56: Commonly used trough chambers (DESAGA).

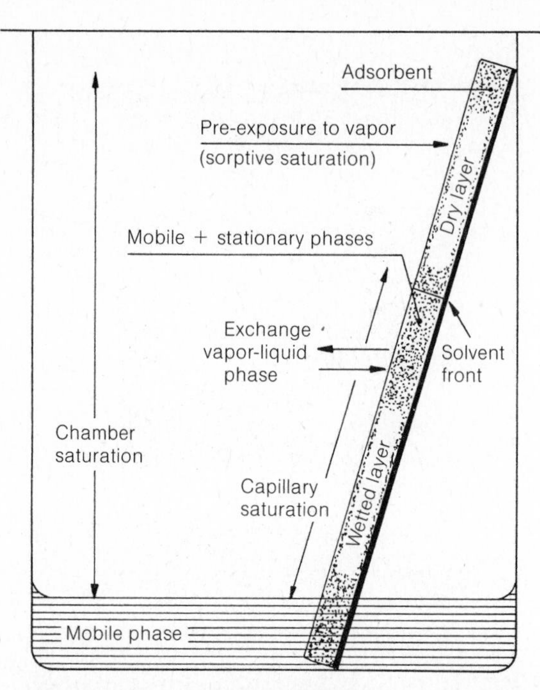

Fig. 57: Schematic representation of the relationships between development, chamber saturation and pre-loading with solvent vapors (acc. [8]).

But there can be no question of chamber saturation if the TLC plate is then placed directly in the chamber. But at least there is a reduction in the evaporation of mobile phase components from the layer. Mobile phase components are simultaneously transported onto the layer (Fig. 57). In the case of multicomponent mobile phases this reduces the formation of β-fronts.

Apart from such trough chambers there are also S-chamber systems (small chambers, sandwich chambers) with deliberately reduced vapor volumes, which are specially suited to adsorptive separations. Such chambers are available for vertical and horizontal development (Fig. 58). Different separation results are naturally obtained in trough and S-chambers [8].

The description of the experiment must, therefore, state what type of chamber was used and whether "chamber saturation" was employed.

The twin-trough chamber (Fig. 43A) was not just developed to economize in mobile phase; it also allows the layer to be impregnated as desired from the vapor

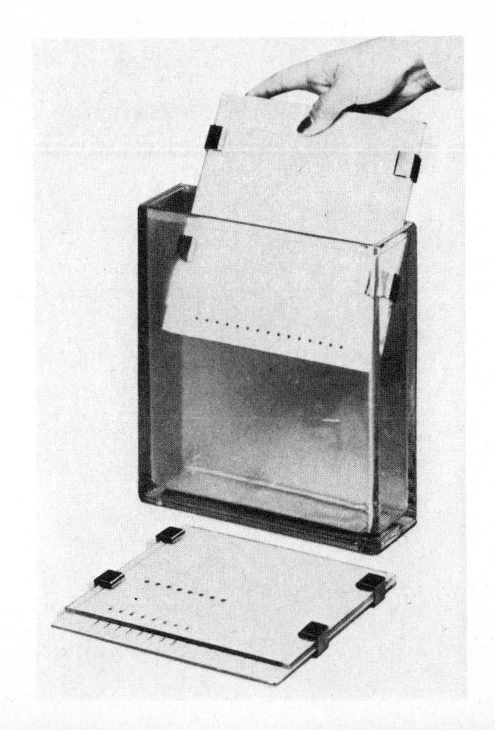

A

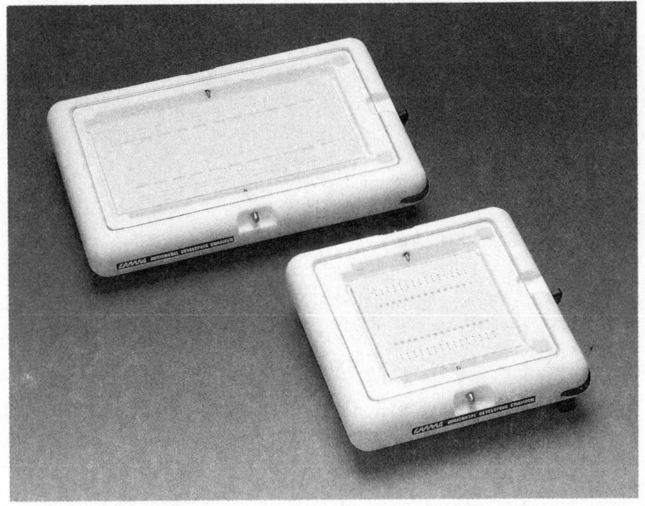

B

Fig. 58: A) S chamber, B) horizontal developing chamber (CAMAG)

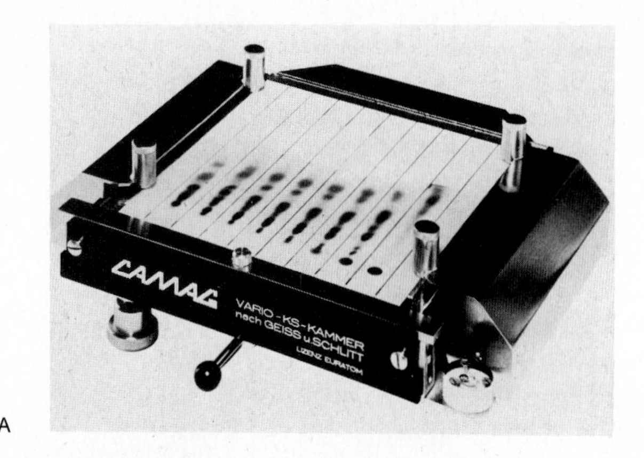

A

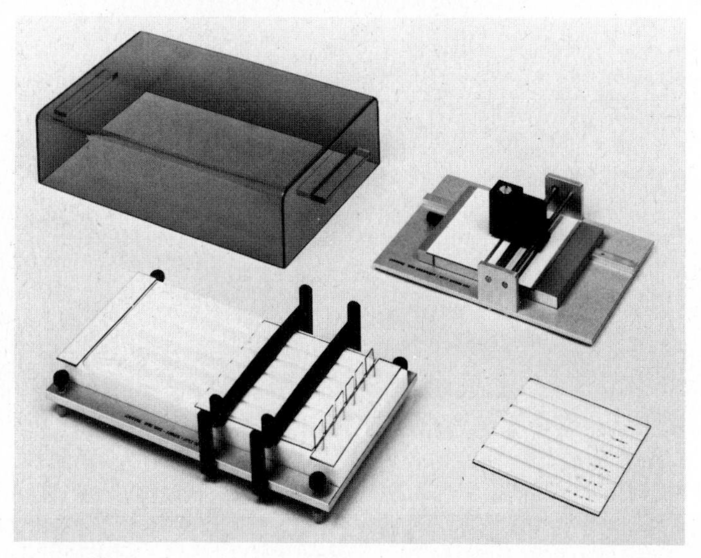

B

Fig. 59: Vario chambers (CAMAG) for plate format 20 cm × 20 cm (A) and 10 cm × 10 cm (B).

phase. In the case of acidic or basic mobile phases a demixing of the mobile phases usually occurs in the lower part of the chromatogram during development. When, however, a vapor such as acetic acid or ammonia has access to the stationary phase it is often possible to substitute an acid- or ammonia-containing mobile phase by a neutral one. The impregnating liquid is placed in one of the troughs, the mobile phase and the TLC/HPTLC in the other one.

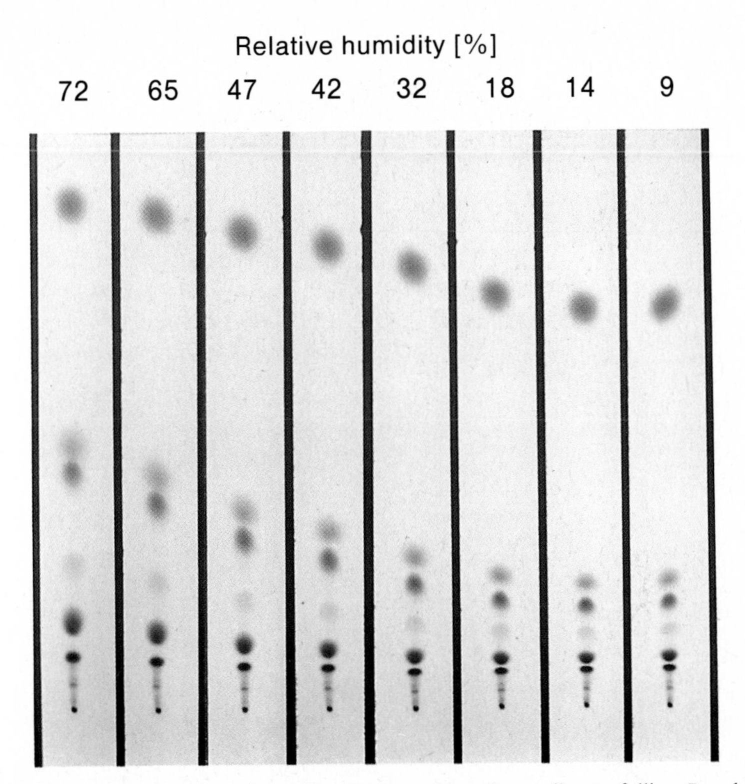

Fig. 60: Chromatogram of a 6 dyestuff mixture made up (according to falling R_f values) of Sudan red 7B, Sudan orange G, Sudan black B, Sudan yellow, Artisil blue 2RP and Sudan black B (2 components) under different humidity conditions. From left to right 72, 65, 47, 42, 32, 18, 14 and 9% relative humidity; layer: silica gel 60; mobile phase: toluene.

A precise mastery of the chromatographic process also requires that the relative humidity be controlled. There are sufficient examples demonstrating that reproducible development is only possible if temperature and relative humidity are maintained constant. The influence of the latter on chromatographic behavior can be investigated using the Vario KS chamber (Fig. 59). When the relative humidity is altered it is possible that not only the zone behavior will be changed but also the order of the zones on the chromatogram (Fig. 60).

It is possible to control the relative humidity with sulfuric acid solutions of particular concentrations (Tab. 26) or with saturated salt solutions in contact with excess salt (Tab. 27). These liquids are placed in a conditioning chamber

Table 26: Relative humidities over various concentrations of sulfuric acid at 20 °C.

Sulfuric acid concentration [%]	Relative humidity [%]	Manufacture: ml conc. sulfuric acid + ml water
98	0	100 + 0
62.5	9	100 + 60
55	15	100 + 82
50	20	100 + 100
40	32	68 + 100
36	42	57 + 100
33	47	50 + 100
28	58	39.5 + 100
25	65	34 + 100
22	72	27.5 + 100

Table 27: Relative humidities over saturated salt solutions in contact with undissolved salts [8].

Salt	Rel. humidity [%]	Salt	Rel. humidity [%]
$ZnCl_2 \cdot 1^1/_2\ H_2O$	10	$NaNO_2$	66
$LiCl \cdot H_2O$	15	$CuCl_2 \cdot 2\ H_2O$	67
$CaBr_2$	16.5	$NH_4Cl + KNO_3$	72.6
$K(HCOO)$	21.3	$NaClO_3$	75
$K(CH_3COO)$	22.7	$NaCl$	75.7
$NiBr_2$	27.1	$H_2C_2O_4 \cdot 2\ H_2O$	76
$NaCl + KNO_3 + NaNO_3$ (16 °C)	30.5	$Na(CH_3COO) \cdot 3\ H_2O$	76
		$Na_2S_2O_3 \cdot 5\ H_2O$	78
KF	30.5	NH_4Cl	79.2
$MgBr_2$	31.8	$(NH_4)_2SO_4$	79.5
$CaCl_2 \cdot 6\ H_2O$	32.3	KBr	84
$MgCl_2 \cdot 6\ H_2O$	32.4	KCl	85
$NaCl + KNO_3$ (16 °C)	32.6	$KHSO_4$	86
CrO_3	35	$BaCl_2 \cdot 2\ H_2O$	88
$NaSCN$	35.7	K_2CrO_4	88
$NaCl + KClO_3$ (16 °C)	36.6	$ZnSO_4 \cdot 7\ H_2O$	90
NaI	38.4	$Na_2CO_3 \cdot 10\ H_2O$	90
$Zn(NO_3)_2 \cdot 6\ H_2O$	42	$NaBrO_3$	92
$K_2CO_3 \cdot 2\ H_2O$	44	K_2HPO_4	92
KNO_2	45	$Na_2SO_4 \cdot 10\ H_2O$	93
$KSCN$	47	$NH_4H_2PO_4$	93
$NaHSO_4 \cdot H_2O$	52	KNO_3	93.2
$Na_2Cr_2O_7 \cdot 2\ H_2O$	52	$Na_2HPO_4 \cdot 12\ H_2O$	95
$Ca(NO_3)_2 \cdot 4\ H_2O$	54	$Na_2SO_3 \cdot 7\ H_2O$	95
$Mg(NO_3)_2 \cdot 6\ H_2O$	54	K_2SO_4	97.2
$FeCl_2 \cdot 4\ H_2O$	55	$CaSO_4 \cdot 5\ H_2O$	98
$NaBr \cdot 2\ H_2O$	58	$Pb(NO_3)_2$	98

(Fig. 43 B) or in one of the troughs of a twin-trough chamber (Fig. 43 A). If a conditioning chamber is employed the equilibrated TLC plate must then be transferred to the chromatography tank without delay. When a twin-trough chamber is employed the chromatography can be started after equilibration has taken place merely by adding the mobile phase to the second trough.

Such details must also be documented in order to make it possible for others to repeat the experiment.

4.2 Documentation on the Chromatogram

One of the great advantages offered by thin-layer chromatography is that several samples (the same or different) can be developed on the same TLC/HPTLC plate together with the appropriate reference substances. A horizontal chamber (Fig. 58B) can be used to generate up to 70 chromatograms simultaneously, with development taking place antiparallel from two opposing edges of the plate and requiring less than 20 ml of mobile phase.

Samples and reference substances should be dissolved in the same solvents to ensure that comparable substance distribution occurs in all the starting zones. In order to keep the size of the starting zones down to a minimum (diameter TLC: 2 to 4 mm, HPTLC: 0.5 to 1 mm) the application volumes are normally limited to a maximum of 5 µl for TLC and 500 nl for HPTLC when the samples are applied as spots. Particularly in the case of adsorption-chromatographic systems layers with concentrating zones offer another possibility of producing small starting zones. Here the applied zones are compressed to narrow bands at the solvent front before the mobile phase reaches the active chromatographic layer.

The use of application schemes and labelling each single chromatogram track can avoid mistakes as far as the order of application and positioning are concerned. Each sample and reference solution is applied twice with a displacement of half a plate width in the data-pair method [9], that has been in use since the 1960s for the purpose of reducing errors of application and chromatography (Fig. 61). The position of application must naturally be chosen so that the starting zone does not dip into the mobile phase.

A soft pencil can be used to write on the chromatogram. But the chromatogram should not be marked *below* the starting point, because the layer could be damaged

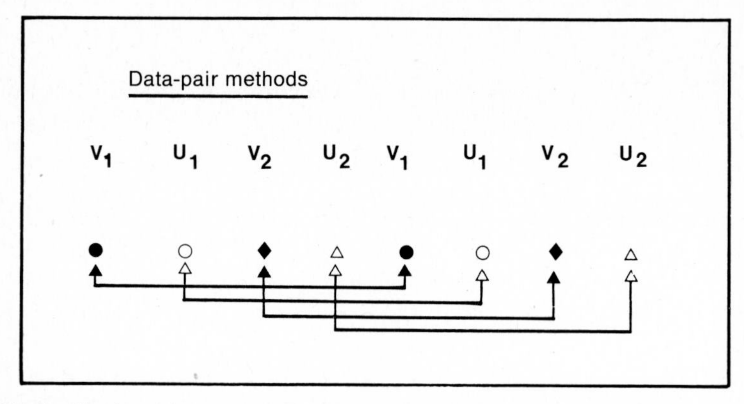

Fig. 61: Application scheme according to the data pair method. – V_1 to V_3 = standard solutions, U_1 to U_3 = sample solutions.

with the result that the substance distribution in the chromatographic zones could be affected. This would lead to errors in direct photometric analysis [10]. It is best to label the chromatogram above the level of the solvent front when the chromatogram is later developed. Immediately after development is completed the level of the solvent front should be marked at the left and right-hand edges of the plate to allow calculation of the R_f values. This should be done even if a solvent front "monitor" is employed [11] or the development is controlled by means of an AMD [12–15] or PMD system [16–18]. The practice of cutting a scratch across the whole adsorbent layer for this purpose is now obsolete – especially doing it before development.

The type of plate, chamber system, composition of mobile phase, running time and detection reagent used must naturally all be recorded. The sample protocol illustrated in Figure 62 can be employed.

When preparing mobile phase mixtures each individual component should be measured out separately and only then placed in the mixing vessel. This prevents not only contamination of the solvent stock by vapors from the already partially filled mixing vessel (e.g. ammonia!) but also volumetric errors caused by volume expansions or contractions on mixing.

All details in the reagent monographs concerning mobile phases are given in parts by volume unless it is specifically stated otherwise.

Test: .... Test No. ....

....

.... Name: ....

.... Date: ....

Chromatographic conditions:

Method: ....

Stationary phase: ....

Mobile phase: ....

Migration distance: .... Migration time: ....

Detection: a) ....

b) ....

c) ....

Evaluation: ....

and ....

comments ....

Preparation of sample: ....

Volume applied: ....

Preparation of standards: ....

Volume applied: ....

Fig. 62: Example of a protocol form

4.3 Fixing the Visual Appearance of the Chromatogram

A chromatogram is produced by developing a TLC/HPTLC plate, but it may be necessary to employ one of the reagents described to make the positions, structures and sizes of the chromatogram zones apparent so that they can be recorded. If the R_f values are the same a comparison of the sizes of the zones of the sample and standard substances gives an indication for estimating the amounts. If, as a result of matrix effects, the R_f values of sample and standard are not the same then their zone size and distribution of substance on the chromatogram will also differ on account of different diffusion patterns. These facts also belong in the protocol of a chromatographic separation.

It is usually only possible to store the original chromatogram if TLC foils are employed or if the adsorbent layer is fixed and removed from the plate as a whole. This can be achieved by treating the chromatogram with collodium [19], "adhesive" [20] or plastic dispersions and exercising a little patience.

4.3.1 Preserving with Neatan

Chromatograms can be made "handleable" and storable by treatment with plastic dispersions, based on polyacrylic ester, polyvinyl chloride or polyvinyl propionate, such as, for example, Neatan (MERCK) [21–23]. In order to avoid clogging the spray head with plastic dispersion residue it is recommended that it be rinsed through immediately after use with tetrahydrofuran or that disposable jets be employed [24].

After drying at a temperature below 50 °C (higher temperatures lead to yellowing!) an adhesive film is rolled over the layer and the covered chromatogram cut down to the glass plate with a razor blade within 2 to 3 mm of its edges and raised at one corner. The "polymerized" layer can then be detached from the glass plate by soaking in water (methanol in the case of polyamide layers [23]) and after drying stuck into the laboratory notebook and labelled (take care when aggressive reagents have been used!).

This method of chromatogram preservation has lost a great deal of its importance with the increasing perfection of photographic methods, particularly since true-colored, instant, paper and slide positives have become available. Photography is the more rapid method of documentation if suitable photographic equipment is available.

4.3.2 Documentation by Sketching, Photocopying or Photographing

The phenomenological analysis results can be recorded by sketching or tracing on translucent paper and colored with crayons or pens to reproduce the impression of color. It is usual to mark fluorescent zones with a "ray-like" edge; phosphorescence-inhibiting zones are marked with a broken line. Stippling and hatching etc. can also be employed by a working group to convey additional information.

Direct copying on Ozalid or Ultrarapid blue print paper has also been employed for documentation [25–27]. For this purpose the chromatograms are laid layer-side up on a light box and, if necessary, covered with a sheet of 1 mm thick glass. The Ozalid paper is then laid on with the yellow-coated side down and covered with an appropriately shaped piece of wood (to keep out the daylight and to maintain contact). The exposure time is 8 to 10 minutes depending on the chromatographic background; it can be shortened to 3 to 4 minutes for weakly colored zones or lengthened appropriately for heavily colored backgrounds. The Ozalid paper is exposed to ammonia vapor after exposure. The chromatogram zones are recorded as red-violet spots on a pale background. AGFA-Copyrapid CpP or CpN papers can also be employed [37].

The simplest method of recording chromatograms nowadays is by means of conventional photocopying (Fig. 63).

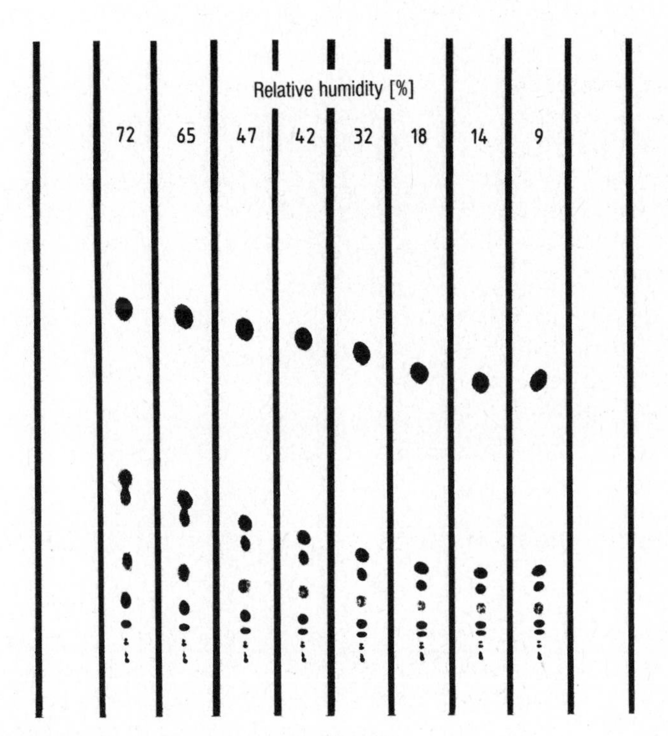

Fig. 63: Photocopy of the chromatogram in Fig. 60.

The chromatograms can also be recorded onto normal photographic paper (e.g. Ilfospeed 2.1 M, glossy 2, Ilford, Essex [28] or Kodak paper [29]) by transmitted light. Here too, as in the case of copying onto Ozalid paper, a sandwich is prepared from chromatogram and photographic paper (dark room!); this is then illuminated for about 1 s with a 25 watt incandescent bulb at a distance of 1.5 m. The negative image on the paper is then developed.

Color reproduction of the chromatograms can be achieved by color photography – the best, but also the most expensive method of documenting thin-layer chromatograms. It can be used not only to produce true-color reproductions of colored zones but also – with the aid of a Reprostar (Fig. 64) or a UVIS analysis lamp (Fig. 6) – of fluorescent or fluorescence-quenched zones. When photograph-

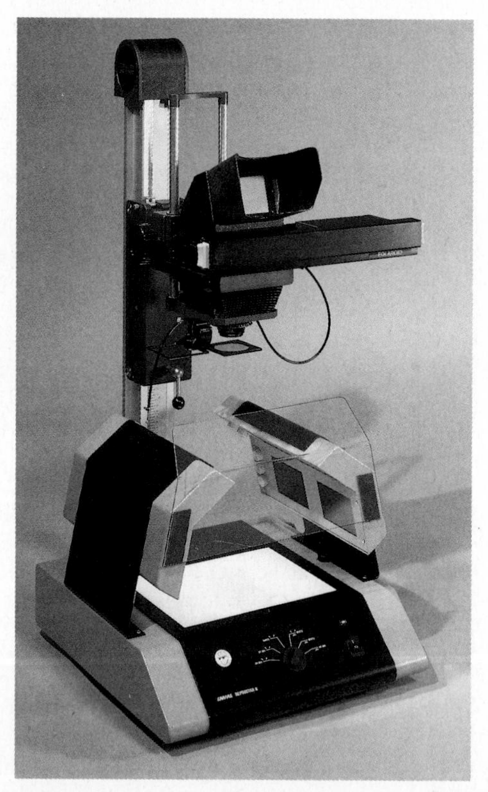

Fig. 64: Reprostar (CAMAG).

ing fluorescent zones it is necessary to ensure that the filter transparency of the UV lamp is optimal so that the chromatogram illumination is even (Sec. 2.2.2 and Fig. 8). It must also be ensured that no exciting radiation reaches the film [31 – 33].

When fluorescence-quenched zones are being photographed (excitation at λ = 254 nm) it is often sufficient to place a Type GG 435 glass filter (SCHOTT, thickness 4 mm) in front of the camera lens. On the other hand, a Wratten Type 2 E filter is recommended for recording fluorescent zones (excitation at λ = 365 nm) [34]. The slight blue coloration of the film caused by light passing the black-light filter of the UV lamp can be avoided by a yellow or pale orange filter.

The required exposure times are difficult to estimate. They are best found by trial and error. Documentation of fluorescence quenching at λ = 254 nm usually only requires one trial. The exposure time found to be adequate here is normally suitable for all following exposures of fluorescence quenching if the exposure conditions are maintained constant (camera type, film type, distance of objective and lamp, aperture etc.). The exposure time required for fluorescent chromatograms is primarily dependent on the intensity of the fluorescence and, therefore, has to be optimized for each chromatogram. It is best to operate with a range of exposure times, e.g. aperture 8 with exposures of 15, 30, 60, 120 and 240 seconds. Experience has shown that one exposure is always optimal.

Favorable settings are given in Table 28 for the employment of Polaroid color film Type 669 in a CU-5 Polaroid-Land camera with a 127 mm objective. Further

Table 28: Setting conditions for a Polaroid Land CU 5 camera with supplementary lens (0.5 dioptres) for photographing TLC plates of 20 × 20 cm format on Type 669 Polaroid color film.

Subject	Aperture	Exposure	Color correction
Colored zones, white transmitted *or* incident light	11	1/8 s	no
Colored zones, white transmitted *and* incident light	16	1/8 s	no
Fluorescent zones, incident light at λ = 365 nm	8	30 s	yes
Fluorescence quenching zones, incident light at λ = 254 nm	8	30 s	yes

detailed data have been reported in the literature for other types of cameras and films [30–32, 35, 36]. The production of autoradiograms will be dealt with in Volume 2.

The color plates for the chromatograms reproduced in the reagent monographs were produced using an Olympus OM-4 camera with 50 mm lens combined with a copying stand with TTL-Makroblitz T 28 (OLYMPUS).

4.4 Documentation by Means of in situ Evaluation by Computer

The methods of documentation of thin-layer chromatograms described until now depend on the photographic recording of the visually detectable chromatograms. The film negatives or slides can then be subjected to densitometric quantitation [26, 38, 39] and included in the protocol. This indirect method of TLC chromatogram quantitation with its intermediate photographic step has been rendered completely obsolete with the coming of the computer-controlled, chromatogram spectrometer.

Particularly elegant documentation is achieved by storing the quantitative TLC data on diskettes. They are then available for years and complement the qualitative record in an excellent manner. In addition they are always available for statistical analysis and, thus, contribute to comprehensive documentation.

References

[1] Vavruch, J.: *CLB Chem. Labor Betr.* **1984,** *35*, 536–543.
[2] Reichardt, C.: *Lösungsmittel-Effekte in der organischen Chemie.* Verlag Chemie, Weinheim 1979.
[3] Hampel, B., Maas, K.: *Chem. Ztg.* **1971,** *95*, 316–325.
[4] Koch, F., Plein, G.: *GIT Fachz. Lab. Supplement 3 „Chromatographie"* **1988,** 52–56.
[5] Stahl, E.: *Dünnschicht-Chromatographie, ein Laboratoriumshandbuch.* 2nd Ed., Springer, Berlin–Heidelberg–New York 1967.
[6] E. MERCK: Company literature *„Wasserfeste, wasserbenetzbare DC-Fertigschichten Kiesel-gel 60 W"*, Darmstadt 1985.
[7] MACHEREY-NAGEL: Company literature *„Fertigprodukte für die Dünnschicht-Chromatographie"*, 1985.

[8] Geiss, F.: *Fundamentals of Thin Layer Chromatography (Planar Chromatography)*. Hüthig-Verlag, Heidelberg-Basel-New York 1987.
[9] Bethke, H., Santi, W., Frei, R. W.: *J. Chromatogr. Sci.* **1974**, *12*, 392-397.
[10] Jork, H., Wimmer, H.: *Quantitative Direktauswertung von Dünnschicht-Chromatogrammen* (TLC-Report). GIT-Verlag, Darmstadt 1982ff.
[11] Omori, T.: *J. Planar Chromatogr.* **1988**, *1*, 66-69.
[12] Burger, K.: *Fresenius Z. Anal. Chem.* **1984**, *318*, 228-233.
[13] Burger, K.: *GIT Fachz. Lab. Supplement „Chromatographie"* **1984**, 29-31.
[14] Jänchen, D. E.: *Proc. Int. Symp. Instrum. High Perform. Thin-Layer Chromatogr. (HPTLC), 3rd, Würzburg,* 1985, 71-82.
[15] Jork, H.: *Proc. Int. Symp. Instrum. High Perform. Thin-Layer Chromatogr., 4th, Selvino* 1987, 193-239.
[16] Jupille, T. H., Perry, J. A.: *Science* **1976**, *194*, 288-293.
[17] Issaq, H. J., Barr, E. W.: *Anal. Chem.* **1977**, 83A-97A.
[18] Jupille, T. H., Perry, J. A.: *J. Chromatogr.* **1975**, *13*, 163-167.
[19] Barrollier, J.: *Naturwissenschaften* **1961**, *48*, 404.
[20] Lüdy-Tenger, F.: *Schweiz. Apoth. Ztg.* **1967**, *105*, 197-198.
[21] Lichtenberger, W.: *Fresenius Z. Anal. Chem.* **1962**, *185*, 111-112.
[22] Foner, H. A.: *Analyst (London)* **1965**, *91*, 400-401.
[23] E. MERCK: Company literature *„Neatan"*, 1975.
[24] Stuart-Thomson, J.: *J. Chromatogr.* **1975**, *106*, 423-424.
[25] Sprenger, H. E.: *Fresenius Z. Anal. Chem.* **1964**, *204*, 241-245.
[26] Rasmussen, H.: *J. Chromatogr.* **1967**, *27*, 142-152.
[27] Abbott, D. C., Blake, K. W., Tarrant, K. R., Thomson, J.: *J. Chromatogr.* **1967**, *30*, 136-142.
[28] Engström, N., Hellgren, L., Vincent, J.: *J. Chromatogr.* **1980**, *189*, 284-285.
[29] Steinfeld, R.: *KODAK-Bulletin* **1947**, *57*, (1), 1-3.
[30] Scholtz, K. H.: *Dtsch. Apoth. Ztg.* **1974**, *114*, 589-592.
[31] Heinz, D. E., Vitek, R. K.: *J. Chromatogr. Sci.* **1975**, *13*, 570-576.
[32] Eggers, J.: *Photogr. Wiss.* **1961**, *10*, 40-43.
[33] Zimmer, H.-G., Neuhoff, V.: *GIT Fachz. Lab.* **1977**, *21*, 104-105.
[34] Rulon, P. W., Cardone, M. J.: *Anal. Chem.* **1977**, *49*, 1640-1641.
[35] Michaud, J. D., Jones, D. W.: *Am. Lab.* **1980**, *12*, 104-107.
[36] Romel, W. C., Adams, D., Jones, D. W.: Private communication.
[37] Gänshirt, H.: *Arch. Pharmaz.* **1963**, *296*, 73-79.
[38] Andreev, L. V.: *J. Liq. Chromatogr.* **1982**, *5*, 1573-1582.
[39] Colarow, L., Pugin, B., Wulliemier, D.: *J. Planar Chromatogr.* **1988**, *1*, 20-23.
[40] Weiss, M., Jork, H.: *GIT Arbeitsblatt 093*, GIT-Verlag, Darmstadt 1982.

Part II

Reagents in Alphabetical Order

Alizarin Reagent

Reagent for:

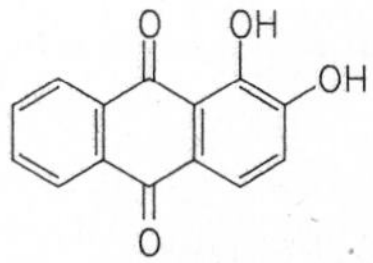

- Cations [1 – 7]

$C_{14}H_8O_4$
$M_r = 240.22$

Preparation of the Reagent

Dipping solution	Dissolve 100 mg alizarin in 100 ml ethanol.
Spray solution	A 0.25% [3] or saturated solution of alizarin [4] in ethanol.
Storage	The solution may be kept for several days at room temperature.
Substances	Alizarin Ethanol Ammonia solution (25%)

Reaction

Metal cations yield colored complexes with alizarin:

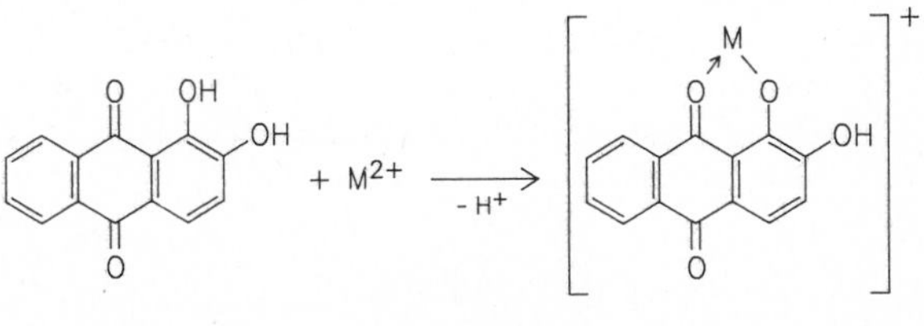

$$+ \; M^{2+} \xrightarrow{-H^+}$$

Method

The chromatogram is dried for 10 min in warm air and either immersed in the dipping solution (1 s) or sprayed evenly with the spray solution. The still-moist TLC plates are placed in the empty part of a twin-trough chamber filled with ammonia solution (25%) for 1 min. In a few minutes red-violet zones appear on a violet background. If the plate is then dipped in either 0.1 – 1% acetic acid in diethyl ether or in a 1% solution of boric acid in methanol-water (9 + 1) the background turns yellow and most of the chromatogram zones appear as red to violet spots. The following cations (arranged as in the periodic table) can be detected:

I	II	III	IV	V	VI	VII	VIII
Li	Be			NH_4			
	Mg	Al					
Cu	Ca Zn	Ga Sc	Ti	As V	Se Cr	Mn	Fe Co Ni
Ag	Sr Cd	In	Sn Zr	Sb			Pd
Au	Ba Hg	La	Pb	Bi			Pt

and Ce, Th, U and other rare earths.

Note: The reagent can be employed on silica gel layers, which may also be impregnated, for example, with 8-hydroxyquinoline or dibenzoylmethane [3] or with 2,2′-dipyridyl or iminodiacetic acid [4] or on cellulose layers.

Rather than dipping the chromatogram in acid solution it is preferable to heat it to 100 °C for 2 – 5 min (fume cupboard!) in order to evaporate the ammonia and turn the background yellow. By this means it is possible to increase the sensitivity of detection for some of cations e.g. Sr^{2+} and Ba^{2+}. However, these zones fade after some time, so that it is necessary to quantify the chromatogram immediately after heating.

Procedure Tested

Nickel, Copper and Beryllium nitrates; Calcium, Magnesium, Strontium and Barium chlorides [5, 7]

Method One-dimensional ascending development in a HPTLC chamber with chamber saturation.

Layer HPTLC plates Cellulose (MERCK). The plates were pre-washed with mobile phase and dried for 10 min in a stream of warm air before use.

Mobile phase
1. Ni, Cu and Be cations: acetone – nitric acid (25%) (35 + 15).
2. Alkaline earth cations [5]: methanol-hydrochloric acid (25%) (80 + 20).

Migration distance 5 – 7 cm

Running time 20 min

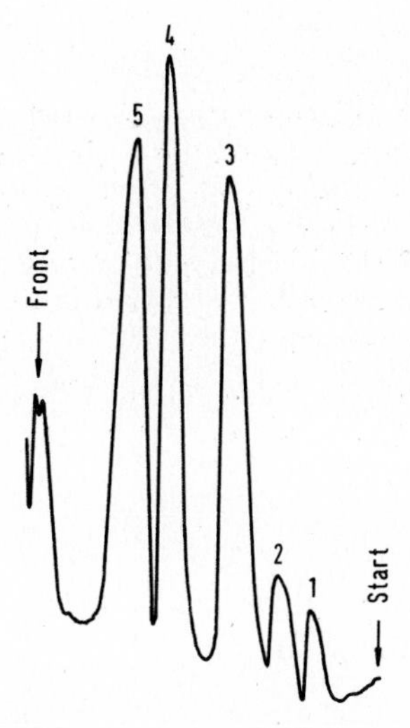

Fig. 1: Absorption scanning curve of the alizarin complexes of barium (1), strontium (2), calcium (3), magnesium (4) and beryllium cations (5). The amounts applied were 2 µg in each case.

Detection and result: The separation was adequate in the systems given. The following hR_f values were obtained after heating for 2–5 min to 100 °C.

Mobile phase 1: Ni^{2+} (blue-violet), hR_f: 45–50; Cu^{2+} (violet), hR_f : 55–60; Be^{2+} (violet), hR_f: 80–85.
Mobile phase 2: Ba^{2+} (pale blue-violet), hR_f: 15–20; Sr^{2+} (pale blue-violet, slightly tailing), hR_f: 30–35; Ca^{2+} (violet), hR_f: 50–55; Mg^{2+} (violet), hR_f: 75–80; Be^{2+} (violet), hR_f: 93–98.
The visual limit of detection was between 30 and 50 ng per chromatogram zone for nickel and copper, a factor of ten worse for the alkaline earths. There was sometimes an "impurity" front in the same hR_f range as beryllium.

In situ quantitation: The photometric determination was made in reflectance mode at $\lambda = 550$ nm (Fig. 1).

References

[1] De Vries, G., Schütz, G. P., Van Dalen, E.: *J. Chromatogr.* **1964**, *13,* 119-127.
[2] Hammerschmidt, H., Müller, M.: *Papier Darmstadt* **1963**, *17,* 448-450.
[3] Srivastava, S. P., Bhushan, R., Chauhan, R. S.: *J. Liq. Chromatogr.* **1984**, *7,* 1341-1344.
[4] Srivastava, S. P., Bhushan, R., Chauhan, R. S.: *J. Liq. Chromatogr.* **1985**, *8,* 571-574.
[5] Gagliardi, E., Likussar, W.: *Mikrochim. Acta (Vienna)* **1965,** 765-769.
[6] Bhushan, R., Srivastava S. P., Chanhan, R. S.: *Anal. Letters* **1985**, *18,* 1549-1553.
[7] Kany, E., Jork, H.: GDCH-Workshop Nr. 300 „Einführung in die Dünnschicht-Chromatographie", Universität des Saarlandes, Saarbrücken 1987.

Aluminium Chloride Reagent

Reagent for:

- Unsaturated 1,2- and 1,3-dihydroxyketones, e.g. flavonoids [1–4]
- Mycotoxins: zearalenone [5, 6, 10]; ochratoxin [5]; sterigmatocystine [7–11]; citrinine [12]
- Trichothecenes: deoxynivalenol (vomitoxin) [13]
- Cholesterol and its esters [14]
- Phospholipids and triglycerides [14]

$AlCl_3 \cdot 6H_2O$
$M_r = 241.45$

Preparation of the Reagent

Solution I Dissolve 0.2 to 1 g aluminium chloride in 100 ml ethanol.

Solution II Dissolve 20 g aluminium chloride in 100 ml ethanol.

Storage The solutions can be stored for long periods in the refrigerator.

Substances Aluminium chloride hexahydrate
Ethanol absolute

Reaction

Aluminium chloride forms, for example, fluorescent complexes with flavonoids:

$$\text{5-hydroxyflavone} + Al^{3+} \xrightarrow{-H^{+}} \text{Al complex} \quad (1/3\ Al)$$

Method

The developed chromatograms are briefly immersed in or evenly sprayed with the appropriate reagent solution. Solution I is employed for flavonoids [1, 3] and solution II for mycotoxins [5, 8, 12], phospholipids, triglycerides and cholesterol [14].

After the dipped or sprayed chromatogram has been dried in a stream of cold air long-wave UV light ($\lambda = 365$ nm) reveals fluorescent yellow zones (flavonoids). Sterigmatocystine, which can be detected without derivatization on account of its red intrinsic fluorescence (detection limit 0.5 µg), also fluoresces pale yellow after being heated to 80 °C [9] or 100 °C [13] for 10 min; on the other hand, citrinine, zearalenone and vomitoxin fluoresce blue.

The detection limits are ca. 20 ng per chromatogram zone.

Note: The reagent can be employed on silica gel, kieselguhr, polyamide, RP, CN, NH_2 and cellulose layers.

The colors of the fluorescing zones can depend on the concentration of the aluminium chloride solution.

Hypericin which is a hydroxyanthraquinone and the antibiotic nystatin also yield fluorescent zones. A higher fluorescence intensity is frequently obtained by heating to 88 °C for 2–5 min instead of simply allowing to dry at room temperature.

Procedure Tested

Flavonoids (Quercetin, Rutin, Hyperoside, Quercitrin) [15]

Method One-dimensional, ascending development in a trough chamber with chamber saturation.

Layer HPTLC plates Silica gel 60 (MERCK) pre-washed by a single development with chloroform-methanol (50+50). The layers were preconditioned for 30 min over water after the sample had been applied.

Mobile phase Ethyl acetate – formic acid (98–100%) – water (85+10 +15).

Migration distance 6 cm

Running time 18 min

Detection and result: The chromatogram was freed from mobile phase and dipped for 1 s in solution I and after drying for 1 min in a stream of cold air it was dipped in a solution of liquid paraffin – *n*-hexane (1 + 2) in order to stabilize and increase the intensity of fluorescence by a factor of 1.5–2.5. The derivatives which were pale yellow in daylight after drying fluoresce pale blue to turquoise in long-wave

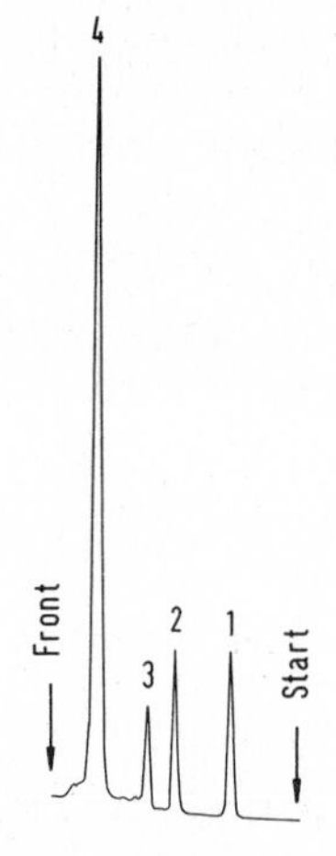

Fig. 1: Separation of flavonoids: fluorescence scanning curve of rutin (1), hyperoside (2), quercitrin (3) and quercetin (4).

UV light ($\lambda = 365$ nm): rutin (hR_f: 25–30), hyperoside (hR_f: 45–50), quercitrin (hR_f: 60–65) and quercetin (hR_f 85–90). The detection limits were 10 ng substance per chromatogram zone.

In situ quantitation: The in situ fluorescence measurement was carried out at λ_{exc} = 436 nm and λ_{fl} = 546 nm (monochromate filter M 546).

References

[1] Gage, T. G., Douglas, C. H., Wender, S. H.: *Anal. Chem.* **1951,** *23* 1582–1585.
[2] Zaprjanowa, A. Z, Angelowa, M. K.: *Mikrochim Acta (Vienna)* **1976,** *II,* 481–486.
[3] Förster, H., Ziege, M.: *Fortschr. Med.* **1971,** *89,* 672–675.
[4] Willuhn, G., Röttger, P. M.: *Dtsch. Apoth. Ztg.* **1980,** *120,* 1039–1042.
[5] Fonseca, H., Nogueira, J., Graner, M.: Proc. of the 6th Int. Congress of Food Science and Technology Dublin, **1983,** *3,* 53–54.
[6] Takeda, Y., Isohata, E., Amano, R., Uchiyama, M.: *J. Assoc. off. Anal. Chem.* **1979,** *62,* 573–578.
[7] Majerus, P., Woller, R., Leevivat, P., Klintrimas, T.: *Fleischwirtschaft* **1985,** *65,* 1155–1158.
[8] Kiermeier, F., Kraus, P. V.: *Z. Lebensm.-Unters. Forsch.* **1980,** *170,* 421–424.
[9] Johann, J., Dose, K.: *Fresenius Z. Anal. Chem.* **1983,** *314,* 139–142.
[10] Josefsson, B. G., Möller, T. E.: *J. Assoc. off. Anal. Chem.* **1977,** *60,* 1369–1371.
[11] Egmond, H. P. van, Paulsch, W. E., Deijll, E.: *J. Assoc. off. Anal. Chem.* **1980,** *63,* 110–114.
[12] Gimeno, A.: *J. Assoc. off. Anal. Chem.* **1979,** *62,* 579–585; **1980,** *63,* 182–186: **1984,** *67,* 194–196.
[13] Eppley, M. R., Trucksess, M. W., Nesheim, S., Thorpe, C. W., Wood, G. F., Pohland, A. E.: *J. Assoc. off. Anal. Chem.* **1984,** *67,* 43–45.
[14] Segura, R., Navarro, X.: *J. Chromatogr.* **1981,** *217,* 329–340.
[15] Gilles, F.: Thesis, Universität Gießen, Fachbereich Agrarwissenschaft, 1986.

4-Aminoantipyrine – Potassium Hexacyanoferrate(III) Reagent
(Emerson Reagent)

Reagent for:

- Arylamines [1, 2]
- Phenols [3 – 5]
- Salithion [6]
- Sympathicomimetics

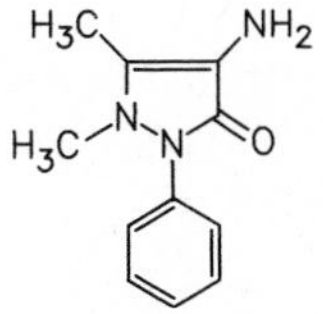

$C_{11}H_{13}N_3O$

$M_r = 203.25$

4-Aminoantipyrine

$K_3[Fe(CN)_6]$

$M_r = 329.26$

Preparation of the Reagent

Dipping solution I Dissolve 1 g 4-aminoantipyrine (4-aminophenazone; 4-amino-2,3-dimethyl-1-phenyl-3-pyrazolin-5-one) in 100 ml 80% ethanol.

Dipping solution II Dissolve 4 g potassium hexacyanoferrate(III) in 50 ml water and make up to 100 ml with ethanol.

Spray solution I Dissolve 2 g 4-aminoantipyrine in 100 ml 80% ethanol.

Spray solution II See dipping solution II.

Storage The dipping solution and spray solution I can be stored in the refrigerator for about 1 week.

Substances 4-Amino-2,3-dimethyl-1-phenyl-3-pyrazolin-5-one
Potassium hexacyanoferrate(III)
Ammonia solution (25%)
Ethanol

Reaction

4-Aminoantipyrine forms with aniline, for instance, a colored diimine derivative under the oxidative influence of iron(III) ions.

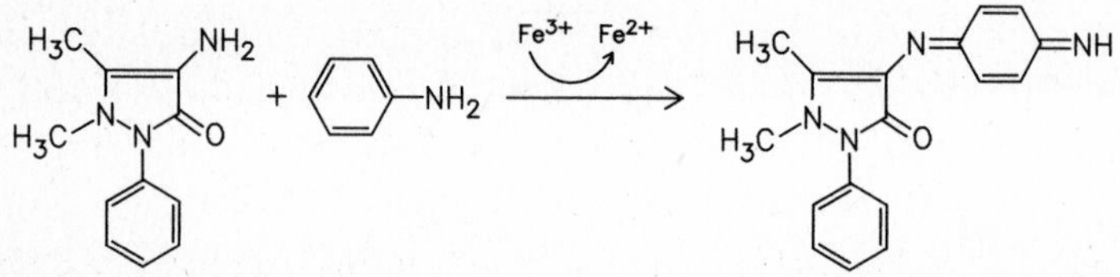

4-Aminoantipyrine Aniline Quinone diimine derivative

Method

The chromatogram is freed from mobile phase and immersed in dipping solution I for 1 s or sprayed with spray solution I, dried in warm air for 5 min and then immersed for 1 s in dipping solution II or sprayed with spray solution II. After redrying the background is decolorized by placing the chromatogram in a twin-trough chamber, one of whose troughs contains 25% ammonia solution. Red-orange colored zones are produced on a pale yellow background. The color intensity of the chromatogram zones is also increased in the case of phenols, since these only react in alkaline medium [4].

Note: The reagent can be just as successfully employed on silica gel, kieselguhr, aluminium oxide and polyamide layers as it can with RP and NH_2 phases. The final treatment with ammonia vapor to decolorize the background can be omitted in the last case.

An iodine solution can be employed as oxidizing agent in place of potassium hexacyanoferrate(III). 4-Aminoantipyrine also produces colored zones with 1- and 1,4-unsaturated 3-ketosteroids (pregnadienediol derivatives) in the absence of oxidizing agents.

In the case of sympathicomimetics it should be checked whether the reaction in solution (PFEIFER, S., MANNS, O.: Pharmazie **1957**, *12*, 401-408) is applicable to TLC.

Procedure Tested

Eugenol, Carvacrol and Thymol in Essential Oils [7]

Method Ascending, one-dimensional development in a trough chamber with chamber saturation. A double development (5 min intermediate drying in a stream of warm air) was required for the separation of carvacrol/thymol.

Layer TLC plates Silica gel 60 W or HPTLC plates Silica gel 60 WRF_{254s} (MERCK); SIL G-25 or Nano-Sil-20 (MACHEREY-NAGEL).

Mobile phase TLC: chloroform; HPTLC: toluene (for carvacrol/thymol separation).

Migration distance TLC: 10 cm; HPTLC: 6 cm

Running time TLC: 20 min; HPTLC: 12 min

Detection and result: The following hR_f values (TLC, chloroform) were obtained: carvacrol and thymol (red) 30 – 35, eugenol (salmon pink) 40 – 45; here the background was pale yellow colored. The visual detection limit was 200 – 500 ng per chromatogram zone. The colored zones faded after 1 to 2 hours.

References

[1] Eisenstaedt, E.: *J. Org. Chem.* **1938**, *3*, 153-165
[2] Mordovina, L. L., Korotkova, V. I., Noskov, V. V.: *Zh. Anal. Khim.* **1974**, *29*, 580-583.
[3] Gabel, G., Müller, K. H., Schonknecht, J.: *Dtsch. Apoth. Ztg.* **1962**, *102*, 293-295.
[4] Emerson, E.: *J. Org. Chem.* **1943**, *8*, 417-428.
[5] Thielemann, H.: Fresenius *Z. anal. Chem.* **1974**, *269*, 125-126; Pharmazie **1977**, *32*, 244.
[6] Murano, A., M. Nagase, S. Yamane: *Japan Analyst* **1971**, *20*, 565-569.
[7] Zentz, V.: Thesis, Universität des Saarlandes, Fachbereich 14, Saarbrücken 1978.

4-Aminobenzoic Acid Reagent

Reagent for:

- Carbohydrates (sugars)

  e.g. monosaccharides [1 – 5]
  disaccharides [2, 3]
  uronic acids [1 – 3]

$C_7H_7NO_2$

$M_r = 137.14$

Preparation of the Reagent

Dipping solution Dissolve 1 g 4-aminobenzoic acid in 18 ml glacial acetic acid and add 20 ml water and 1 ml 85% phosphoric acid; immediately before use dilute with acetone in the ratio 2 + 3 [2].

Storage The reagent may be stored for 1 week in the dark at room temperature.

Substances 4-Aminobenzoic acid
Glacial acetic acid
Orthophosphoric acid (85%)
Acetone

Reaction

Sugars react with the reagent probably with the formation of SCHIFF's bases:

$$R^1R^2C{=}O + H_2N{-}C_6H_4{-}COOH \xrightarrow{-H_2O} R^1R^2C{=}N{-}C_6H_4{-}COOH$$

Method

The chromatogram is freed from mobile phase and dipped in the reagent for 2 s or uniformly sprayed with it, dried for several minutes in a stream of cold air and heated to 100 °C for 10 – 15 min. The result is reddish-brown chromatogram zones on a colorless to pale brown background.

Note: The reagent can be employed on cellulose layers. Sodium acetate-buffered kieselguhr layers are less suitable [6]. Only a few sugars are detectable and those with lower sensitivity if acid is not added to the reagent [7].

Procedure Tested

Sugar in Diabetic Chocolate [8]

Method Ascending, one-dimensional development in a trough chamber with chamber saturation.

Layer TLC plates Silica gel 60 (MERCK).

Mobile phase Dichloromethane – methanol – glacial acetic acid – water (50 + 50 + 25 + 10).

Migration distance 15 cm

Running time 120 min

Detection and result: The chromatogram was freed from mobile phase and immersed in the reagent solution for 2 s and placed in a drying cupboard while still moist. After heating to 120 °C for 15 min red-brown (fructose) and grey-blue (lactose) chromatogram zones were produced, which fluoresced turquoise under long-wavelength UV light ($\lambda = 365$ nm). The detection limits in visible light were 200 – 300 ng substance per chromatogram zone. The detection limits were appreciably lower with less than 5 ng per chromatogram zone on fluorimetric analysis.

In situ quantitation: Quantitation was performed fluorimetrically ($\lambda_{exc} = 365$ nm, $\lambda_{fl} > 560$ nm). The baseline structure was most favorable under the chosen conditions (Fig. 1).

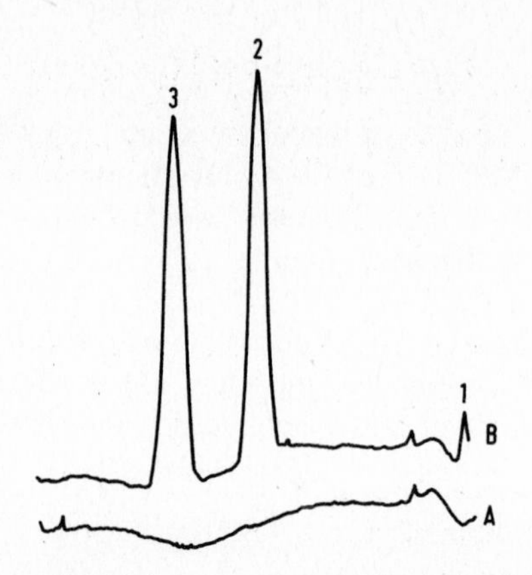

Fig. 1: Fluorescence plot of a blank (A) and a chromatogram track of a diabetic chocolate extract (B). Circa 50 ng lactose and fructose were applied. Start (1), lactose (2), fructose (3).

References

[1] Metraux, J. P.: *Chromatogr.* **1982,** *237,* 525-527.
[2] Ersser, R. S., Andrew, B. C.: *Med. Lab. Technol.* **1971,** *28,* 355-359.
[3] Damonte, A., Lombard, A., Tourn, M. L., Cassone, M. C.: *J. Chromatogr.* **1971,** *60,* 203-211.
[4] Menzies, I. S., Mount, J. N.: *Med. Lab. Technol.* **1975,** *32,* 269-276.
[5] Kröplin, U.: *J. agric. Food Chem.* **1974,** *22,* 110-116.
[6] Bell, D. J., Talukder, M. Q.-D.: *J. Chromatogr.* **1970,** *49,* 469-472.
[7] Roy, J. K.: *Analyst* (London) **1960,** *85,* 294-295.
[8] Müller, E., Jork, H.: GDCh-training course Nr. 302 „Möglichkeiten der Quantitativen Auswertung von Dünnschicht-Chromatogrammen“, Universität des Saarlandes, Saarbrücken 1987.

2-Aminodiphenyl – Sulfuric Acid Reagent

Reagent for:

- Carbonyl compounds [1, 2]
  e.g. aliphatic aldehydes from C-8,
  glycol aldehyde [5], glyoxalic acid,
  2,3-pentanedione
- Vitamin B_6 (pyridoxal) [2]
- Sugars [3, 4]

NH_2

$C_{12}H_{11}N$ $M_r = 169.23$

H_2SO_4 $M_r = 98.08$

Preparation of Reagent

Solution I Dissolve 1 g 2-aminodiphenyl (biphenyl-2-ylamine) in 100 ml ethanol.

Solution II 20% sulfuric acid.

Dipping solution Carefully mix equal volumes of solutions I and II immediately before the dipping process.

Spray solution For sugars [3]: dissolve 3 g 2-aminodiphenyl in 100 ml glacial acetic acid and add 1.5 ml 85% orthophosphoric acid.

Storage Solution I may be stored in the refrigerator for several days.

Substances Biphenyl-2-ylamine
Sulfuric acid (25%)
Ethanol
Glacial acetic acid
Orthophosphoric acid (85%)

Reaction

2-Aminodiphenyl reacts with carbonyl compounds to form colored or fluorescent SCHIFF's bases with the elimination of water:

$$R^1R^2C{=}O + \text{2-aminodiphenyl}\ (NH_2) \xrightarrow{-H_2O} \text{2-biphenylyl--}N{=}CR^1R^2$$

Method

The chromatogram is freed from mobile phase and immersed for 1 s in the freshly prepared reagent solution and then heated to 105 to 110 °C for 5 to 10 min. Green, blue or purple fluorescence appears on a dark background under long-wavelength UV light ($\lambda = 365$ nm).

Note: The dipping solution, which can also be used as a spray solution, can be employed with silica gel, kieselguhr, cellulose, RP, NH_2 and CN phases. Sugars (exceptions include, for example, fructose, melezitose and raffinose) yield brilliantly colored zones on an almost colorless background when the spray solution is employed. Aldohexoses appear brown, aldopentoses bright red and hexuronic acids orange in color [3]. The detection limit differs for different substances; it ranges from 10 ng (pyridoxal) over 100 ng (cinnamaldehyde) up to 2 µg (citral).

Procedure Tested

Citral, Citronellal, Cinnamaldehyde [6]

Method Ascending, one-dimensional development in a trough chamber.

Layer HPTLC plates Silica gel 60 F_{254} (MERCK).

Mobile phase Toluene – ethyl acetate – glacial acetic acid (90 + 5 + 5).

Migration distance 5 cm

Running time 20 min

Detection and result: The chromatogram was freed from mobile phase and immersed for 1 s in the freshly prepared dipping solution and then heated to 105 to 110 °C for 5 to 10 min. Citral (hR_f 60) and citronellal (hR_f 80) produced brown zones on a light brown background in visible light, the zones had a purple fluorescence under long-wavelength UV light ($\lambda = 365$ nm). Cinnamaldehyde acquired an intense yellow color but did not fluoresce.

In situ quantitation: This could be made under long-wavelength UV light ($\lambda_{exc} =$ 365 nm; $\lambda_{fl} > 560$ nm). However, it was not very sensitive.

References

[1] Nakai, T., Demura, H., Koyama, M.: *J. Chromatogr.* **1972**, *66*, 87–91.
[2] Nakai, T., Ohta, T., Takayama, M.: *Agric. Biol. Chem.* **1974**, *38*, 1209–1212.
[3] Timell, T. E., Glaudemans, C. P. J., Currie, A. L.: *Anal. Chem.* **1956**, *28*, 1916–1920.
[4] McKelvy, J. F., Scocca, J. A.: *J. Chromatogr.* **1970**, *51*, 316–318.
[5] Nakai, T., Ohta, T., Wanaka, N., Beppu, D.: *J. Chromatogr.* **1974**, *88*, 356–360.
[6] Kany, E., Jork, H.: GDCh-training course Nr. 301, „Dünnschicht-Chromatographie für Fortgeschrittene“, Universität des Saarlandes, Saarbrücken 1986.

4-Aminohippuric Acid Reagent

Reagent for:

- Sugars (monosaccharides) [1, 2]

$H_2N-C_6H_4-C(=O)-NH-CH_2-COOH$

$C_9H_{10}N_2O_3$
$M_r = 194.19$

Preparation of the Reagent

Dipping solution Dissolve 0.5 g 4-aminohippuric acid in 50 ml ethanol and make up to 100 ml with toluene.

Spray solution Dissolve 0.5 g 4-aminohippuric acid in 100 ml ethanol.

Storage Both solutions are stable for several days.

Substances 4-Aminohippuric acid
Ethanol
Toluene

Reaction

Sugars react with the reagent, probably with the production of SCHIFF's bases:

$$R^1R^2C{=}O + H_2N-C_6H_4-C(=O)-NH-CH_2-COOH \xrightarrow{-H_2O} R^1R^2C{=}N-C_6H_4-C(=O)-NH-CH_2-COOH$$

Carbonyl compound — 4-Aminohippuric acid — Reaction product

Method

The chromatogram is freed from mobile phase and immersed in the reagent solution for 1 s or homogeneously sprayed with the spray solution and then heated to 140°C for 8 min [2]. Hexoses and pentoses yield orange-colored zones on an almost colorless background, the zones fluoresce blue under long-wavelength UV light ($\lambda = 365$ nm).

Note: The layers on which the reagent can be employed include silica gel, cellulose and polyamide.

Procedure Tested

Hexoses and Pentoses [1]

Method	Ascending, one-dimensional development in a HPTLC trough chamber with chamber saturation.
Layer	HPTLC plates Silica gel 60 (MERCK).
Mobile phase	2-Propanol – 0.75% aqueous boric acid – glacial acetic acid (40 + 5 + 1).
Migration distance	8 cm
Running time	80 min

Detection and result: The chromatogram was freed from mobile phase and immersed in the reagent solution for 1 s and heated to 140°C for 10 min. The following appeared as blue fluorescent zones under long-wavelength UV light ($\lambda =$ 365 nm): lactose (hR_f 25–30), fructose (hR_f 30–35), arabinose (hR_f 45–50), xylose (hR_f 55–60) and rhamnose (hR_f 60–65) (Fig. 1A).

Immersion in a liquid paraffin – *n*-hexane (1 + 3) did not lead to an appreciable increase in fluorescence intensity.

In situ quantitation: The fluorimetric analysis was made in UV light ($\lambda_{exc} = 313$ nm, $\lambda_{fl} > 460$ nm; Fl 46 filter). The signal-noise ratio was better above $\lambda = 460$ nm than when a Fl 39 filter is employed.

The detection limits per chromatogram zone were 10–20 ng for xylose, 50–100 ng for fructose, arabinose and rhamnose and 1–2 µg for lactose (Fig. 1B).

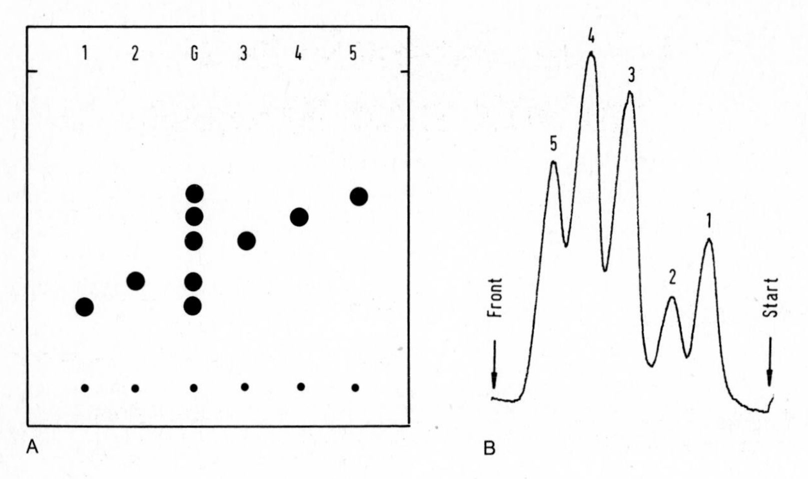

Fig. 1: Schematic diagram of the chromatographic separation (A) and the fluorescence scan (B) of a sugar mixture containing 1 µg substance per chromatogram zone. Lactose (1), fructose (2), arabinose (3), xylose (4), rhamnose (5), mixture (G).

References

[1] Jork, H., Kany, E.: GDCh-training course Nr. 300 „Einführung in die Dünnschicht-Chromatographie", Universität des Saarlandes, Saarbrücken 1986.

[2] Sattler, L., Zerban, F. W.: *Anal. Chem.* **1952**, *24,* 1862.

4-Aminohippuric Acid – Phthalic Acid Reagent

Reagent for:

- Sugars (mono- and disaccharides) [1 – 3]

  e.g. hydrolysates of tragacanth [4]

$C_8H_6O_4$	$C_9H_{10}N_2O_3$
$M_r = 166.13$	$M_r = 194.19$
Phthalic acid	4-Aminohippuric acid

Preparation of the Reagent

Dipping solution Dissolve 0.2 g 4-aminohippuric acid and 3 g phthalic acid in 50 ml ethanol and dilute with 50 ml toluene.

Spray solution Dissolve 0.3 g 4-aminohippuric acid and 3 g phthalic acid in 100 ml ethanol.

Storage Both solutions are stable for several days.

Substances 4-Aminohippuric acid
Phthalic acid
Ethanol
Toluene

Reaction

Sugars react with the reagent probably with the formation of SCHIFF's bases:

$$R^1R^2C{=}O + H_2N{-}C_6H_4{-}CO{-}NH{-}CH_2{-}COOH \xrightarrow{-H_2O} R^1R^2C{=}N{-}C_6H_4{-}CO{-}NH{-}CH_2{-}COOH$$

Method

The chromatogram is dried in a stream of warm air and immersed for 10 s in the reagent solution or the spray solution is applied to it homogeneously and it is then heated to 115–140 °C for 8–15 min [2, 4]. Yellow to orange-red zones are produced on an almost colorless background; these emit an intense blue fluorescence under long-wavelength UV light ($\lambda = 365$ nm).

Note: Subsequent immersion of the chromatogram in a mixture of liquid paraffin – *n*-hexane (1 + 2) leads to an increase in the fluorescence by a factor of 2.5 to 4.5 for some carbohydrates.

The visual limit of detection, when irradiating with UV light, is 250 ng per chromatogram zone; the zones are only detectable in visible light when the amounts are 4 to 5 times greater [2].

The reagent can be employed on silica gel, kieselguhr, Si 50 000, NH_2, cellulose and polyamide layers.

Procedure Tested

Glucose, Fructose, Maltose [1]

Method Ascending, one-dimensional development in a trough chamber without chamber saturation.

Layer HPTLC plates NH_2 F_{254s} (MERCK).

Mobile phase Acetonitrile – water – phosphate buffer (pH = 5.9) (80 + 15 + 10).
Preparation of the phosphate buffer:
Dissolve 680 mg potassium dihydrogen phosphate in 50 ml water and add 4.6 ml 0.1 M caustic soda; then dilute 1 + 9 with water.

Migration distance 7 cm

Running time 20 min

Detection and result: The chromatogram was freed from the mobile phase, immersed in the reagent solution for 10 s and heated to 150 °C for 8 min. Maltose (hR_f 10–15), glucose (hR_f 20–25) and fructose (hR_f 25–30) appeared under long-wavelength UV light ($\lambda = 365$ nm) as pale blue fluorescent zones on a weakly bluish fluorescent background.

The fluorescence intensity of the chromatogram zones could be stabilized and increased by a factor of 2.5 to 3.5 by subsequent immersion in liquid paraffin – *n*-hexane (1 + 2).

In situ quantitation: The in situ fluorimetric analysis was made under long-wavelength UV light ($\lambda_{exc} = 365$ nm, $\lambda_{fl} > 560$ nm) and is illustrated in Figure 1. The detection limits for maltose, glucose and fructose were ca. 10 ng substance per chromatogram zone.

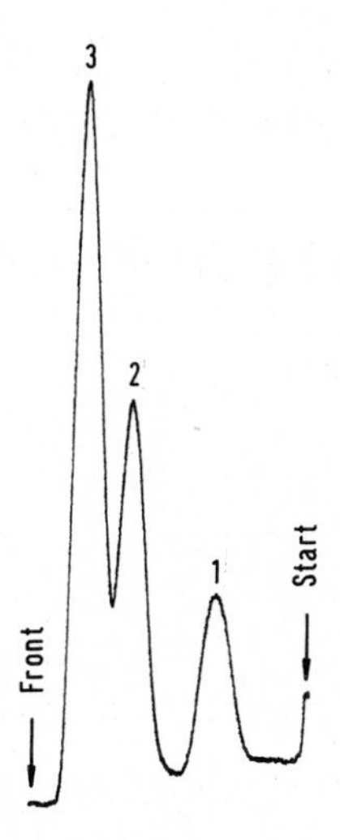

Fig. 1: Fluorescence plot for a sugar mixture containing 50 ng substance per chromatogram zone. Maltose (1), glucose (2), fructose (3).

References

[1] Patzsch, K.: Thesis, Fachhochschule Gießen, Fachbereich Technisches Gesundheitswesen, 1986.
[2] Sattler, L., Zerban, F. W.: *Anal. Chem.* **1952,** *24,* 1862.
[3] Kubelka, W., Eichhorn-Kaiser, S.: *Pharm. Acta Helv.* **1970,** *45,* 513–519.
[4] Stahl, E., Tugrul, L.: *Dtsch. Apoth. Ztg.* **1981,** *121,* 1409–1413.

Ammonia Vapor Reagent

Reagent for:

- Alkaloids
  e.g. morphine, heroin [1, 2]
  6-monoacetylmorphine [2]
- Mycotoxins
  e.g. ochratoxin A [3 – 5]
- Flavonoids, flavonoid glycosides [6 – 9]
- Sennosides [10]
- Naphthoquinone glucosides [11]
- Valepotriates [12]
- Antibiotics
  e.g. penicillic acid [13]
  rifamycin [14]
  tetracyclins [15]
- Anthracene derivatives [16]
- Homogentisic acid [17]

NH_3
$M_r = 17.03$

Preparation of Reagent

Solution Ammonia solution (25%).

Storage The reagent may be stored over an extended period.

Substances Ammonia solution (25%)

Reaction

Morphine and heroin form fluorescent oxidation products on heating in the presence of ammonia [1].

Method

The chromatograms are dried in a stream of cold air (alkaloids: 110–120 °C for 25 min in drying cupboard) and placed for 15 min in a twin-trough chamber – in the case of alkaloids while still hot – whose second trough contains 10 ml 25% ammonia solution.

Valepotriates are detected by placing the chromatogram 0.3 mm from a TLC plate sprayed with conc. ammonia solution (sandwich configuration layer to layer), fastening with clips and heating to 110 °C in a drying cupboard for 10 min [12].

The result is usually chromatogram zones that fluoresce yellow, green or blue on a dark background under long-wavelength UV light ($\lambda = 365$ nm), in some cases colored zones are detectable in visible light (e.g. homogentisic acid [17], sennosides [10], rifamycin [14]).

Note: The natural fluorescence colors of some flavonoids [7, 9] and anthracene derivatives [16] are altered by the ammonia treatment. This makes possible differentiation on the basis of color. Detection limits per chromatogram zone have been reported of 2 ng for morphine and heroin [2], 6 ng for ochratoxin A [5] and 1 µg for penicillic acid [13].

The reagent can be employed on silica gel, kieselguhr, Si 50 000, polyamide, RP and cellulose layers.

Procedure Tested

Alkaloids [2]

Method Ascending, one-dimensional development in a trough chamber without chamber saturation.

Layer HPTLC plates Silica gel 60 F_{254s} with concentrating zone (MERCK), which had been prewashed by developing once with chloroform – methanol (50 + 50) and then dried at 110 °C for 30 min.

Application The samples were applied to the concentrating zone as bands in the direction of chromatography. The zones were concentrated by brief development in the mobile phase described below almost to the junction between the concentrating zone and the chromatographic layer, followed by drying for 5 min in a stream of warm air. The actual chromatographic separation was then carried out.

Mobile phase Methanol – chloroform – water (12 + 8 + 2).

Migration distance 4.5 cm

Running time 20 min

Detection and result: The chromatogram was heated in the drying cupboard to 110 – 120 °C for 25 min and immediately placed – while still hot – in a twin-trough chamber, whose second trough contained 10 ml 25% ammonia solution, for 15 min. The chromatogram was then immersed for 2 s in a solution of liquid paraffin – *n*-hexane (1 + 2).

Morphine (hR_f 20 – 25), 6-monoacetylmorphine (hR_f 35 – 40) and heroin (hR_f 50 – 55) appeared as blue fluorescent zones on a dark background under long-

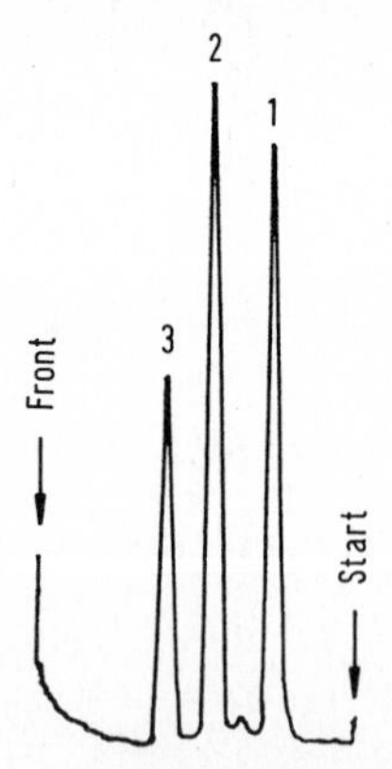

Fig. 1: Fluorescence scan of a mixture of alkaloids with ca. 50 ng substance per chromatogram zone: morphine (1), 6-monoacetylmorphine (2) and heroin (3).

wavelength UV light ($\lambda = 365$ nm). In each case the detection limits were 2 ng substance per chromatographic zone (Fig. 1).

In situ quantitation: The fluorimetric determination was carried out in UV light ($\lambda_{exc} = 313$ nm, $\lambda_{fl} > 390$ nm).

References

[1] Wintersteiger, R., Zeipper, U.: *Arch. Pharm.* (Weinheim) **1982,** *315*, 657-661.
[2] Patzsch, K., Funk, W., Schütz, H.: *GIT Fachz. Lab. Supplement 3 „Chromatographie"* **1988,** *32*, 83-91.
[3] Majerus, P., Woller, R., Leevivat, P., Klintrimas, T.: *Fleischwirtschaft* **1985,** *65*, 1155-1158.
[4] Takeda, Y., Isohata, E., Amano, R., Uchiyama, M.: *J. Assoc. Off. Anal. Chem.* **1979,** *62*, 573-578.
[5] Asensio, E., Sarmiento, I., Dose, K.: *Fresenius Z. Anal. Chem.* **1982,** *311*, 511-513.
[6] Nilsson, E.: *Acta Chem. Scand.* **1969,** *23*, 2910-2911.
[7] Ulubelen, A., Kerr, K. M., Mabry, T.: *Phytochemistry* **1980,** *19*, 1761-1766; **1982,** *21*, 1145-1147.
[8] Henning, W., Herrmann, K.: *Phytochemistry* **1980,** *19*, 2727-2729.
[9] Theodor, R., Zinsmeister, H. D., Mues, R.: *Phytochemistry* **1980,** *19*, 1695-1700.
[10] Kobashi, K., Nishimura, T., Kusaka, M.: *Planta Med.* **1980,** *40*, 225-236.
[11] Steinerova, N., Cludlin, J., Vanek, Z.: *Collect. Czech. Chem. Commun.* **1980,** *45*, 2684-2687.
[12] Rücker, G., Neugebauer, M., El Din, M. S.: *Planta Med.* **1981,** *43*, 299-301.
[13] Vesely, D., Vesela, D.: *Chem. Listy* (CSSR) **1980,** *74*, 289-290.
[14] Jankova, M., Pavlova, A., Dimov, N., Boneva, V., Chaltakova, M.: *Pharmazie* **1981,** *36*, 380.
[15] Urx, M., Vondrackova, J., Kovarik, L.: *J. Chromatogr.* **1963,** *11*, 62-65.
[16] Sims, P.: *Biochem. J.* **1972,** *130*, 27-35; **1973,** *131*, 405-413.
[17] Treiber, L. R., Örtengren, B., Lindstein, R.: *J. Chromatogr.* **1972,** *73*, 151-159.

Ammonium Thiocyanate – Iron(III) Chloride Reagent

Reagent for:

- Phosphates and phosphonic acids in detergents* [1]
- Organic acids and phosphate esters [2] e.g. sugar phosphates

NH_4SCN $M_r = 76.12$

$FeCl_3$ $M_r = 162.22$

Preparation of the Reagent

Dipping solution I Dissolve 1 g ammonium thiocyanate (ammonium rhodanide) in 100 ml acetone.

Dipping solution II Dissolve 50 mg iron(III) chloride in 100 ml acetone.

Storage The solutions are stable – when stored in the dark – for one month at room temperature [2].

Substances Ammonium thiocyanate
Iron(III) chloride anhydrous
Acetone

Reaction

Iron(III) thiocyanate is not formed to any extent in the chromatogram zones. The result is white zones on a pink-colored background:

$$Fe^{3+} + 3SCN^- \not\longrightarrow Fe(SCN)_3$$

*) Rüdt, U.: Private communication, Chemische Untersuchungsanstalt, Stuttgart, 1984

Method

The chromatogram is dried for 10 min in a stream of warm air and immersed in solution I for 1 s. It is then dried for 5 min in a stream of warm air and finally immersed in solution II for 1 s. White zones result on a pink background.

Note: It is necessary to remove acid mobile phases completely, since the color reaction only occurs in neutral to weakly acid medium. This is often difficult when cellulose layers are employed so that interference can occur.

Some acids such as cinnamic, lactic, oxalic and quinaldic acid yield yellow zones. Maleic, fumaric and *o*-phthalic acids turn red in color, salicylic acid grey and 4-aminobenzoic acid bluish.

Dipping solution I alone is a sensitive detection reagent for phosphate esters. Combination with dipping solution II does increase the sensitivity limit for organic acids but it is still insufficient for sensitive detection. Inorganic ions that form complexes with iron(III) ions can interfere with detection. It is necessary to replace dipping solution II after each dipping procedure because accumulated impurities (e.g. NH_4SCN from I) can discolor the background.

The reagent can be employed on silica gel, kieselguhr, polyamide and cellulose layers. Only dipping solution I can be employed on amino phases.

Procedure Tested

Inorganic Phosphates and Phosphonic Acids in Detergents [1]

Method Ascending, one-dimensional development in a trough chamber with chamber saturation.

Layer HPTLC plates Cellulose (MERCK).

Mobile phase Dioxan – trichloroacetic acid solution (70 + 30).
Preparation of the trichloroacetic acid solution: 16 g trichloroacetic acid were dissolved in 50 ml water. 0.8 ml 33% ammonia solution was (carefully!) added (fume cupboard) and made up to 100 ml with water.

Migration distance 6 cm

Running time 30 min

Detection and result: The chromatogram was dried for 10 min in a stream of warm air and immersed in solution I for 1 s, it was then dried for 5 min in a stream of warm air and immersed in solution II for 1 s.
The phosphates and phosphonic acids appeared as white zones on a pink background (Fig. 1A). Figure 1B is a reproduction of the reflectance plots ($\lambda = 480$ nm). Detection limits of 50 ng have been found for PO_4^{3-} and $P_2O_7^{4-}$. In the case of $P_3O_9^{3-}$ and $P_3O_{10}^{5-}$ the detection limits were 125 ng per chromatogram zone.

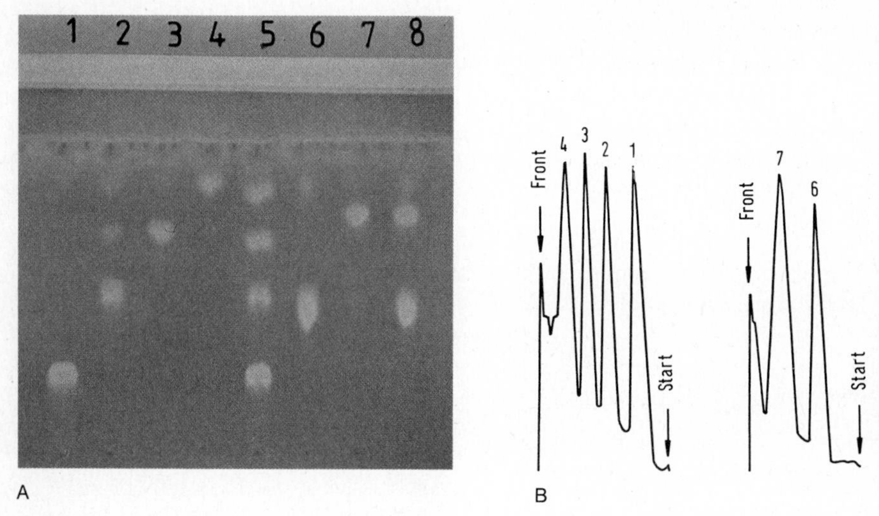

Fig. 1: A) Separation of phosphates and phosphonic acids; B) absorption plots of track 5 and track 8. $P_3O_9^{3-}$ (1), $P_3O_{10}^{5-}$ (2), $P_2O_7^{4-}$ (3), PO_4^{3-} (4), mixture I (5), aminotrimethylene-phosphonic acid (6), 1-hydroxyethane-1,1-diphosphonic acid (7), mixture II (8).

References

[1] Mörsdorf, W.: Thesis, Universität des Saarlandes, Saarbrücken 1986.
[2] Firmin, J. L., Gray D. O.: *J. Chromatogr.* **1974**, *94*, 294–297.

Amylose – Potassium Iodate/Iodide Reagent

Reagent for:

- Nonvolatile aromatic and aliphatic carboxylic acids [1]
- Ampicillin [2]
- Phenylethylamines e.g. ephedrine [3]

KI	KIO_3	Amylose
$M_r = 166.01$	$M_r = 214.00$	$M_r \approx 10^6$

Preparation of Reagent

Solution I Dissolve 1 g amylose with warming in 100 ml water.

Solution II Dissolve 2 g potassium iodate in 100 ml water.

Solution III Dissolve 8 g potassium iodide in 100 ml water.

Dipping solution Combine 10 ml each of solutions I to III immediately before use and dilute to 100 ml with water.

Spray solution Mix equal volumes of solutions I to III immediately before spraying.

Storage Solutions I to III are stable for several weeks in the refrigerator.

Substances Amylose
Potassium iodide
Potassium iodate

Reaction

Iodide and iodate ions react under the influence of protons to yield iodine molecules which react with amylose to yield a blue clathrate complex:

$$IO_3^- + 5I^- + 6\,H^+ \longrightarrow 3I_2 + 3H_2O \xrightarrow{\text{Amylose}} \text{blue complex.}$$

Method

The well-dried chromatogram (1 h at 105 °C if acidic or basic eluents have been employed) is immersed in the dipping solution for 1 s or homogeneously sprayed with the spray solution and then dried in a stream of cold air. Acids yield blue zones on a colorless or pale blue background [1] which gradually darkens.

For the detection of ampicillin it is necessary to add acetic acid to the dipping or spray solution. Ampicillin then yields pale zones on a blue background [2].

Note: The reagent can be employed on silica gel and cellulose layers. Starch can also be employed in place of amylose [2]. The blue coloration of the amylose complex turns brown after a short time.

Saccharin and the three diphenols, pyrocatechol, resorcinol and hydroquinone, react only weakly or not at all. The same is true of picric acid. On the other hand, cyclohexanesulfamic acid and bis-(2-ethylhexyl)-phosphoric acid are readily detected [1].

The detection limit for ampicillin is 50 ng per chromatogram zone.

Procedure Tested

Organic Acids [4]

Method Ascending, one-dimensional development in a trough chamber with chamber saturation.

Layer TLC plates Cellulose (MERCK).
HPTLC plates Silica gel WRF_{254s} (MERCK).
SIL G-25 (MACHEREY-NAGEL).

Mobile phase *a) for cellulose layers*
1. 1-butanol – ethanol (96%) – ammonia solution (25%) – water (60 + 60 + 60 + 15)

2. ethanol (96%) – ammonia solution (25%) (112 + 16).

b) for M & N SIL G-25 *and* MERCK HPTLC *plates*
3. diisopropyl ether – formic acid – water (90 + 7 + 3).

Migration distance 6 cm

Running time 25–30 min

Detection and result: The chromatograms had to be freed from mobile phase before they were immersed; otherwise a blue background was produced. After it had been dipped the chromatogram was dried in a stream of cold air. Zones appearing on an initially pale background were first brown and then turned blue. The background, however, darkened so much that after 5 min it was scarcely possible to discern the zones. Table 1 lists some hR_f values.

Table 1: hR_f values of some carboxylic acids

Acid	hR_f value	
	Cellulose	Silica gel
4-Aminobenzoic acid	55–60	–
2-Aminobenzoic acid	60–65	–
Fumaric acid	–	80–85
Benzoic acid	70–75	–
Malic acid	–	15–20
Tartaric acid	5–10	0
Phthalic acid	15–20	60–65
Adipic acid	–	55–60
Salicylic acid	80–85	–

Note: The reagent was not particularly sensitive for acids. On cellulose layers the detection limit was ca. 1 µg (salicylic acid ≥ 5 µg) and on silica gel layers it was 5 µg (fumaric acid ca. 1 µg).

In situ quantitation: The reagent was not suitable for a sensitive, direct, photometric analysis.

References

[1] Chafetz, L, Penner M. H.: *J. Chromatogr.* **1970**, *49*, 340–342.
[2] Larsen, C., Johansen, M.: *J. Chromatogr.* **1982**, *246*, 360–362.
[3] Chafetz, L.: *J. pharmac. Sci.* **1971**, *60*, 291–294.
[4] Klein, I., Jork, H.: GDCh-training course Nr. 300 „Einführung in die Dünnschicht-Chromatographie", Universität des Saarlandes, Saarbrücken 1985.

Aniline – Aldose Reagent

Reagent for:

- Organic acids [1–4]
- N-Acylglycine conjugates [7]

NH2

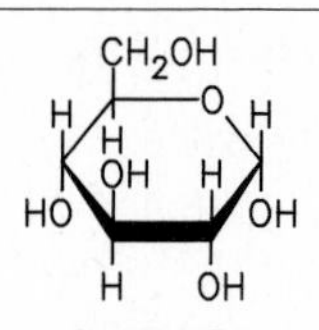

C_6H_7N
$M_r = 93.13$
Aniline

$C_6H_{12}O_6$
$M_r = 180.16$
Glucose

Preparation of the Reagent

Solution I Mix 2 ml freshly distilled aniline with 18 ml ethanol.

Solution II Dissolve 2 g of an aldose (e.g. glucose) in 20 ml water.

Dipping solution Mix 20 ml each of solutions I and II immediately before use and make up to 100 ml with 1-butanol.

Storage Solutions I and II are stable for a long period in the refrigerator.

Substances D(+) Glucose
Aniline
Ethanol
1-Butanol

Reaction

Furfural is produced from glucose under the influence of acid and this reacts with aniline to yield a colored product [5].

$$\text{Glucose} \xrightarrow[-\,3\,H_2O,\;-\,HCHO]{H^+} \text{Furfural}$$

$$\xrightarrow{2\,C_6H_5NH_2} \left[C_6H_5\text{-}\overset{+}{N}H_2\text{-}CH{=}CH\text{-}CH{=}C(OH)\text{-}CH{=}N\text{-}C_6H_5\right]^+$$

Method

The chromatogram is freed from mobile phase and immersed in the dipping solution for 3 s or the solution is sprayed on homogeneously; the chromatogram is then heated to 90 – 140 °C for 5 – 10 min. Brown zones are produced on a beige-grey background.

Note: Aldoses other than glucose can also be used e.g. arabinose [1], xylose [2, 3, 7] or ribose [4]. The background color is least on cellulose layers; when cellulose acetate, aluminium oxide 150, silica gel, RP, NH_2 or polyamide layers are employed the background is a more or less intense ochre. The detection limit of carboxylic acids on cellulose layers is ca. 0.5 µg substance per chromatogram zone.

It has not been possible to ascertain why this reagent is occasionally referred to as SCHWEPPE reagent.

Procedure Tested

Organic Acids [6]

Method Ascending, one-dimensional development in a trough chamber with chamber saturation.

Layer TLC plates Sil G-25 (MACHEREY-NAGEL).

Mobile phase Diisopropyl ether – formic acid – water (90 + 7 + 3).

Migration distance 10 cm

Running time 30 min

Detection and result: The chromatogram was freed from mobile phase and immersed for 3 s in the dipping solution and heated to 125 °C for 5–10 min. The carboxylic acids: terephthalic acid (hR_f 5), succinic acid (hR_f 50–55), phthalic acid (hR_f 55–60), suberic acid (hR_f 60–65), sebacic acid (hR_f 65–70), benzoic acid (hR_f 75–80) and salicylic acid (hR_f 80–85) yielded brown zones on a light brown background. The detection limit was 2 µg acid per chromatogram zone.

Note: If a dipping solution was employed for detection whose concentration was reduced to 1/10th that given above the acids appeared as white zones on a light brown background.

In situ quantitation: The reagent was not suitable for a sensitive, direct, photometric analysis.

References

[1] Bourzeix, M., Guitraud, J., Champagnol, F.: *J. Chromatogr.* **1970**, *50* 83–91.
[2] Köhler, F.: *J. Chromatogr.* **1972**, *68*, 275–279.
[3] Lin, L., Tanner, H.: *J. High Resolut. Chromatogr. Chromatogr. Commun.* **1985**, *8*, 126–131.
[4] Beaudoin, A. R., Moorjani, S., Lemonde, A.: *Can. J. Biochem.* **1973**, *51*, 318–320.
[5] Kakáč, B., Vejdělek, Z. J.: *Handbuch der photometrischen Analyse photometrischer Verbindungen.* Weinheim: Verlag Chemie, 1974.
[6] Jork, H., Klein, I: GDCh-training course Nr. 300 „Einführung in die Dünnschicht-Chromatographie" Unversität des Saarlandes, Saarbrücken 1987.
[7] Berg, H. van den, Hommes, F. A.: *J. Chromatogr.* **1975**, *104*, 219–222.

Aniline – Diphenylamine – Phosphoric Acid Reagent

Reagent for:

- Sugars
  Mono- and disaccharides [1–5]
  Oligosaccharides [5–7]
  Starch hydrolysates [8–12]
- Thickening agents [13, 14]
- Glycosides [15, 18]
  e.g. arbutin, prunasin, amygdalin, rutin

C_6H_7N	$C_{12}H_{11}N$	H_3PO_4
$M_r = 93.13$	$M_r = 169.23$	$M_r = 98.00$
Aniline	Diphenylamine	Orthophosphoric acid

Preparation of the Reagent

Dipping solution Dissolve 2 g diphenylamine and 2 ml aniline in 80 ml acetone. Carefully add 15 ml phosphoric acid and make up to 100 ml with acetone [2, 8].

Spray solution Dissolve 1 to 2 g diphenylamine and 1 to 2 ml aniline in 80 ml methanol or ethanol. After addition of 10 ml phosphoric acid make up to 100 ml with methanol [3, 9] or ethanol [7].

Storage Both reagents can be stored in the dark at 4°C for up to 14 days. It is recommended that the reagent be prepared daily for in situ quantitation.

Substances Diphenylamine
Aniline
Orthophosphoric acid

Reaction (according to [16])

Heating the sugar with strong acid yields furfural derivatives. Aldohexoses can eliminate water and formaldehyde under these conditions yielding furfural. This adehyde reacts with amines according to I to yield colored SCHIFF's bases. Ketohexoses condense with diphenylamine in acid medium with simultaneous oxidation according to II to yield the condensation product shown.

I

$$\text{Furfural} + 2\ C_6H_5{-}NH_2 \xrightarrow[-\ H_2O]{H^+} \left[C_6H_5{-}\overset{+}{N}H_2{-}CH{=}CH{-}CH{=}\overset{OH}{C}{-}CH{=}N{-}C_6H_5\right]^+$$

II

$$\text{2-Ketohexose} \xrightarrow{H^+} 2\ OHC{-}C_4H_2O{-}CH_2OH\ (1) \xrightarrow{H_2O} OHC{-}C_4H_2O{-}CH_2{-}O{-}CH_2{-}C_4H_2O{-}CHO\ (2)$$

$$(2) + 4\ C_6H_5{-}NH{-}C_6H_5\ (3) \xrightarrow[-\ 2\ H_2O]{+\ 2\ H^+}$$

$$\left[(C_6H_5{-}\overset{+}{N}H{=}C_6H_4{=})(C_6H_5{-}NH{-}C_6H_4{-})C{-}C_4H_2O{-}CH_2{-}O{-}CH_2{-}C_4H_2O{-}C(C_6H_4{-}NH{-}C_6H_5)({=}C_6H_4{=}\overset{+}{N}H{-}C_6H_5)\right]^{2+}\ (4)$$

Method

The chromatogram is freed from mobile phase and evenly sprayed with the spray solution or immersed for 1 s in the dipping solution. After drying the TLC plate is heated to 85–120 °C normally for 10 to 15 min but in exceptional cases for 60 min. It is advisable to observe the chromatogram during the reaction period, because the temperature and duration of heating strongly affect color development.

Aldohexoses are stained dark grey to grey-blue. 1,4-Linked aldohexose-oligosaccharides appear as blue zones on a pale background if they are derived from maltose. Sucrose goes brown, lactose pale blue and raffinose grey-green. Ketohexoses yield olive green zones while pentoses (rhamnose) yield grey-green to pale green ones. Glucuronic acid (very dark grey) and galacturonic acid (pale violet) also react. Trehalose and sugar alcohols do not react.

Note: For in situ quantitation the scanning should begin 20 min after applying the reagent [7]. Suitable stationary phases include silica gel [2, 8] – also buffered with 0.02 M sodium acetate solution [9] – kieselguhr, Si 50000 [10] and, in particular, NH_2 layers [5]. An adequate resolution is often obtained on amino layers after a single development (cf. Procedure Tested).

To detect glycosides heat the chromatograms to 130–150 °C for 15 min. Blue-grey zones are produced (detection limit prunasin: 0.3–0.5 µg [18]). Flavonoids are better detected with a modified reagent of the following composition: phosphoric acid (85%) – acetic acid – aniline – diphenylamine (20 ml + 100 ml + 5 ml + 5 g).

Absorption plots of oligosaccharide separations are reproduced in Figure 1 (A: maltodextrin, B: glucose syrup), those of mono- and disaccharide separations in Figure 2.

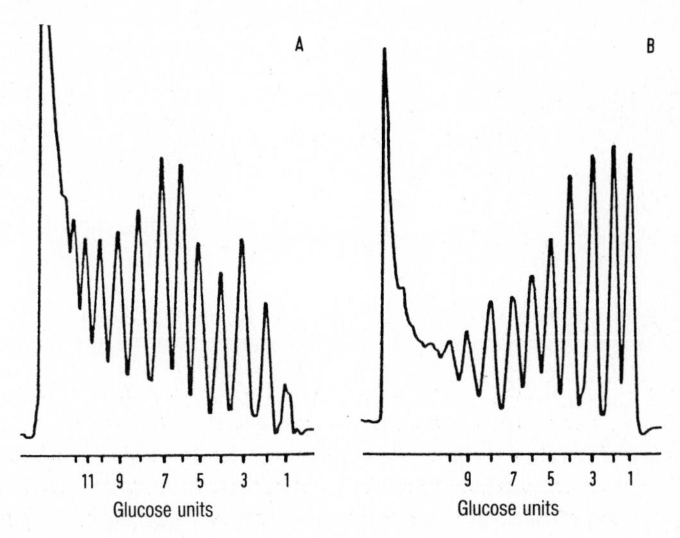

Fig. 1: Absorption plot (A) of a maltodextrin and (B) of a glucose syrup [8].

Procedure Tested

Glucose, Fructose, Maltose or Lactose, Sucrose and Raffinose [17]

Two noninterchangeable methods of procedure are reported below under the headings A and B.

Method

A) Ascending, one-dimensional, double development in a trough chamber (5 min drying in warm air between developments) with chamber saturation.
B) Ascending, one-dimensional development at 20 °C in a trough chamber.

Layer

A) HPTLC plates Si 50000 (MERCK); develop once in chloroform – methanol (1 + 1) to prewash and then dry at 110 °C for 30 min.
B) HPTLC plates NH_2 F_{254s} (MERCK).

Mobile phase

A) Acetonitrile – water (17 + 3).
B) Acetonitrile – water – phosphate buffer (16 + 3 + 2)
Preparation of the phosphate buffer: Make up 10 ml of a mixture of 50 ml potassium dihydrogen phosphate solution ($c = 0.1$ mol/l) and 4.6 ml caustic soda solution ($c = 0.1$ mol/l) to 100 ml with water.

Migration distance

A) 2 × 7 cm
B) 7 cm

Running time

A) 2 × 10 min
B) 15 – 20 min

Detection and result: A) and B): The chromatogram was dried for 3 min in a stream of warm air, immersed in the dipping solution for 9 s and then heated to 105–110 °C for 15 min and finally immersed for 2 s in a solution of liquid paraffin – *n*-hexane (1 + 2) to stabilize and increase the fluorescence intensity (factor 2 to 3).

Grey-green zones on a white background resulted; they exhibited weak red fluorescence under long-wavelength UV light ($\lambda = 365$ nm); glucose and fructose exhibited pale blue fluorescence in method B. Some hR_f values are listed in Table 1.

Table 1: hR_f values of some sugars

Sugar	hR_f value	
	Method A	Method B
Fructose	45–50	25–30
Glucose	35–40	20–25
Sucrose	25–30	15–20
Lactose	15–20	–
Maltose	–	10–15
Raffinose	10–15	5–10

In situ quantitation: Quantitative analysis (Figs. 2 and 3) could be performed both absorption-photometrically with long-wavelength UV light ($\lambda = 365$ nm) or fluorimetrically ($\lambda_{exc} = 436$ nm; $\lambda_{fl} = 546$ nm [monochromation filter M 546] or $\lambda_{fl} > 560$ nm).
The detection limit for fluorimetric quantitation was 10 ng substance per chromatogram zone.

Note: If the chromatogram developed by method B was exposed to ammonia vapors for 10 min before being immersed in liquid paraffin–*n*-hexane (1 + 2) the fluorescence of the chromatogram zones became deep red. Glucose and fructose also appeared red.

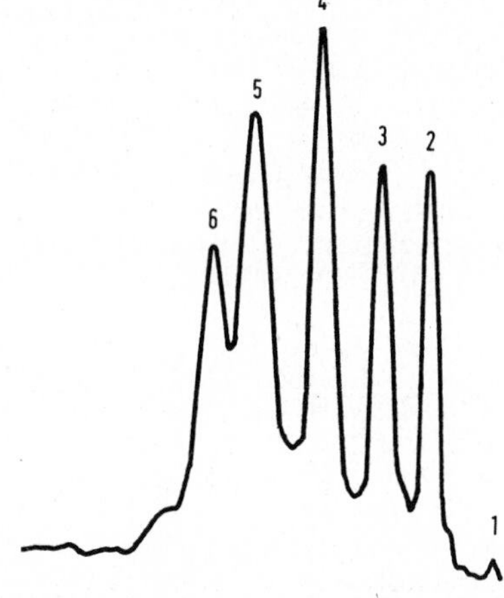

Fig. 2: Fluorescence plots of the sugars after separation on a Si-50 000 layer without ammonia-vapor treatment. Start (1), raffinose (2), lactose (3), sucrose (4), glucose (5), fructose (6).

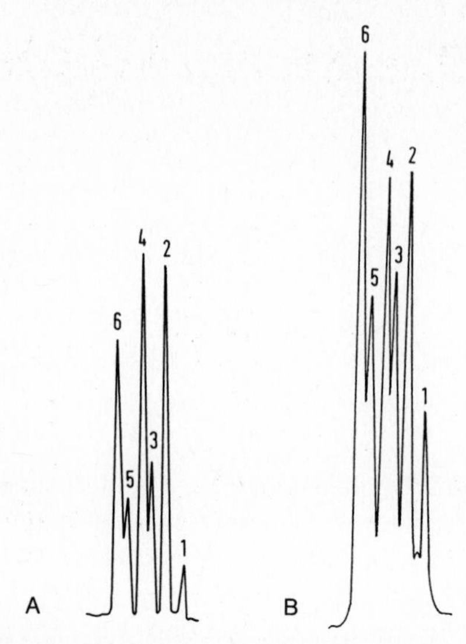

Fig. 3: Absorbance plots (A) and fluorescence plots of the sugars without ammonia-vapor treatment (B) after separation on a NH_2 layer. Start (1), raffinose (2), maltose (3), sucrose (4), glucose (5), fructose (6).

References

[1] Bailey, R. W., Bourne E. J.: *J. Chromatogr.* **1960**, *4*, 206–213.
[2] Lee, K. Y., Nurok, D., Zlatkis, A.: *J. Chromatogr.* **1979**, *174*, 187–193.
[3] Kreuzig, F.: *J. Liq. Chromatogr.* **1983**, *6*, 1227–1238.
[4] Martinez-Castro, I., Olano, A.: *Chromatographia* **1981**, *14*, 621–622.
[5] Doner, L. W., Biller, L. M.: *J. Chromatogr.* **1984**, *287*, 391–398.
[6] Buffa, M., Congiu, G., Lombard, A., Tourn, M. L.: *J. Chromatogr.* **1980**, *200*, 309–312.
[7] Mansfield, C. T., McElroy, H. G. jr.: *Anal. Chem.* **1971**, *43*, 586–587.
[8] Schweizer, T. F., Reimann, S.: *Z. Lebensm. Unters. Forsch.* **1982**, *174*, 23–28.
[9] Conway, R. L., Hood, L. F.: *Die Stärke* **1976**, *28*, 341–343; *J. Chromatogr.* **1976**, *129*, 415–419.
[10] Koizumi, K., Utamura, T., Okada, Y.: *J. Chromatogr.* **1985**, *321*, 145–157.
[11] Stefanis, V. A., Ponte, J. G. jr: *J. Chromatogr.* **1968**, *34*, 116–120.
[12] Würsch, P., Roulet, P.: *J. Chromatogr.* **1982**, *244*, 177–182.
[13] Scherz, H., Mergenthaler, E.: *Z. Lebensm. Unters. Forsch.* **1980**, *170*, 280–286.
[14] Friese, P.: *Fresenius Z. Anal. Chem.* **1980**, *301*, 389–397.
[15] Wolf, S. K., Denford, K. E.: *Biochem. Syst. Ecol.* **1984**, *12*, 183–188.
[16] Kakáč, B., Vejdělek, Z. J.: *Handbuch der photometrischen Analyse organischer Verbindungen.* Weinheim, Verlag Chemie, 1974.
[17] Patzsch, K.: Thesis, Fachhochschule Gießen, Fachbereich Technisches Gesundheitswesen, 1986.
[18] Jork, H.: Private communication, Universität des Saarlandes, Fachbereich 14, Saarbrücken 1986

Aniline – Phosphoric Acid Reagent

Reagent for:

- Carbohydrates (sugars)
  e.g. monosaccharides [1]
- Glucosides
  e.g. aryl- and thioglucosides [2]

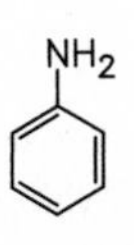

C_6H_7N
$M_r = 93.13$
Aniline

H_3PO_4
$M_r = 98.00$
Orthophosphoric acid

Preparation of the Reagent

Solution I Make 15 ml aniline up to 100 ml with 1-butanol in a volumetric flask.

Solution II Carefully make 30 ml orthophosphoric acid (88–90%) up to 100 ml with 1-butanol in a volumetric flask.

Dipping solution Immediately before use add 20 ml solution I to 50 ml solution II and mix well to redissolve the precipitate that is produced. If this does not go back into solution it should be filtered off.

Storage Solutions I and II are stable for several days.

Substances Aniline
Orthophosphoric acid
1-Butanol

Reaction (after [3])

Heating the sugars with strong acid yields furfural derivatives. Under these conditions aldohexoses can eliminate formaldehyde and water to yield furfural. This aldehyde reacts with amines to yield colored SCHIFF's bases.

$$\text{Furfural} + 2\ C_6H_5-NH_2 \xrightarrow[-\ H_2O]{H^+,\ \Delta T} \left[C_6H_5-\overset{+}{N}H_2-CH{=}CH-CH{=}\underset{}{\overset{OH}{C}}-CH{=}N-C_6H_5 \right]^+$$

Method

The chromatogram is freed from mobile phase in the drying cupboard (10 min, 120 °C) and immersed for 1 s in the reagent solution or sprayed homogeneously with it until the plate starts to appear transparent; it is then dried briefly in a stream of warm air and heated to 125–130 °C for 45 min.

Many aryl- and also thio-β-D-glucosides produce yellow fluorescent chromatogram zones on a dark violet background under long-wavelength UV light (λ = 365 nm), the zones are also sometimes recognizable in visible light as grey-brown zones on a white background [2]. Monosaccharides produce dark brown chromatogram zones on a white background [1].

Note: The detection limit for aryl- and thioglucosides is 100–200 ng substance per chromatogram zone [2]. Reduction of the proportion of phosphoric acid in the reagent leads to loss of sensitivity [2].

The reagent can be employed on silica gel, RP-18, CN and NH_2 layers. Cellulose and polyamide layers are not suitable.

Procedure Tested

Thioglucosides [4]

Method Ascending, one-dimensional development in a trough chamber with chamber saturation.

Layer HPTLC plates Silica gel 60 F_{254} (MERCK).

Mobile phase 1-Butanol – 1-propanol – glacial acetic acid – water (30 + 10 + 10 + 10).

Migration distance 5 cm

Running time 45 min

Detection and result: The chromatogram was freed from mobile phase in the drying cupboard (10 min, 125 °C) and immersed for 1 s in the reagent solution, then heated to 120 °C for 45 min.

Sinigrin (hR_f 35 – 40) appeared as a yellow fluorescent chromatogram zone on a dark background under long-wavelength UV light (λ = 365 nm). The detection limit was 25 – 50 ng substance per chromatogram zone.

In situ quantitation: Excitation at $\lambda_{exc} = 365$ nm and measurement at $\lambda_{fl} > 560$ nm were employed for fluorimetric quantitation (Fig. 1).

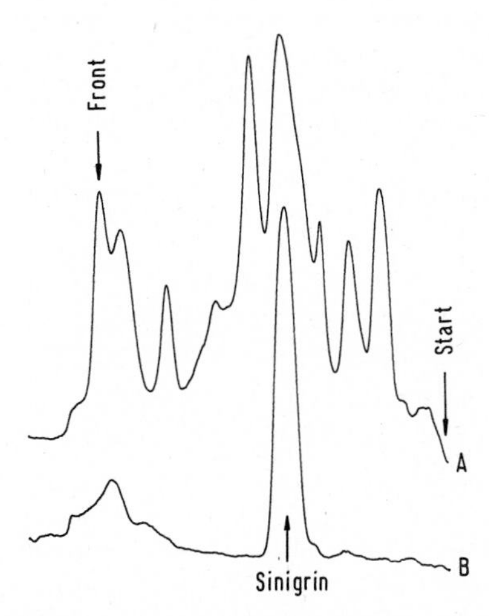

Fig. 1: Fluorescence plot (A) of the chromatogram track of an "unpurified" extract of sinapis seed (application: 2 µl of a 1% solution in methanol) and (B) of a reference track with 1 µg sinigrin per chromatogram zone.

References

[1] Ogan, A. U.: *Planta Med.* **1972**, *21*, 431-434.
[2] Garraway, J. L., Cook, S. E.: *J. Chromatogr.* **1970**, *46*, 134-136.
[3] Kakăc, B., Vejdělek, Z. J.: *Handbuch der photometrischen Analyse organischer Verbindungen.* Weinheim: Verlag Chemie, 1974.
[4] Kany, E., Jork, H.: GDCh-training course Nr. 300 „Einführung in die DC" Universität des Saarlandes, Saarbrücken 1987.

Aniline – Phthalic Acid Reagent
(Aniline phthalate Reagent)

Reagent for:

- Halogen oxyacids [1]
  e.g. chlorate, chlorite, perchlorate, bromate, bromite, iodate
- Reducing carbohydrates (sugars)
  e.g. monosaccharides [2–8]
  oligosaccharides [2, 8]
  oligouronic acids [9]
  methyl sugars [10, 11]

NH_2 $COOH$ $COOH$

$C_5H_5NH_2$	$C_8H_6O_4$
$M_r = 93.13$	$M_r = 166.13$
Aniline	Phthalic acid

Preparation of Reagent

Dipping solution Dissolve 0.9 ml aniline and 1.66 g phthalic acid in 100 ml acetone.

Spray solution Dissolve 0.9 ml aniline and 1.66 g phthalic acid in 100 ml water-saturated 1-butanol [1].

Storage The reagent solutions may be stored for several days.

Substances Aniline
Phthalic acid
1-Butanol
Acetone

Reaction (according to [12])

Furfural derivatives are produced when sugars are heated with acids (see Aniline–Diphenylamine–Phosphoric Acid Reagent), these condense with aniline to SCHIFF's bases. Chlorate, chlorite, perchlorate etc. oxidize aniline to 1,4-quinone, which then reacts with excess aniline to yield indophenol:

Aniline —Oxidation→ 1,4-benzoquinone + aniline $\xrightarrow[-H_2O]{\Delta T}$ Indophenol

Method

The chromatograms are freed from mobile phase, immersed in the reagent solution for 1 s or homogeneously sprayed with it, dried briefly in a stream of warm air and heated to 80 – 130 °C for 20 – 30 min.

Variously colored chromatogram zones are formed on an almost colorless background [1, 4], some of which fluoresce after irradiation with long-wavelength UV light ($\lambda = 365$ nm) [5].

Note: The dipping solution can also be employed as spray reagent. The detection limits per chromatogram zone are reported to be 1 – 5 µg substance [1] for the oxyacids of halogens and ca. 10 µg substance for reducing sugars [4].

The reagent may be employed on silica gel, kieselguhr, aluminium oxide, cellulose and polyamide layers.

Procedure Tested

Halogen Acids and Halogen Oxyacids [13]

Method Ascending, one-dimensional development in a trough chamber with chamber saturation.

Layer HPTLC plates Silica gel 60 F_{254} (MERCK).

Mobile phase 2-Propanol – tetrahydrofuran – ammonia solution (32%) (50 + 30 + 20).

Migration distance 5 cm

Running time 25 min

Detection and result: The chromatogram was freed from mobile phase in a stream of warm air, immersed in the reagent solution for 1 s and heated to 120 °C for 20 min. Intense yellow to brown zones of various hues were produced; these appeared as dark zones on a fluorescent background under long-wavelength UV light ($\lambda = 365$ nm).

While chlorate and bromate yielded strong brown zones, iodate did not react at all. Iodide yielded an intense yellow zone, the colors produced by chloride, bromide and perchlorate were weak (Fig. 1).

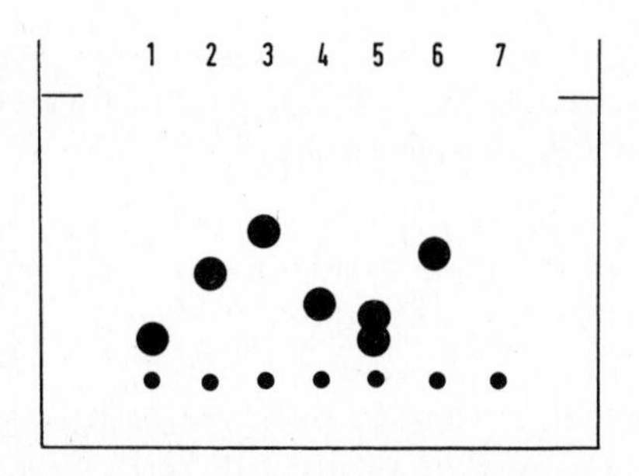

Fig. 1: Schematic diagram of a chromatogram of halogen acids and halogen oxyacids: Chloride (1), chlorate (2), perchlorate (3), bromide (4), bromate (5), iodide (6), iodate (7).

References

[1] Peschke, W.: *J. Chromatogr.* **1965,** *20*, 572–579.
[2] Weicker, H., Brossmer, R.: *Klin. Wochenschrift* **1961,** *39*, 1265–1266.
[3] Scott, R. W.: *J. Chromatogr.* **1970,** *49*, 473–481.
[4] Young, D. S., Jackson, A. J.: *Clin. Chem.* **1970,** *16*, 954–959.
[5] Grafe, I., Engelhardt, H.: *Chromatographia* **1972,** *5*, 307–308.
[6] David, J., Wiesmeyer, H.: *Biochim. Biophys. Acta* **1970,** *208*, 68–76.
[7] Wolfrom, M. L., De Lederkremer, R. M., Schwab, G.: *J. Chromatogr.* **1966,** *22*, 474–476.
[8] Sinner, M., Dietrichs, H. H., Simatupang, M. H.: *Holzforschung* **1972,** *26*, 218–228.
[9] Markovic, O., Slezarik, A.: *J. Chromatogr.* **1984,** *312*, 492–496.
[10] Mergenthaler, E., Scherz, H.: *Z. Lebensm. Unters. Forsch.* **1978,** *166*, 225–227.
[11] Tschöpe, G.: *Hoppe-Seyler's Z. Physiol. Chem.* **1971,** *352*, 71–77.
[12] Kacáč, B., Vejdělek, Z. J.: *Handbuch der photometrischen Analyse organischer Verbindungen.* Weinheim: Verlag Chemie, 1974.
[13] Kany, E., Jork, H.: GDCh-training course „Einführung in die Dünnschicht-Chromatographie", Universität des Saarlandes, Saarbrücken, 1986.

8-Anilinonaphthalene-1-sulfonic Acid Ammonium Salt Reagent
(ANS Reagent)

Reagent for:

- Lipids, phospholipids [1 – 8]
  e.g. fatty acids or their methylesters,
  fatty alcohols [9]
- Cholesterol and cholesteryl esters [1, 2]
- Steroids [3]
- Detergents [3]
- Hydrocarbons [3]

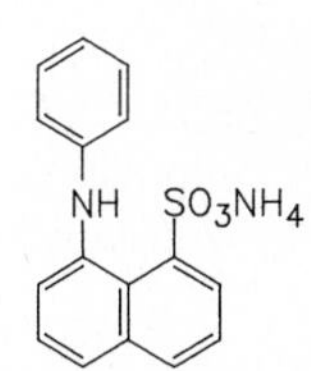

$C_{16}H_{16}N_2O_3S$

$M_r = 316.38$

Preparation of the Reagent

Dipping solution Dissolve 100 mg of 8-anilinonaphthalene-1-sulfonic acid ammonium salt in a mixture of 40 ml caustic soda solution ($c = 0.1$ mol/l) and 57 ml of an aqueous solution containing 21 g citric acid monohydrate and 8 g sodium hydroxide per liter.

Spray solution Dissolve 100 mg of 8-anilinonaphthalene-1-sulfonic acid ammonium salt in 100 ml water [1].

Storage The solutions are stable for at least 3 months if stored in a refrigerator in the dark [3].

Substances 8-Anilinonaphthalene-1-sulfonic acid ammonium salt
Citric acid monohydrate
Sodium hydroxide pellets
Caustic soda (0.1 mol/l)

Reaction

8-Anilinonaphthalene-1-sulfonic acid ammonium salt, which scarcely fluoresces in aqueous solution, is stimulated to intense fluorescence by long-wavelength UV light ($\lambda = 365$ nm) if it is dissolved in nonpolar solvents or adsorptively bound to nonpolar molecular regions [3].

Method

The developed chromatogram is freed from mobile phase by heating to 110 °C for 10 min in the drying cupboard. It is allowed to cool and immersed for 1 s in or sprayed homogeneously with the reagent; the plate is then examined (while still moist).

Yellow-green fluorescent zones are easily visible against a dark background under long-wavelength UV light ($\lambda = 365$ nm).

Note: The developed chromatogram must be completely freed from nonpolar solvents before derivatization, otherwise an intense fluorescence will be stimulated over the whole plate. The fluorescence intensity of the chromatogram zones remains stable for ca. 40 min; it decreases slowly as the layer dries out and can be returned to its original intensity by renewed immersion in the reagent solution or in water.

The reagent can be employed on silica gel, kieselguhr and Si 50 000 layers and, if necessary, on RP-2 and RP-8 phases. It cannot be used on RP-18 layers [9] because here the whole plate is fluorescent.*)

Procedure Tested

Cholesterol [2]

Method Ascending, one-dimensional development in a twin-trough chamber.

---

*) Jork, H.: Private communication, Universität des Saarlandes, 66 Saarbrücken, 1987

Layer HPTLC plates Silica gel 60 WRF$_{245s}$ (MERCK). After application of the samples the plate was preconditioned for 30 min at 0% relative humidity.

Mobile phase Cyclohexane – diethyl ether (1 + 1).

Migration distance 6 cm

Running time 15 min

Detection and result: The developed chromatogram was freed from mobile phase by drying for 10 min at 110 °C, allowed to cool and immersed for 1 s in the reagent solution. The plate was evaluated as rapidly as possible while it was moist since the fluorescent background increased in intensity as the plate dried out. Cholesterol appeared as a yellow-green fluorescent zone (hR_f 20–25).

In situ quantitation: Fluorimetric analysis was made with long-wavelength UV light (λ_{exc} = 365 nm, λ_{fl} > 430 nm). The detection limit on HPTLC plates that were analyzed in a moist state was 25 ng cholesterol per chromatogram zone (Fig. 1).

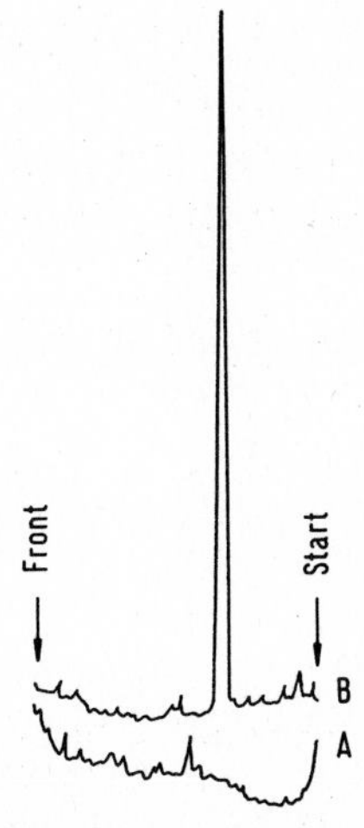

Fig. 1: Fluorescence plot of a blank track (A) and of a cholesterol standard with 200 ng substance per chromatogram zone (B).

References

[1] Vinson, J. A., Hooyman, J. E.: *J. Chromatogr.* **1977** *135,* 226-228.
[2] Zeller, M.: Thesis, Fachhochschule Gießen, Fachbereich Technisches Gesundheitswesen, 1986.
[3] Gitler, C.: *Anal. Biochem.* **1972,** *50,* 324-325.
[4] Gitler, C., in: *Biomembranes;* Manson, L., (Ed.). New York: Plenum, 1971; Vol. *2*, p. 41-47.
[5] Larsen, H. F., Trostmann, A. F.: *J. Chromatogr.* **1981,** *226*, 484-487.
[6] Blass, G., Ho, C. S.: *J. Chromatogr.* **1981,** *208,* 170-173.
[7] Lichtenthaler, H., Boerner, K.: *J. Chromatogr.* **1982,** *242*, 196-201.
[8] Ozawa, A., Jinbo, H., Takahashi, H., Fujita, T., Hirai, A., Terano, T., Tamura, Y., Yoshida, S.: *Bunseki Kagaku (Japan Anal. Chem.)* **1985,** *34,* 707-711.
[9] Hüttenhain, S. H., Balzer, W.: *Fresenius Z. Anal. Chem.* **1989,** *334*, 31–33.

Anisaldehyde – Sulfuric Acid Reagent

Reagent for:

- Antioxidants [1]
- Steroids [2 – 4]
  e.g. estrogens [2, 3], androgens [3]
  sterols, bile acids [3]
- Prostaglandins [5]
- Carbohydrates (sugars) [4, 6]
- Phenols [4, 7]
  e.g. salicyl alcohol, salicylsalicin [8]
- Glycosides
  e.g. cardiac glycosides [4, 12]
  diterpene glycosides [17]
- Sapogenins
  e.g. polygalaic acid [8], diosgenin
  tigogenin, gitogenin [9]
- Essential oil components or terpenes [4, 10, 11]
  e.g. from Hedeoma pulegioides [10],
  Melissae folium [11]
- Antibiotics [13, 14]
  e.g. macrolide antibiotics [13]
  heptaene antibiotics [14]
  tetracyclines [15]
- Mycotoxins (trichothecenes) [16]

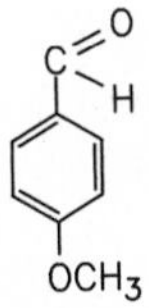

$C_8H_8O_2$	H_2SO_4
$M_r = 136.15$	$M_r = 98.08$
Anisaldehyde	Sulfuric acid

Preparation of the Reagent

Dipping solution	Dissolve 1 ml 4-methoxybenzaldehyde (anisaldehyde) and 2 ml conc. sulfuric acid in 100 ml glacial acetic acid.
Spray solution	Carefully add 8 ml conc. sulfuric acid and 0.5 ml anisaldehyde under cooling with ice to a mixture of 85 ml methanol and 10 ml glacial acetic acid [1, 4, 8, 11, 16].
Storage	The reagents are stable for several weeks in the refrigerator [4].
Substances	4-Methoxybenzaldehyde Sulfuric acid Acetic acid Methanol Ethanol

Reaction

The mechanism of reaction with steroids has not been elucidated. Various nonquantitative reactions occur simultaneously. Cyclopentenyl cations have been postulated as intermediates which condense with anisaldehyde to yield colored compounds [4]. It is probable that triphenylmethane dyes are also formed with aromatic compounds.

Method

The chromatogram is freed from mobile phase in a stream of warm air, immersed for 1 s in the dipping solution or sprayed homogeneously with the spray solution until the layer begins to become transparent and then heated to 90–125 °C for 1–15 min.

Variously colored chromatogram zones result on an almost colorless background, they are often fluorescent under long-wavelength UV light ($\lambda = 365$ nm) (e.g. prostaglandins, salicylsalicin).

Note: Anisaldehyde – sulfuric acid is a universal reagent for natural products, that makes color differentiation possible [6]. The background acquires a reddish coloration if the heating is carried out for too long; it can be decolorized again by interaction with water vapor. The dipping solution can be modified by the addition of *n*-hexane for the detection of glycosides [17]. The detection limits are 50 ng substance per chromatogram zone for prostaglandins and sugars [5, 6]. Bacitracin, chloramphenicol and penicillin do not react [15].

The reagent can be employed on silica gel, kieselguhr, Si 50 000 and RP layers.

Procedure Tested

Essential Oils [18]

Method Ascending, one-dimensional development in a trough chamber with chamber saturation.

Layer HPTLC plates Silica gel 60 F_{254} (MERCK).

Mobile phase Toluene – chloroform (10 + 10).

Migration distance 2 × 6 cm with intermediate drying in a stream of cold air.

Running time 2 × 10 min

Detection and result: The chromatogram was freed from mobile phase and immersed for 1 s in the dipping solution and then heated to 100 °C for 10 min.

Menthol (hR_f 15) and menthyl acetate (hR_f 55) yielded blue chromatogram zones; caryophyllene (hR_f 90) and caryophyllene epoxide (hR_f 20–25) appeared red-violet and thymol (hR_f 40–45) appeared brick red in color.

In situ quantitation: Absorption photometric recording in reflectance was performed at a medium wavelength $\lambda = 500$ nm (Fig. 1).

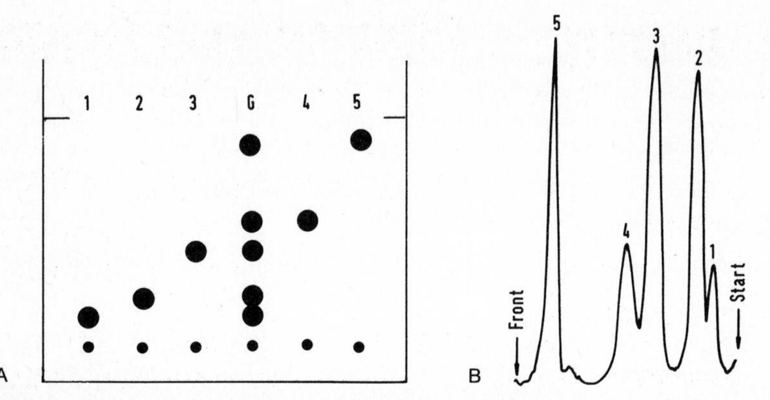

Fig. 1: Schematic sketch (A) of the separation of essential oil components (ca. 500 ng of each component) and reflectance scan of the mixture (B). Menthol (1), caryophyllene epoxide (2), thymol (3), menthyl acetate (4), caryophyllene (5), mixture (G).

References

[1] Van der Heide, R. F.: *J. Chromatogr.* **1966**, *24*, 239–243.
[2] Jarc, H., Ruttner, O., Krocza, W.: *Fleischwirtschaft* **1976**, *9*, 1326–1328.
[3] Kritchefsky, D., Tepper, S. A.: *J. Chromatogr.* **1968**, *37*, 361–362.
[4] Stahl, E., Glatz, A.: *J. Chromatogr.* **1982**, *240*, 518–521; *243*, 139–143.
[5] Ubatuba, F. B.: *J. Chromatogr.* **1978**, *161*, 165–177.
[6] Stahl, E., Kaltenbach, U.: *J. Chromatogr.* **1961**, *5*, 351–355.
[7] Sancin, P.: *Planta Med.* **1971**, *20*, 153–155.
[8] Genius, O.-B.: *Dtsch. Apoth. Ztg.* **1980:** , *120*, 1417–1419 **1980**, *120*, 1739–1740.
[9] Dawidar, A. M., Fayez, M. B. E.: *Fresenius Z. Anal. Chem.* **1972**, *259*, 283–285.
[10] Sleckman, B. P., Sherma, J., Mineo, L. C.: *J. Liq. Chromatogr.* **1983**, *6*, 1175–1182.
[11] Kloeti, F., Christen P., Kapetanidis, I.: *Fresenius Z. Anal. Chem.* **1985**, *321*, 352–354.
[12] Bulger, W. H., Talcott, R. E., Stohs, S. J.: *J. Chromatogr.* **1972**, *70*, 187–189.
[13] Kibwage, I. O., Roets, E., Hoogmartens, J.: *J. Chromatogr.* **1983**, *256*, 164–171.
[14] Thomas, A. H., Newland, P.: *J. Chromatogr.* **1986**, *354*, 317–324.
[15] Langner, H. J., Teufel, U.: *Fleischwirtschaft* **1972**, *52*, 1610–1614.
[16] Martin, P. J., Stahr, H. M., Hyde, W., Domoto, M.: *J. Liq. Chromatogr.* **1986**, *9*, 1591–1602.
[17] Mätzel, U., Maier, H. G.: *Z. Lebensm. Unters. Forsch.* **1983**, *176*, 281–284.
[18] Kany, E., Jork, H.: GDCh-training course Nr. 300 „Einführung in die Dünnschicht-Chromatographie", Universität des Saarlandes, Saarbrücken 1987.

p-Anisidine – Phthalic Acid Reagent

Reagent for:

- Carbohydrates
  e.g. monosaccharides [1 – 7, 12]
  oligosaccharides [2, 6]
  uronic acids [1, 4, 6]

p-Anisidine	Phthalic acid
C_7H_9NO	$C_8H_6O_4$
$M_r = 123.16$	$M_r = 166.13$

Preparation of the Reagent

Solution I Dissolve 1.25 g *p*-anisidine (4-methoxyaniline) in 25 ml methanol and add 25 ml ethyl acetate.

Solution II Dissolve 1.5 g phthalic acid in 25 ml methanol and dilute with 25 ml ethyl acetate.

Dipping solution Mix equal quantities of solutions I and II immediately before dipping.

Storage The two solutions I and II may be stored for several weeks in the refrigerator.

Substances *p*-Anisidine
Phthalic acid
Methanol
Ethyl acetate

Reaction

Sugars react with the reagent probably with the formation of SCHIFF's bases:

$$R^1R^2C{=}O + H_2N{-}C_6H_4{-}OCH_3 \xrightarrow[-\,H_2O]{} R^1R^2C{=}N{-}C_6H_4{-}OCH_3$$

Method

The chromatogram is freed from mobile phase and immersed in the dipping solution for 1 s or uniformly sprayed with it and then heated to 100–130 °C for 10 min. The result is reddish-brown (pentoses) to brown chromatogram zones on a colorless background, which also becomes brown after a time.

Note: Phosphoric acid [8] and hydrochloric acid [6, 9] have both been suggested in the literature as substitutes for phthalic acid. The addition of sodium dithionite [9] is also occasionally mentioned and sometimes no additives are employed [10]. The alternative reagents offer no advantages over the phthalic acid containing reagent since they usually cause more background coloration. The limits of detection are about 0.1–0.5 µg per chromatogram zone [5].

The reagent can be employed on silica gel, RP, NH_2 and polyamide layers.

Procedure Tested

Monosaccharides [11]

Method Ascending, one-dimensional double development in a HPTLC trough chamber with chamber saturation.

Layer HPTLC plates Silica gel 60 (MERCK).

Mobile phase 2-Propanol – boric acid (2% aqueous solution) – glacial acetic acid (40 + 5 + 1).

Migration distance 2 × 8 cm

Running time 2 × 90 min

Detection and result: The chromatogram was freed from mobile phase and immersed in the reagent solution for 1 s and then heated to 130 °C for 10 min. Rhamnose (hR_f 35–40) and fructose (hR_f 70–75) yielded brown and xylose (hR_f 45–50) and arabinose (hR_f 60–65) red-brown chromatogram zones on a pale background. The detection limit for the pentoses was 0.1 µg and for fructose it was 0.5 µg substance per chromatogram zone.

In situ quantitation: Absorption photometric scanning was carried out in reflectance at $\lambda = 480$ nm (Fig. 1).

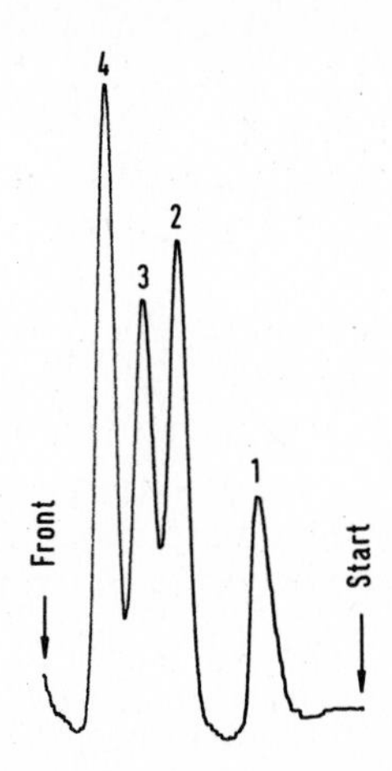

Fig. 1: Absorption curve of a chromatogram track with 4 μg of each substance per chromatogram zone. Rhamnose (1), xylose (2), arabinose (3), fructose (4).

References

[1] Friese, P.: *Fresenius Z. Anal. Chem.* **1980**, *301*, 389–397.
[2] Damonte, A., Lombard, A., Tourn, M. L., Cassone M. C.: *J. Chromatogr.* **1971**, *60*, 203–211.
[3] Trachtenberg, S., Mayer, A. M.: *Phytochemistry* **1981**, *20*, 2665–2668.
[4] Metraux, J. P.: *J. Chromatogr.* **1982**, *237*, 525–527.
[5] Schweiger, A.: *J. Chromatogr.* **1962**, *9*, 374–376.
[6] Petre, R., Dennis, R., Jackson, B. P., Jethwa, K. R.: *Planta Med.* **1972**, *21*, 81–83.
[7] Dobson, R. L., Cooper, M. F.: *Biochim. Biophys. Acta* **1971**, *254*, 393–401.
[8] Niemann, G. J.: *J. Chromatogr.* **1979**, *170*, 227.
[9] Bell, D. J., Talukder, M. Q.-K.: *J. Chromatogr.* **1970**, *49*, 469–472.
[10] Loub, W. D., Fong, H. H. S., Theiner, M., Farnsworth, N. R.: *J. Pharm. Sci.* **1973**, *62*, 149–150.
[11] Jork, H., Kany, E.: GDCh-training course Nr. 300 „Einführung in die Dünnschicht-Chromatographie", Universität des Saarlandes, Saarbrücken 1986
[12] Hartley, R. D., Jones, E. C., Wood, T. M.: *Phytochemistry* **1976**, *15*, 305–307.

Anthrone Reagent

Reagent for:

- Ketoses [1]
- Glycolipids, gangliosides [2–7]
- Cyclodextrins [8]

$C_{14}H_{10}O$

$M_r = 194.24$

Preparation of Reagent

Dipping solution *For ketoses:* Dissolve 300 mg anthrone in 10 ml acetic acid and add in order 20 ml ethanol, 3 ml 85% phosphoric acid and 1 ml water [1].

Spray solution *For glycolipids:* Carefully add 72 ml conc. sulfuric acid to 28 ml water with cooling. Dissolve 50 mg anthrone in the mixture with gentle warming [5].

Storage The dipping solution may be stored in the refrigerator for several weeks, the spray solution should be freshly prepared each day.

Substances Anthrone
Sulfuric acid (95–97%)
Acetic acid (96%)
Ethanol
Orthophosphoric acid (85%)

Reaction

Carbonyl derivatives react with anthrone in acidic medium to yield condensation products of types 1 (pentoses) or 2 (hexoses) [9]:

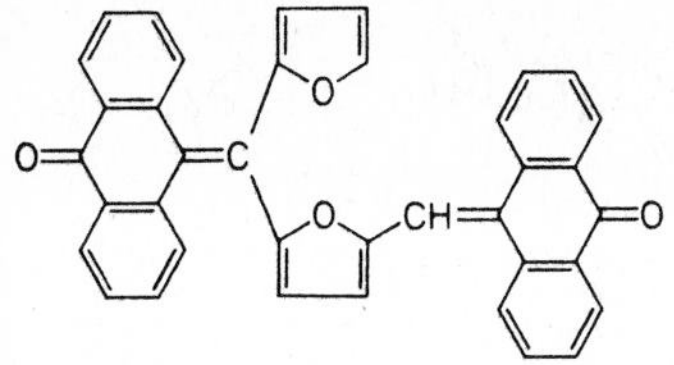

(1)

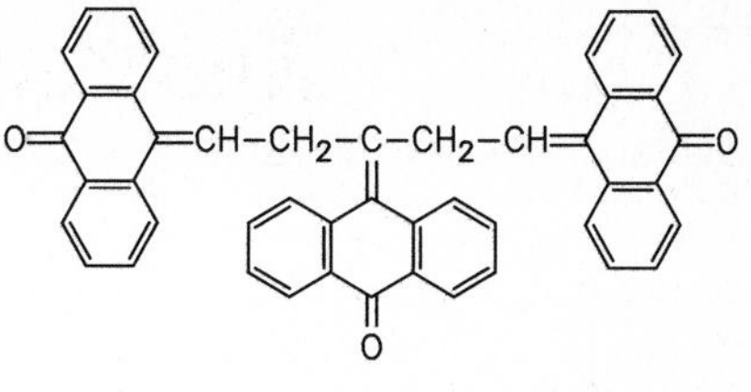

(2)

Method

The chromatogram is freed from mobile phase in a stream of warm air and immersed for 4 s in the dipping solution or evenly sprayed with it until the layer begins to be transparent (the spray solution is employed for glycosides) and then heated to 105 – 120 °C for 5 – 15 min.

Colored chromatogram zones appear on an almost colorless background; ketoses, for example, are yellow in color [1].

Note: Aldoses do not react or only react with greatly reduced sensitivity. The reagent can be employed with silica gel, kieselguhr and Si 50 000 layers. Paraffin-impregnated silica gel layers may also be employed [8].

Procedure Tested

Raffinose, Sucrose, Fructose [10]

Method Ascending, one-dimensional double development in a trough chamber with chamber saturation.

Layer HPTLC plates Si 50 000 (MERCK); before application of the samples the layer was developed once in chloroform – methanol (50 + 50) to precleanse it and dried at 110 °C for 30 min.

Mobile phase Acetonitrile – water (85 + 15).

Migration distance 2 × 7 cm with 5 min intermediate drying in a stream of warm air.

Running time 2 × 10 min

Detection and result: The chromatogram was dried for 5 min in a stream of warm air, immersed in the reagent for 4 s and then heated to 110 °C for 8 min. After cooling to room temperature it was immersed for 2 s in a 20% solution of dioctylsulfosuccinate in chloroform.

In visible light raffinose (hR_f 10 – 15), sucrose (hR_f 30 – 35) and fructose (hRf 45 – 50) produced yellow zones on a light background, in long-wavelength UV light ($\lambda = 365$ nm) the zones had a red fluorescence on a pale blue background. The detection limits were less than 10 ng substance per chromatogram zone (Fig. 1).

In situ quantitation: The absorption photometric scan in reflectance was made at $\lambda = 435$ nm (detection limit 20 – 30 ng per chromatogram zone). Fluorimetric scanning was performed at $\lambda_{exc} = 436$ nm and $\lambda_{fl} > 560$ nm (detection limit < 10 ng per chromatogram zone).

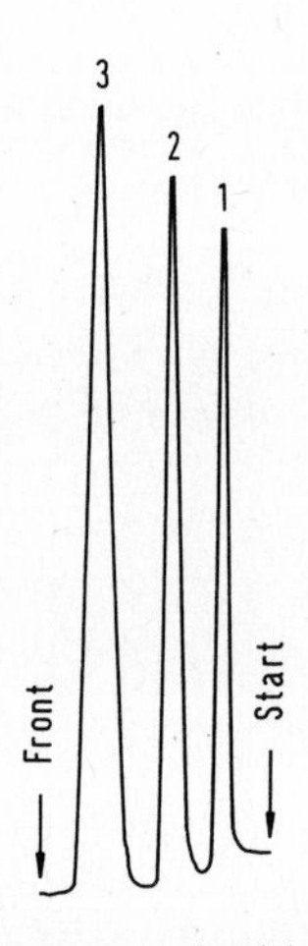

Fig. 1: Fluorescence scan of a chromatogram track with 100 ng substance per chromatogram zone. Raffinose (1), sucrose (2), fructose (3).

References

[1] MERCK E.: Company literature *"Dyeing Reagents for Thin-Layer and Paper Chromatography"*, Darmstadt 1980.
[2] Kasai, N., Sillerud, L. O., Yu, R. K.: *Lipids* **1982**, *17*, 107–110.
[3] Patton, S., Thomas, A. J.: *J. Lipid Res.* **1971**, *12*, 331–335.

[4] Stoffel, W., Hanfland, P.: *Hoppe-Seylers Z. Physiol. Chem.* **1973**, *354*, 21–31.
[5] Taketomi, T., Kawamura, N.: *J. Biochem.* **1972**, *72*, 791–798: **1972**, *72*, 799–806.
[6] Saito, S., Tamai, Y.: *J. Neurochem.* **1983**, *41*, 737–744.
[7] Koenig, F.: *Z. Naturforsch.* **1971**, *26b*, 1180–1187.
[8] Cserhati, T., Szente, L., Szeijtli, J.: *J. High Resolut. Chromatogr. Chromatogr. Commun.* **1984**, *7*, 635–636.
[9] Kakáč, B., Vejdělek, Z. J.: *Handbuch der photometrischen Analyse organischer Verbindungen.* Weinheim: Verlag Chemie, 1974.
[10] Patzsch, K., Netz, S., Funk, W.: *J. Planar Chromatogr.* **1988**, *1*, 39–45.

Antimony(III) Chloride Reagent
(Carr-Price Reagent)

Reagent for:

- Vitamins A and D [1, 2]
- Carotinoids [1, 2]
- Terpenes, triterpenes [1 – 5]
- Sterols [6]
  e.g. *β*-sitosterol [7]
- Steroids, steroid hormones [8 – 10]
  e.g. contraceptives [10]
  steroid alkaloids [11]
- Bile acids [12]
- Sapogenins, steroid sapogenins [2, 13 – 16]
- Glycosides [1]
  e.g. steroid glycosides [2]
  digitalis glycosides [17, 18]
- Flavonoids [3]
- Phospholipids [19]

$SbCl_3$

$M_r = 228.11$

Preparation of Reagent

Dipping solution Dissolve 2 g antimony(III) chloride in 50 ml chloroform.

Spray solution Dissolve 10 g antimony(III) chloride in 50 ml chloroform or carbon tetrachloride.

Storage The solution should always be freshly made up.

Substances Antimony(III) chloride
Chloroform
Carbon tetrachloride

Reaction

Antimony(III) chloride forms colored π-complexes with double bond systems (e.g. vitamin A).

$$\text{>C=C<} + SbCl_3 \longrightarrow \text{>C=C<} \rightarrow SbCl_3$$

Double bond system π-Complex

Method

The chromatograms are freed from mobile phase, immersed in the dipping solution for 1 s or sprayed evenly with the spray solution and then heated to 110–120°C for 5–10 min.

Variously colored chromatogram zones are sometimes produced even before heating; they often fluoresce in long-wavelength UV light ($\lambda = 365$ nm)

Note: The reagent is also occasionally employed in hydrochloric acid [5, 14–16] or acetic acid [10] solution or with the addition of acetic anhydride [7] or sulfuric acid [12]. The solvents employed should be as anhydrous as possible.

The reagent can be employed on silica gel, kieselguhr, Si 50 000 and aluminium oxide layers.

Procedure Tested

Vitamin D_3 [20]

Method Ascending, one-dimensional development in a trough chamber with chamber saturation.

Layer HPTLC plates Silica gel 60 (MERCK). Before sample application the layers were prewashed by developing once with chloroform – methanol (50 + 50) and dried at 110 °C for 30 min.

Mobile phase Cyclohexane – diethyl ether (40 + 20).

Migration distance 6 cm

Running time 15 min

Detection and result: The developed chromatogram was dried in a stream of cold air for 15 min, then immersed in the dipping solution for 1 s and, after brief drying in a stream of warm air, heated to 110 °C for 10 min. Vitamin D_3 (hR_f 15–20, detection limit ca. 1 ng per chromatogram zone) appeared in visible light as a grey chromatogram zone on a white background and in long-wavelength UV light (λ = 365 nm) as a bright red fluorescent zone on a blue background.

The fluorescence intensity was stabilized and enhanced (ca. 2-fold) by immersion of the chromatogram for 1 s in a mixture of liquid paraffin – *n*-hexane (1 + 2).

In situ quantitation: The fluorimetric analysis was made at λ_{exc} = 436 nm and $\lambda_{fl} > 560$ nm (Fig. 1).

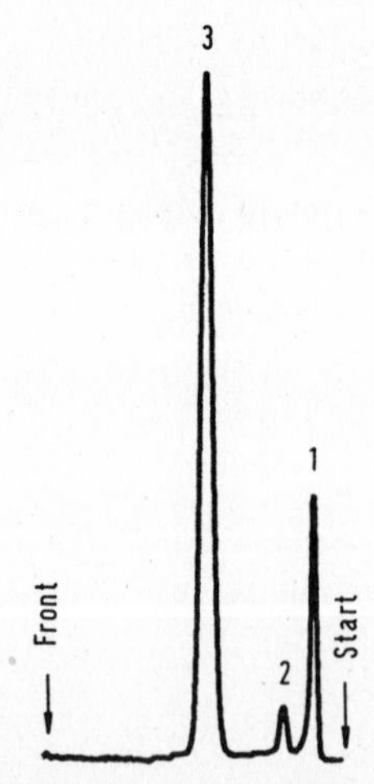

Fig. 1: Fluorescence scan of a chromatogram track with 50 ng vitamin D_3 per chromatogram zone: by-products (1, 2), vitamin D_3 (3).

References

[1] Hörhammer, L., Wagner, H., Hein, K.: *J. Chromatogr.* **1964,** *13*, 235-237.
[2] Ikan, R., Kashman, J., Bergmann, E. D.: *J. Chromatogr.* **1964,** *14*, 275-279.
[3] Stahl, E.: *Chem.-Ztg.* **1958,** *82*, 323-329.
[4] Griffin, W. J., Parkin, J. E.: *Planta Med.* **1971,** *20*, 97-99.
[5] Auterhoff, H., Kovar, K.-A.: *Identifizierung von Arzneistoffen.* 4th Ed., Stuttgart: Wissenschaftliche Verlagsgesellschaft, 1981; p. 69.
[6] Mermet-Bouvier, R.: *J. Chromatogr.* **1971,** *59*, 226-230.
[7] Elghamry, M. I., Grunert, E., Aehnelt, E.: *Planta Med.* **1971,** *19*, 208-214.
[8] Vaedtke, J., Gajewska, A.: *J. Chromatogr.* **1962,** *9*, 345-347.
[9] Abraham, R., Gütte, K.-F., Hild, E., Taubert, H.-D.: *Clin. Chim. Acta* **1970,** *28*, 341-347.
[10] Szekacs, I., Klembala, M.: *Z. Klin. Chem. Klin. Biochem.* **1970,** *8*, 131-133.
[11] Wagner, H., Seegert, K., Sonnenbichler, H., Ilyas, M., Odenthal, K. P.: *Planta Med.* **1987,** *53*, 444-446.
[12] Anthony, W. L., Beher, W. T.: *J. Chromatogr.* **1964,** *13*, 567-570.
[13] Takeda, K., Hara, S., Wada, A., Matsumoto, N.: *J. Chromatogr.* **1963,** *11*, 562-564.
[14] Blunden, G., Yi, Y., Jewers, K.: *Phytochemistry* **1978,** *17*, 1923-1925.
[15] Hardman, R., Fazli, F. R. Y.: *Planta Med.* **1972,** *21*, 131-138.
[16] Blunden, G., Hardman, R.: *J. Chromatogr.* **1968,** *34*, 507-514.
[17] Thieme, H., Lamchav, A.: *Pharmazie* **1970,** *25*, 202-203.
[18] Eder, S. R.: *Fette, Seifen, Anstrichm.* **1972,** *74*, 519-524.
[19] Horvath, P., Szepesi, G., Hoznek, M., Vegh, Z.: *Proc. Int. Symp. Instrum. High Perform. Thin-Layer Chromatogr. (HPTLC), 1st.* Bad Dürkheim, IfC-Verlag, 1980, p. 295–304.
[20] Netz, S., Funk, W.: Private communication, Fachhochschule Gießen, Fachbereich Technisches Gesundheitswesen, 1987.

Antimony(V) Chloride Reagent

Reagent for:

- Esters of phenoxyalkanecarboxylic acids e.g herbicides [1]
- Components of essential oils [2–4]
- Terpenes, triterpenes [2, 5]
- Steroids [6, 7]
- Phenols, phenol ethers [2, 8, 9]
- Antioxidants [9]
- Aromatic hydrocarbons [10]

$SbCl_5$

$M_r = 299.02$

Preparation of Reagent

Dipping solution Mix 2 ml antimony(V) chloride with 8 ml carbon tetrachloride.

Storage The reagent solution should always be freshly prepared.

Substances Antimony(V) chloride
Carbon tetrachloride

Reaction

Antimony(V) chloride forms colored π-complexes with double bond systems.

$$>C=C< + SbCl_5 \longrightarrow >C=C< \rightarrow SbCl_5$$

Double bond system $\qquad$ π-Complex

Method

The chromatograms are freed from mobile phase, immersed in the dipping solution for 1 s or homogeneously sprayed with it until they begin to be transparent and then heated to 105 – 120° for 5 – 10 min.

Variously colored chromatogram zones are produced on a colorless background; some of them fluoresce under long-wavelength UV light ($\lambda = 365$ nm).

Note: The solvents employed should be anhydrous. The esters of phenoxy-alkanecarboxylic acids (detection limits: 500 ng) [1] yield brown to violet, terpenes violet-grey [2] and triterpenes yellow to violet [5] colored chromatogram zones.

The reagent can be employed on silica gel, kieselguhr, Si 50 000 and aluminium oxide layers.

Procedure Tested

Essential Oil of Peppermint

Method	Ascending, one-dimensional development in a HPTLC trough chamber with chamber saturation.
Layer	HPTLC plates Silica gel 60 F_{254} (MERCK).
Mobile phase	Dichloromethane.
Migration distance	5 cm
Running time	8 min

Detection and result: The chromotograms were dried in a stream of warm air, immersed for 1 s in the reagent solution and dried at 120 °C for 5 min. Menthol (hR_f 15 – 20), cineole (hR_f 20 – 25), menthone (hR_f 35 – 40), menthyl acetate (hR_f 45 – 50) and menthofuran (hR_f 80 – 85) yielded grey to brown chromatogram zones on a pale background, they fluoresced yellow under long-wavelength UV light ($\lambda = 365$ nm). The detection limit for menthol was 15 ng substance per chromatogram zone (Fig. 1).

In situ quantitation: The fluorimetric analysis was carried out at $\lambda_{exc} = 365$ nm and $\lambda_{fl} > 560$ nm (Fig. 2).

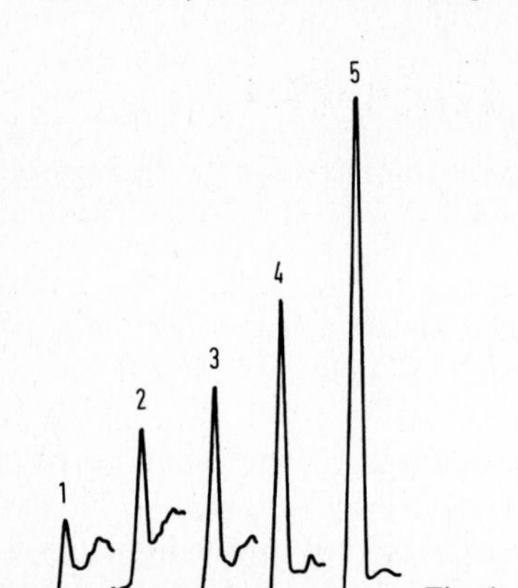

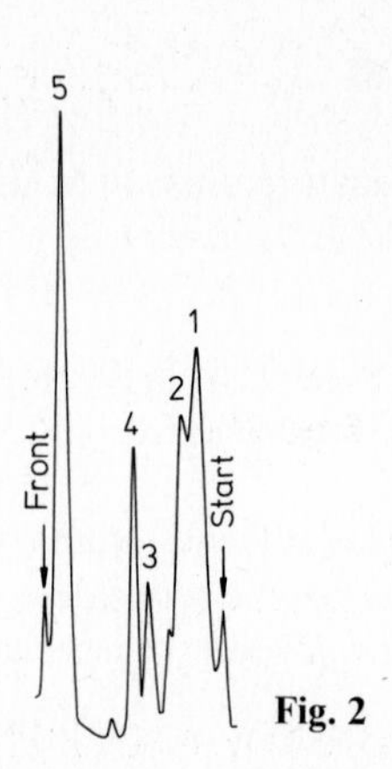

Fig. 1: Fluorescence scans of a series of dilutions for evaluation of the detection limit of menthol: 15 ng (1), 30 ng (2), 75 ng (3), 150 ng (4), 300 ng (5).

Fig. 2: Fluorescence scan of a chromatogram of peppermint oil: menthol (1), cineole (2), menthone (3), menthyl acetate (4), menthofuran/terpenes (5).

References

[1] Henkel, H. G., Ebing, W.: *J. Chromatogr.* **1964,** *14*, 283–285.
[2] Stahl, E.: *Chem.-Ztg.* **1958,** *82*, 323–327.
[3] Szejtli, J., Szente, L., Banky-Elöd, E.: *Acta Chim. Hung.* **1979,** *101*, 27–46.
[4] Lindner, K., Szente, L., Szejtli, J.: *Acta Alimentaria* **1981,** *10*, 175–186.
[5] Ikan, R., Kashman, J., Bergmann, E. D.: *J. Chromatogr.* **1964,** *14*, 275–279.
[6] Neder, A., Uskert, A., Nagy, E., Mehesfalvi, Z., Kuszmann, J.: *Acta Chim. Hung.* **1980,** *103*, 231–240; **1980,** *104*, 123–140..
[7] Neder, A., Pelczer, I., Mehesfalvi, Z., Kuszmann, J.: *Acta Chim. Hung.* **1982,** *109*, 275–285.
[8] Thielemann, H.: *Mikrochim. Acta (Vienna)* **1971,** 717–723.
[9] Van der Heide, R. F.: *J. Chromatogr.* **1966,** *24*, 239–243.
[10] Thielemann, H.: *Mikrochim. Acta (Vienna)* **1971,** 838–840.
[11] Kany, E., Jork, H.: GDCh-training course Nr. 301 „Dünnschicht-Chromatographie für Fortgeschrittene“, Universität des Saarlandes, Saarbrücken 1988.

Berberine Reagent

Reagent for:

- Sterols [1 – 3]
  e.g. sitosterol, stigmasterol,
  campesterol, cholesterol
- Saturated compounds [4]

H3CO
OCH3
N
+
Cl−

$C_{20}H_{18}ClNO_4 \cdot 2H_20$
$M_r = 407.86$

Preparation of Reagent

Dipping solution Dissolve 10 mg berberine chloride in 100 ml ethanol.

Storage The dipping solution may be kept for several days.

Substances Berberine chloride
Ethanol

Reaction

The mechanism of the reaction has not been elucidated. Berberine is probably enriched in the lipophilic chromatogram zones which then fluoresce more intensely than the environment.

Method

The chromatograms are freed from mobile phase, immersed in the reagent solution for 1 s or sprayed homogeneously with it and then dried in a stream of cold air.

Under UV light ($\lambda = 254$ nm or 365 nm) the chromatogram zones appear as pale yellow fluorescent zones on a weakly yellow fluorescent background.

Note: The reagent can also be applied before chromatography e.g. by impregnating the layer, which should preferably be free from fluorescent indicator; it is not eluted by most mobile phases [4].

The reagent can be employed on silica gel, silver nitrate-impregnated silica gel, carboxymethylcellulose-containing silica gel, kieselguhr and Si 50000 layers; RP phases are not suitable.

Procedure Tested

Sterols, Fatty Acids, Triglycerides, Hydrocarbons [5]

Method Ascending, one-dimensional development in a trough chamber with chamber saturation.

Layer HPTLC plates Silica gel 60 F_{254} (MERCK).

Mobile phase *n*-Hexane – diethyl ether – glacial acetic acid (80 + 20 + 1).

Migration distance 5 cm

Running time 7 min

Detection and result: The chromatogram was dried in a stream of cold air for 5 min and immersed in the dipping solution for 1 s. Cholesterol (hR_f 10–15),

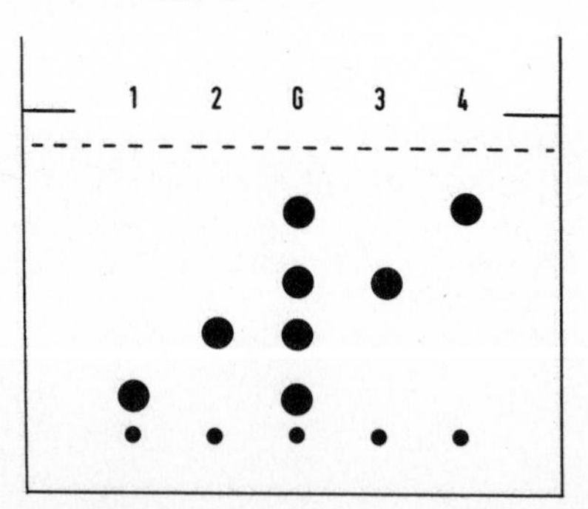

Fig. 1: Sketch of the chromatographic separation of a mixture (G) of cholesterol (1), stearic acid (2), tripalmitin (3) and caryophyllene (4).

stearic acid (hR_f 20–25), tripalmitin (hR_f 45–50) and caryophyllene (hR_f 90–95) appeared as pale yellow fluorescent zones on a less intense yellow fluorescent background. The visual detection limits per chromatogram zone were 10 ng for cholesterol, 50 ng for stearic acid and tripalmitin and 100 ng for caryophyllene (Fig. 1).

In situ quantitation: The reagent is unsuitable for in situ quantitation.

References

[1] Misso, N. L. A., Goad, L. J.: *Phytochemistry* **1984,** *23*, 73-82.
[2] Huang, L.-S., Grunwald, C.: *Phytochemistry* **1986,** *25*, 2779-2781.
[3] Heintz, R., Benveniste, P., Robinson, W. H., Coates, R. M.: *Biochem. Biophys. Res. Commun.* **1972,** *49*, 1547-1553.
[4] Mamlok. L.: *J. Chromatogr. Sci.* **1981,** *19*, 53.
[5] Kany, E., Jork, H.: GDCh-training course Nr. 301 „Dünnschicht-Chromatographie für Fortgeschrittene", Universität des Saarlandes, Saarbrücken 1988.

2,2′-Bipyridine – Iron(III) Chloride Reagent
(Emmerie-Engel Reagent)

Reagent for:

- Phenols [1]
- Vitamin E (tocopherols) [2–4]
- Antioxidants [5]
- Reducing substances
  e.g. ascorbic acid [6]

$C_{10}H_8N_2$

$M_r = 156.19$

2,2′-Bipyridine

$FeCl_3 \cdot 6H_2O$

$M_r = 270.30$

Preparation of the Reagent

Solution I Dissolve 0.1 g iron(III) chloride hexahydrate in 50 ml ethanol.

Solution II Dissolve 0.25 g 2,2′-bipyridine (α, α′-dipyridyl) in 50 ml ethanol.

Dipping solution Mix equal quantities of solutions I and II immediately before use.

Storage Solution I should be stored in the dark.

Substances 2,2′-Bipyridine
Iron(III) chloride hexahydrate
Ethanol

Reaction

Ascorbic acid, for example, is oxidized to dehydroascorbic acid with reduction of the iron(III) ions. The Fe(II) ions so produced react with 2,2′-bipyridine with formation of a colored complex.

Method

The chromatogram is freed from mobile phase in a stream of warm air and immersed in the reagent solution for 1 s and then dried in a stream of cold air. Red to reddish-brown zones are formed on a colorless background. They often appear immediately but sometimes they only appear after some minutes; their color intensity is completely developed after 30 min [1]. They can be employed for quantitative analysis.

Note: The dipping reagent, which can also be applied as a spray reagent, can be employed on cellulose and silica gel layers. A 3% solution of 2,2'-bipyridine in 40% thioglycolic acid can be employed as a specific spray reagent for the detection of iron (red coloration) [7].

Procedure Tested

α-Tocopherol [3]

Method	Horizontal, one-dimensional development in a linear chamber (CAMAG).
Layer	HPTLC plates Silica gel 60 F_{254} (MERCK).

Mobile phase Toluene – chloroform (10 + 10).

Migration distance 5 cm

Running time 10 min

Detection and result: The chromatogram was freed from mobile phase in a stream of warm air, immersed in the freshly prepared reagent mixture for 1 s and then dried in a stream of cold air. After a short time α-tocopherol yielded a red chromatogram zone on a colorless background at hR_f 35–40. The visual detection limit on the HPTLC layer was 20–25 ng per chromatogram zone.

In situ quantitation: The plate was scanned with visible light ($\lambda = 520$ nm) in the reflectance mode (Fig. 1).

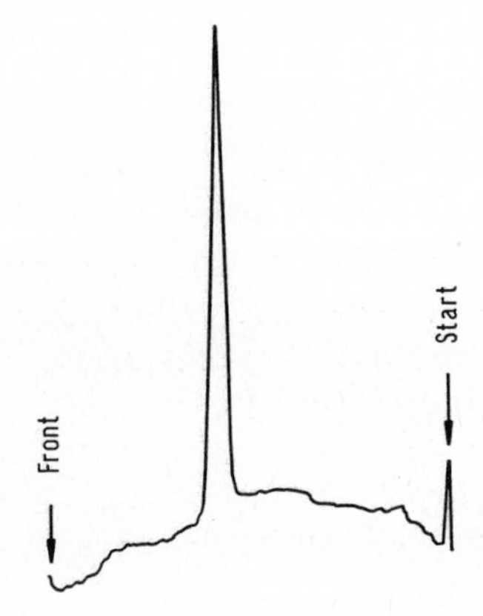

Fig. 1: Reflectance scan of a chromatogram of 200 ng D-α-tocopherol per chromatogram zone.

References

[1] BARTON, G. M.: *J. Chromatogr.* **1965,** *20,* 189.
[2] STROHECKER, R., HENNING, H. M.: *Vitaminbestimmungen.* Weinheim: Verlag Chemie 1963, p. 311–312.
[3] KOOP, R.: Dissertation, Universität Gießen, Fachbereich Ernährungswissenschaften, 1984.
[4] PEREDI, J., BALOGH, A.: *Olaj, Szappan, Kozmet.* **1981,** *30,* 1-5.
[5] HEIDE, R. F. VAN DER: *J. Chromatogr.,* **1966,** *24,* 239–243.
[6] SCHNEKENBURGER, G.: Thesis, Fachhochschule Gießen, Fachbereich Technisches Gesundheitswesen, 1987.
[7] FEIGL, F.: *Fresenius Z. Anal. Chem.* **1956,** *152,* 52–55.

Blue Tetrazolium Reagent

Reagent for:

- Corticosteroids [1–6]
- Reducing steroids [2]
- Carbohydrates [7]

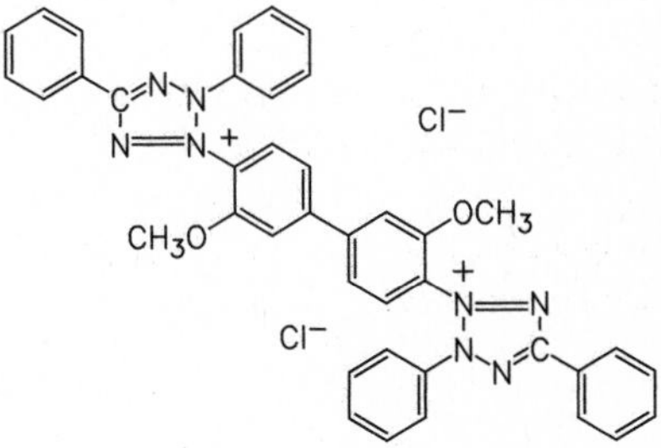

$C_{40}H_{32}Cl_2N_8O_2$
$M_r = 727.66$

Preparation of Reagent

Solution I Dissolve 0.5 g blue tetrazolium (3,3′-(3,3′-dimethoxy-4,4′-biphenylylene)-bis(2,5-diphenyl-2H-tetrazolium)-chloride) in 100 ml methanol.

Solution II Dissolve 24 g sodium hydroxide carefully in 50 ml water and make up to 100 ml with methanol.

Dipping solution Mix equal quantities of solution I and solution II before use.

Spray solution Dilute the dipping solution with methanol in a ratio of 1 + 2.

Storage Each solution may be stored for a longer period of time in the refrigerator.

Substances Blue tetrazolium
Sodium hydroxide pellets
Methanol

Reaction

Blue tetrazolium is transformed into the colored or fluorescent formazan by reducing compounds.

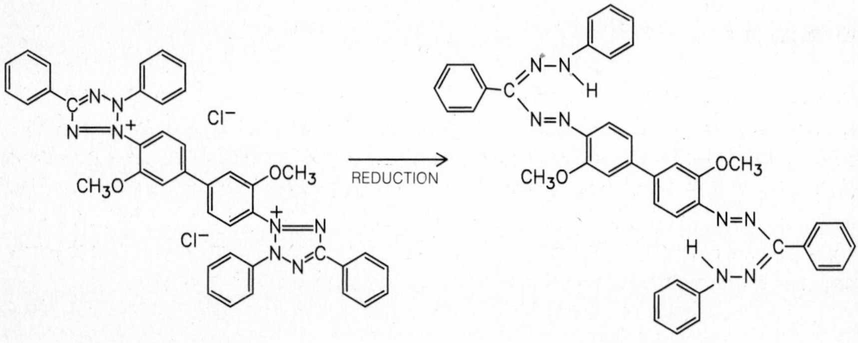

Blue tetrazolium REDUCTION Formazan

Method

The developed chromatograms are either immersed briefly in the dipping solution or homogeneously sprayed with the spray solution. The color reaction occurs immediately at room temperature or on gentle heating. In general violet-colored zones are formed on a light background. If they are heated to 90 °C for 10 to 20 min Δ^4-3-ketosteroids exhibit rather characteristic deep yellow fluorescence under long-wavelength UV light ($\lambda = 365$ nm).

Note: The reagent can be employed on silica gel, alumina, polyamide and cellulose layers. In the case of the latter it is to be recommended that the solutions be diluted 1 + 3 with methanol. The detection limit is reported to be 0.1 to 0.5 µg per chromatogram zone [5].

Procedure Tested

Corticosteroids [8]

Method Ascending, one-dimensional, double development in a trough chamber with chamber saturation (5 min intermediate drying in a stream of warm air).

Layer TLC or HPTLC plates Silica gel 60 F_{254} (MERCK).

Mobile phase Chloroform – methanol (93 + 7).

Migration distance TLC: 2 × 12 cm; HPTLC: 2 × 6 cm.

Running time TLC: 2 × 35 min; HPTLC: 2 × 20 min.

Detection and result: The chromatogram was freed from mobile phase and evenly sprayed with reagent solution. After a short time at room temperature violet-colored chromatogram zones appeared for the corticosteroids tetrahydrocortisol (hR_f 10–15), tetrahydrocortisone (hR_f 15–20), prednisolone (hR_f 15–20), hydrocortisone (hR_f 20–25), prednisone (hR_f 30–35), cortisone (hR_f 35–40), corticosterone (hR_f 45–50), cortexolone (REICHSTEIN S., hR_f 50–55), 11-dehydrocorticosterone (hR_f 60–65), 11-desoxycorticosterone (hR_f 75–80).

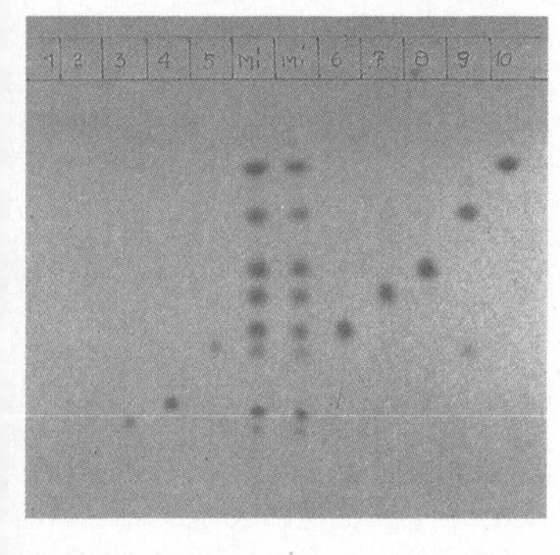

A

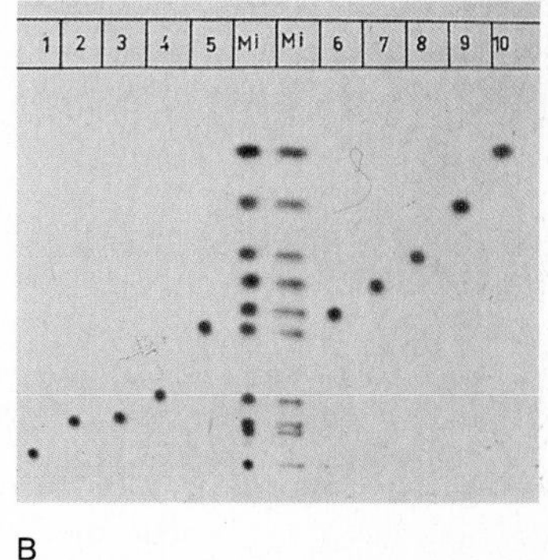

B

Fig. 1: Separation of the corticosteroids. (A) Detection in UV light ($\lambda = 254$ nm, (B) staining with blue tetrazolium. Tetrahydrocortisol (1), tetrahydrocortisone (2), prednisolone (3), hydrocortisone (4), prednisone (5), cortisone (6), corticosterone (7), cortexolone (8), 11-dehydrocorticosterone (9), 11-desoxycorticosterone (10), mixture (Mi).

Note: In the case of HPTLC plates the detection limit for the visual recognition of the violet ($\lambda_{max} = 530$ nm) colored chromatogram zones was 20 ng per chromatogram zone. With the exception of the two tetrahydrosteroids the corticosteriods could be detected on TLC plates with fluorescent indicators by reason of fluorescence quenching (Fig. 1 A). Figure 2 illustrates the absorption scans of the separations illustrated in Figures 1 A and 1 B.

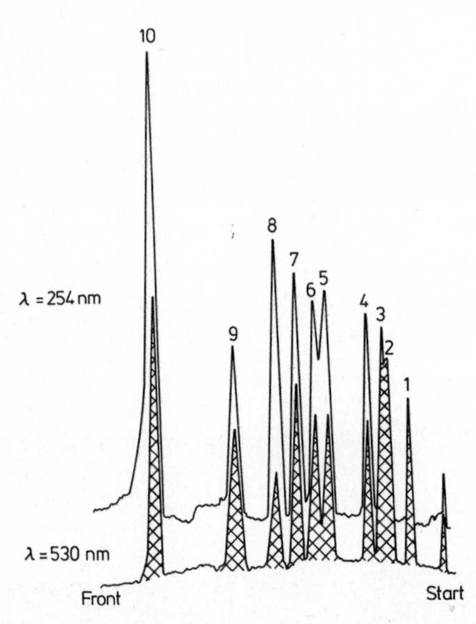

Fig. 2: Absorption scan of the corticosteroid mixture in Figure 1.

References

[1] Adamec, O., Matis, J., Galvanek, M.,: *Steroids* **1963,** *1,* 495–505.
[2] Feher, T.: *Mikrochim. Acta* (Vienna) **1965,** 105–116.
[3] Freimuth, U., Zawta, B., Büchner, M.: *Acta Biol. Med. Ger.* **1964,** *13,* 624–628.
[4] Nishikaze, O., Abraham, R., Staudinger, Hj.: *J. Biochem.* (Tokyo) **1963,** *54,* 427–431.
[5] McCarthy, J. L., Brodsky, A. L., Mitchell, J. A., Herrscher, R. F.: *Anal. Biochem* **1964,** *8,* 164–170.
[6] Nieminen, E., Castren, E.: *Zbl. Pharmaz.* **1970,** *109,* 571–578.
[7] Schmoldt, A., Machut, M.: *Dtsch. Apoth. Ztg.* **1981,** *121,* 1006–1009.
[8] Kany, E., Jork, H.: GDCh-training course Nr. 300 „Einführung in die Dünnschicht-Chromatographie", Universität des Saarlandes, Saarbrücken 1985.

Bratton-Marshall Reagent

Reagent for:

- Primary aromatic amines [1 – 3]
  e.g. benzodiazepines, aminobenzophenones [3 – 5]
- Sulfonamides [6 – 10]
- Urea, carbamate and anilide herbicides [11]
- Folic acid [12]
- Dulcin [13]

—NH—CH₂—CH₂—NH₂ · 2 HCl

$C_{12}H_{16}Cl_2N_2$

$M_r = 259.18$

N-(1-Naphthyl)ethylene-diaminedihydrochloride

Preparation of Reagent

Spray solution I Dissolve 1 g sodium nitrite in 20 ml water and make up to 100 ml with a mixture of 17 ml conc. hydrochloric acid and 83 ml ethanol [11].

Spray solution IIa Dissolve 1 g *N*-(1-naphthyl)ethylenediamine dihydrochloride in 10 ml water and make up to 100 ml with ethanol [11].

Spray solution IIb Dissolve 1 g *N*-(1-naphthyl)ethylenediamine dihydrochloride in 50 ml dimethylformamide and 50 ml hydrochloric acid ($c_{HCl} = 4$ mol/l) with warming. If the cooled solution is not clear it should be filtered. A pale violet coloration does not interfere with the reaction [4].

Storage Spray solutions I and IIa are only stable for a short period of time and, hence, should always be freshly made up [11], spray solution IIb can be stored for several weeks in the refrigerator [4].

Substances

Sodium nitrite
N-(1-Naphthyl)ethylenediamine dihydrochloride
Hydrochloric acid 32%
Ethanol
N,*N*-Dimethylformamide

Reaction

Primary aromatic amines are first diazotized and then coupled to yield azo dyestuffs [4].

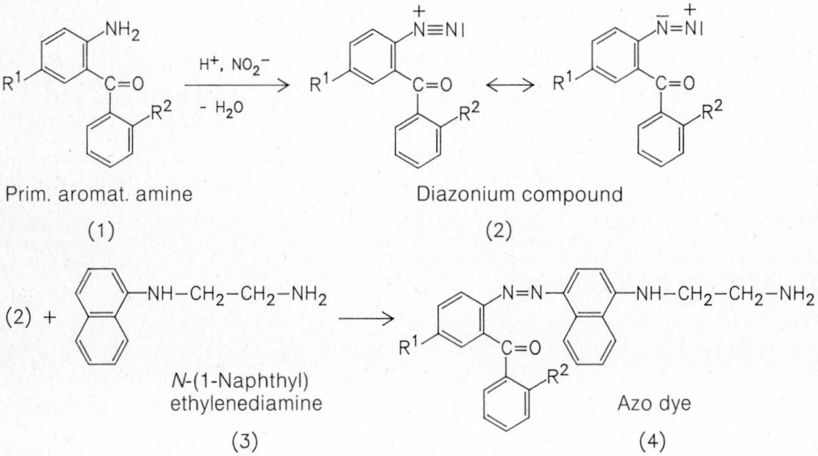

H^+, NO_2^-; $- H_2O$

Prim. aromat. amine

(1)

Diazonium compound

(2)

(2) +

N-(1-Naphthyl)
ethylenediamine

(3)

Azo dye

(4)

Method

The chromatograms are freed from mobile phase and then sprayed with spray solution I until the layer begins to be transparent, then dried in a stream of cold air for 10 min and finally sprayed again to the start of transparency this time with spray solution IIa or IIb and dried in a stream of warm air.

The pink to violet-colored chromatogram zones on a colorless background usually appear immediately.

Note: Note that the diazotization of primary aromatic amines can also be achieved by placing the chromatogram for 3–5 min in a twin-trough chamber containing nitrous fumes (fume cupboard!). The fumes are produced in the empty trough of the chamber by addition of 25% hydrochloric acid to a 20% sodium nitrite solution [2, 4]. *N*-(1-Naphthyl)ethylenediamine can be replaced in the reagent by α- or β-naphthol [10, 14], but this reduces the sensitivity of detection [2]. Spray solutions IIa and IIb can also be used as dipping solutions.

A range of benzodiazepines which do not contain free aromatic amino groups are only able to react with BRATTON-MARSHALL reagent after in situ hydrolysis to the corresponding aminobenzophenones [4]. In such cases the chromatograms are sprayed with 25% hydrochloric acid after being freed from mobile phase and then heated to 110°C for 10 min before diazotization [4]. Some benzodiazepines form secondary amines on hydrolysis which can be photolytically dealkylated to the corresponding primary amines by irradiation with UV light [4].

Urea, carbamate and anilide herbicides, which are reported only to be detectable on layers containing fluorescence indicators [11], also have to be hydrolyzed, for example by spraying the dried chromatogram with a mixture of 60 ml conc. hydrochloric acid and 50 ml ethanol and then heating (10 min 180°C, covering the chromatogram with a glass plate) before reacting with BRATTON-MARSHALL reagent.

Folic acid is detected by irradiating the chromatogram with broad-spectrum UV light for 30 min before reaction with BRATTON-MARSHALL reagent (detection limit: 200 ng) [12].

The detection limits for benzodiazepines, aminobenzophenones and sulfonamides lie in the lower nanogram range.

The reagent can be employed on silica gel, cellulose and RP layers.

Procedure Tested

Hydrolysis Products of Benzodiazepines [15]

Method Ascending, one-dimensional development in a trough chamber without chamber saturation.

Layer HPTLC plates Silica gel 60 F_{254} (MERCK). Before sample application the plate was prewashed by developing once with chloroform – methanol (50 + 50) and then dried at 110°C for 30 min.

Mobile phase Benzene.

Migration distance 6 cm

Running time 15 min

Detection and result: The chromatogram was freed from mobile phase (10 min in a stream of warm air) and placed for 10 min in the empty half of a twin-trough chamber in whose second half nitrous fumes were being generated by the addition of 10 drops 37% hydrochloric acid to 5 ml 20% aqueous sodium nitrite solution. After the nitrous fumes had cleared (3–5 min in air, fume cupboard!) the chromatogram was immersed in solution IIa for 1 s and dried in a stream of cold air.

(2-Amino-5-bromophenyl(pyridin-2-yl)methanone ("ABP", deep violet, hR_f < 5), 2-amino-5-nitrobenzophenone ("ANB", pink, hR_f 15–20), 2-amino-5-chloro-

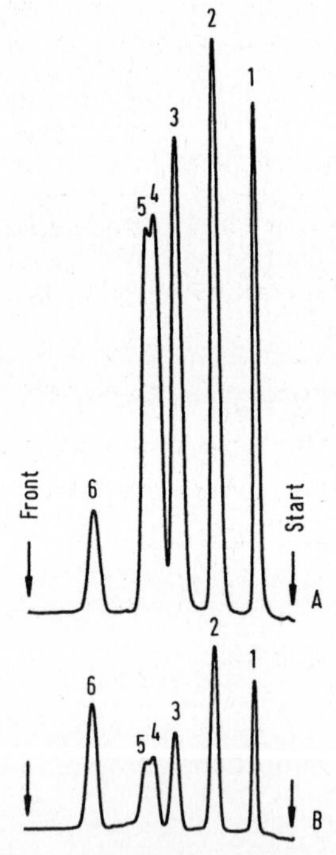

Fig. 1: Reflectance scan (A: scan at $\lambda = 560$ nm, B: scan at $\lambda = 470$ nm) of a chromatogram track of a mixture of 100 ng each of ABP, ANB, ACB, ACFB and ADB and 500 ng MACB per chromatogram zone: ABP (1), ANB (2), ACB (3), ACFB (4), ADB (5), MACB (6).

benzophenone ("ACB", violet, hR_f 30–35), 2-amino-5-chloro-2′-fluorobenzophenone ("ACFB", violet, hR_f 35–40), 2-amino-2′,5-dichlorobenzophenone ("ADB", violet, hR_f 40–50) and 5-chloro-2-(methylamino)benzophenone ("MACB", ochre, hR_f 60–65) appeared as colored zones on a white background. The detection limits per chromatogram zone were 5 ng for ABP, ANB, ACB, ACFB and ADB and ca. 50 ng for MACB.

In situ quantitation: The absorption photometric analysis in reflectance was performed at a mean wavelength of $\lambda = 560$ nm ($\lambda_{max\,(ABP)} = 580$ nm, $\lambda_{max\,(ANB)} = 550$ nm, $\lambda_{max\,(ACB,\,ACFB,\,ADB)} = 560$ nm, $\lambda_{max\,(MACB)} = 470$ nm, see Fig. 1).

References

[1] Bratton, A. C., Marshall, E. K., jr.: *J. Biol. Chem.* **1939,** *128,* 537-550.
[2] Narang, A. S., Choudhury, D. R., Richards, A.: *J. Chromatogr. Sci.* **1982,** *20,* 235-237.
[3] Ebel, S., Schütz, H.: *Dtsch. Apoth. Ztg.* **1977,** *117,* 1605-1609.
[4] Schütz, H.: *Dtsch. Apoth. Ztg.* **1981,** *121,* 1816-1823; *Fresenius Z. Anal. Chem.* **1979,** *294,* 135-139; **1985,** *321,* 359-362; Mitt. VI der Senatskommission für Klinisch-toxikologische Analytik, Verlag Chemie, Weinheim 1986.
[5] Chiarotti, M., De Giovanni, N., Fiori, A.: *J. Chromatogr.* **1986,** *358,* 169-178.
[6] Parks, O. W.: *J. Assoc. Off. Anal. Chem.* **1982,** *65,* 632-634; **1984,** *67,* 566-569; **1985,** *68,* 20-23; **1985,** *68,* 1232-1234.
[7] Heizmann, P., Haefelfinger, P.: *Fresenius Z. Anal. Chem.* **1980,** *302,* 410-412; *Experientia* **1981,** *37,* 806-807.
[8] Klein, S., Kho, B. T.: *J. Pharm. Sci.* **1962,** *51,* 966-970.
[9] Goodspeed, D. P., Simpson, R. M., Ashworth, R. B., Shafer, J. W., Cook, H. R.: *J. Assoc. Off. Anal. Chem.* **1978,** *61,* 1050-1053.
[10] Bican-Fister T., Kajganovic, V.: *J. Chromatogr.* **1963,** *11,* 492-495.
[11] Sherma, J., Boymel, J. L.: *J. Liq. Chromatogr.* **1983,** *6,* 1183-1192.
[12] Nuttall, R. T., Bush, J. E.: *Analyst* (London) **1971,** *96,* 875-878.
[13] Anonymous: MSZ (Hungarian Norm) 14474/1-81, S. 3.
[14] Jones, G. R. N.: *J. Chromatogr.* **1973,** *77,* 357-367.
[15] Netz, S., Funk, W.: *J. Planar Chromatogr.* **1989,** in press.

Bromocresol Green – Bromophenol Blue – Potassium Permanganate Reagent

Reagent for:

- Organic acids

$KMnO_4$	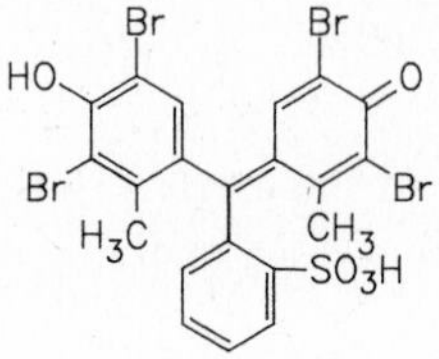 $C_{21}H_{14}Br_4O_5S$	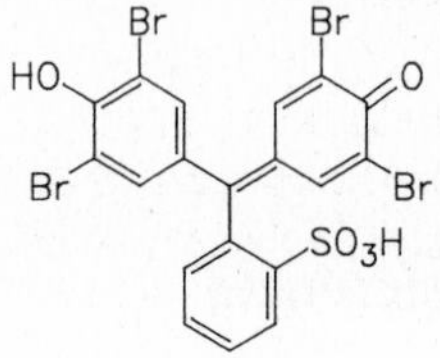 $C_{19}H_{10}Br_4O_5S$
$M_r = 158.04$	$M_r = 698.04$	$M_r = 669.96$
	Bromocresol green	Bromophenol blue

Preparation of the Reagent

Solution I Dissolve 40 mg bromocresol green and 15 mg bromophenol blue in 100 ml ethanol.

Solution II Dissolve 250 mg potassium permanganate and 500 mg sodium carbonate in 100 ml water.

Dipping solution Mix solutions I and II to a ratio of 9 + 1 immediately before use.

Storage Solutions I and II may be stored in the refrigerator for an extended period. The dipping solution must be employed within 5 – 10 min of preparation.

Substances Bromocresol green
Bromophenol blue
Potassium permanganate
Sodium carbonate decahydrate
Ethanol

Reaction

The detection of acids takes place on the basis of the pH-dependent color change of the two indicators bromocresol green (pH range: 3.8 – 5.4) and bromophenol blue (pH range: 3.0 – 4.6) from yellow to blue.

Method

The chromatogram is freed from mobile phase in a stream of warm air or in the drying cupboard (5 min 100 °C), immersed for 1 s in freshly prepared dipping solution or sprayed homogeneously with it and then heated to 100 °C for 5 – 10 min. This usually results in the formation of blue-green colored zones on a grey-blue background.

Note: The reagent can be employed on silica gel, kieselguhr, RP and, with lower sensitivity of detection, on cellulose layers. The color differentiation is probably greater on cellulose layers [1]. A dark blue background is produced on polyamide layers.

Procedure Tested

Organic Acids [2]

Method	Ascending, one-dimensional development with chamber saturation.
Layer	TLC plates Silica gel 60 F_{254} (MERCK).
Mobile phase	Diisopropyl ether – formic acid – water (90 + 7 + 3).
Migration distance	10 cm
Running time	30 min

Detection and result: The chromatogram was freed from mobile phase in the drying cupboard (5 min 100 °C), immersed in the freshly prepared dipping reagent for 1 s and then heated to 100 °C for 10 min.

Even before heating all the acids rapidly appeared as blue zones on a yellow-blue background. After heating tartaric acid (hR_f 2–5) and malic acid (hR_f 5–10) retained their color while lactic acid (hR_f 30–35), succinic acid (hR_f 35–40), pimelic acid (hR_f 50), maleic acid (hR_f 55), suberic acid (hR_f 55–60), benzoic acid (hR_f 80–85), stearic acid (hR_f 85–90) and arachidic acid (hR_f 85–90) appeared as pale yellow zones on a blue-yellow background (Fig. 1). The detection limits lay at 1 to 2 μg substance per chromatogram zone.

In situ quantitation: This reagent was not suitable for direct, precise photometric quantitation.

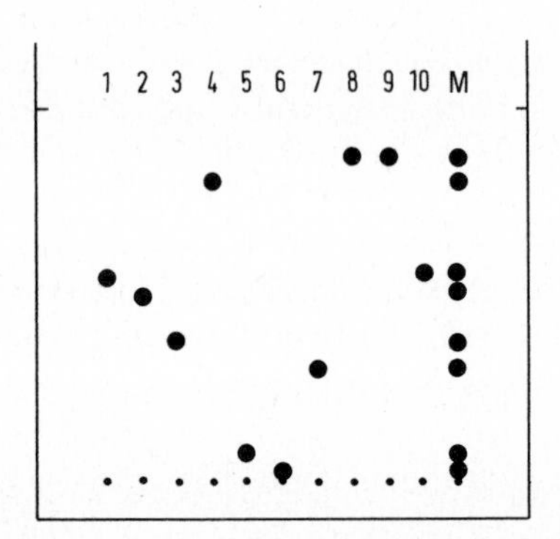

Fig. 1: Schematic representation of the chromatographic separation of carboxylic acids. Maleic acid (1), pimelic acid (2), succinic acid (3), benzoic acid (4), malic acid (5), tartaric acid (6), lactic acid (7), stearic acid (8), arachidic acid (9), suberic acid (10), mixture (M).

References

[1] Paskova, J., Munk, V.: *J. Chromatogr.* **1960,** *4,* 241–243.
[2] Jork, H., Klein, I.: Private communication, Universität des Saarlandes, Fachbereich 14, Saarbrücken, 1987.

Bromocresol Purple Reagent

Reagent for:

- Organic acids [1–4]
- Halogen anions [5]
- Phenols [6]
- 5-Aminodibenzo(a,d)cycloheptane derivatives [7]

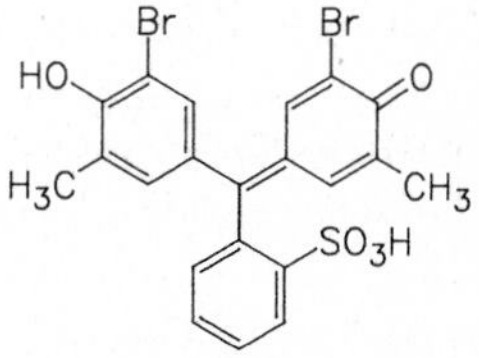

$C_{21}H_{16}Br_2O_5S$

$M_r = 540.24$

Preparation of the Reagent

Dipping solution I *Organic acids:* Dissolve 40 mg bromocresol purple in 100 ml 50% ethanol and adjust to pH = 10.0 (glass electrode) with caustic soda solution ($c = 0.1$ mol/l) [1].

Dipping solution II *Halogen anions:* Dissolve 100 mg bromocresol purple in 100 ml ethanol and add a few drops of 10% ammonia solution until the color changes.

Storage The bromocresol solution keeps well in the refrigerator before basification.

Substances Bromocresol purple
Sodium hydroxide solution (0.1 mol/l)
Ammonia solution (25%)
Ethanol

Reaction

The pH indicator bromocresol purple changes in color from yellow to purple in the pH range 5.2–6.8.

Method

The chromatogram is freed from mobile phase in a stream of warm air or in the drying cupboard (10 min 100 °C) and after cooling dipped for 1 s in dipping solution I or II or sprayed homogeneously with it. Then it is heated to 100 °C for 10 min.

Organic acids yield lemon-yellow zones on a blue background [1]. Halide ions migrate as ammonium salts in ammoniacal mobile phases and are also colored yellow. The colors fade rapidly in the air. This can be delayed for some days by covering the chromatogram with a glass plate.

Note: The background color depends on the pH of the layer, it is, therefore, affected by the efficiency of removal of acidic mobile phase components before staining.

Dicarboxylic acids react more sensitively than do monocarboxylic acids. Fatty acids and amino acids cannot be detected.

The reagent can be employed on silica gel, kieselguhr and Si 50 000 layers (also when they are impregnated with polyethylene glycol [1]) and on cellulose layers.

Procedure Tested

Multibasic Carboxylic Acids [8]

Method Ascending, one-dimensional development in a trough chamber with chamber saturation.

Layer 1. TLC plates Silica gel 60 (MERCK).
2. TLC plates Silica gel 60 (MERCK) impregnated by dipping once in a 10% solution of polyethylene glycol 1000 in methanol.

Mobile phase Diisopropyl ether – formic acid – water (90 + 7 + 3).

Migration distance 10 cm

Running time 30 min (silica gel layer)
50 min (impregnated layer)

Detection and result: The chromatogram was freed from mobile phase in a stream of warm air or in the drying cupboard (10 min 100 °C), immersed in dipping solution I for 1 s and then heated on a hotplate to 100 °C for 15 min. Citric acid (hR_f 0–5), lactic acid (hR_f 30), phthalic acid (hR_f 40–45), sebacinic acid (hR_f 60–65) and salicylic acid (hR_f 80–85) yielded lemon-yellow chromatogram zones on a yellow background on silica gel layers (Fig. 1). Figure 2 illustrates a similar separation of other dicarboxylic acids.

In situ quantitation: The reagent was not suitable for an exact quantitative determination of acids because the background structure did not allow sensitive reproducible scanning.

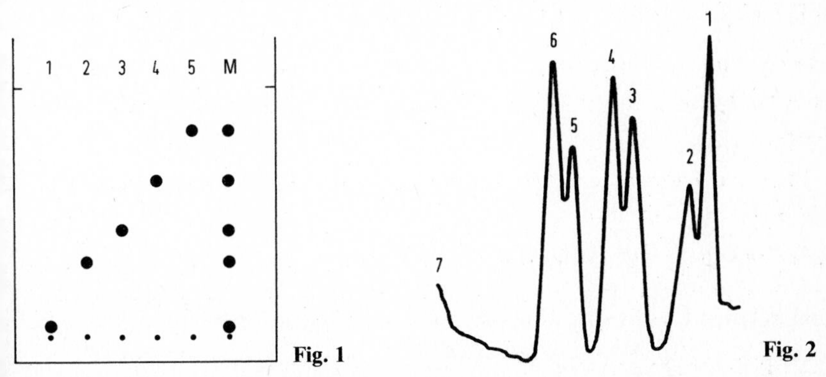

Fig. 1: Separation of carboxylic acids (schematic representation). Citric acid (1), lactic acid (2), phthalic acid (3), sebacinic acid (4), salicylic acid (5), mixture (M).

Fig. 2: Separation of dicarboxylic acids (reflectance scan). Start + tartaric acid (1), malic acid (2), phthalic acid (3), succinic acid (4), adipic acid (5), fumaric acid (6), front (7).

References

[1] Knappe, E., Peteri, D.: *Fresenius Z. Anal. Chem.* **1962,** *188,* 184–189; 352–355; **1962,** *190* 380–386.

[2] Fortnagel, P.: *Biochim. Biophys. Acta,* **1970,** *222,* 290–298.

[3] Tjan, G. H., Konter, T.: *J. Assoc. off. Anal. Chem.* **1973,** *55,* 1223–1225.

[4] Yang Zuying: *Chinese Brew (Zhongguo Niangzao)* **1983,** *2,* 32–34.

[5] Seiler, H., Kaffenberger, T.: *Helv. Chim. Acta,* **1961,** *44,* 1282–1283.

[6] Braun, D., Lee, D. W.: *Kunststoffe* **1972,** *62,* 517–574.

[7] Maulding, H. V., Brusco, D., Polesuk, J., Nazareno, J., Michaelis, A. F.: *J. Pharmac. Sci.* **1972,** *61,* 1197–1201

[8] Jork, H., Klein, I.: Private communication, Universität des Saarlandes, Fachbereich 14, Saarbrücken 1987.

tert-Butyl Hypochlorite Reagent

Reagent for:

- Fatty acids, triglycerides, amino acids, sugars, steroids [1]
- Peptides, nucleotides [2]
- Vitamin B_1 [3]
- Alkaloids [5]
- Carbamazepine [4]

$H_3C-C(CH_3)_2-OCl$

C_4H_9ClO

$M_r = 108.57$

Preparation of the Reagent

Dipping solution I Dissolve 1 ml *tert*-butyl hypochlorite in 100 ml carbon tetrachloride or cyclohexane.

Dipping solution II Mix chloroform, paraffin oil and triethanolamine in the ratio of 6 + 1 + 1.

Storage Dipping solution I may be stored in the refrigerator for a few days. *tert*-Butyl hypochlorite should be stored cool in the dark under an atmosphere of nitrogen.
Caution: *tert*-Butyl hypochlorite is very reactive, e.g. it reacts violently with rubber seals particularly under the influence of light.

Substances *tert*-Butyl hypochlorite
Carbon tetrachloride
Paraffin viscous
Chloroform
Triethanolamine

Reaction

The reaction mechanism has not been elucidated.

Method

The developed chromatogram is freed from mobile phase in a stream of warm air and then either immersed in dipping solution I for 1 s or sprayed homogeneously with it or exposed to its vapors for 15 min in a twin-trough chamber in whose second chamber 5 ml of dipping solution I has been placed ca. 10 min previously. The chromatograms are then immersed in dipping solution II for 1 s and dried in a stream of hot air. If derivatization is carried out by dipping, the chromatograms should be dried for ca. 1 min in a stream of hot air and allowed to cool to room temperature before the second treatment.

There then appear, in long-wavelength UV light ($\lambda = 365$ nm), yellow to violet fluorescent zones on a dark background; these can be quantitatively analyzed.

Note: The reagent can be employed on silica gel and cellulose layers. When derivatization is carried out from the vapor phase the detection limit for morphine is 10 ng and that for papaverine 1 ng per chromatogram zone [5]. In some cases it has been recommended that ammonium sulfate be added to the layer with subsequent heating to 150–180 °C [1] after derivatization. It is also possible to spray afterwards with an aqueous solution of potassium iodide (1%) and starch (1%) [2].

Procedure Tested

Vitamin B_1 [3]

Method Ascending, one-dimensional development in a twin-trough chamber with chamber saturation with the layer being preconditioned in the solvent-free half of the trough for 15 min after application of the sample.

Layer HPTLC plates Silica gel 60 (Merck), which had been prewashed by single development of the plate with chloroform – methanol (50 + 50) and then dried at 110 °C for 30 min.

Mobile phase Methanol – ammonia solution (25%) – glacial acetic acid (8 + 1 + 1).

Migration distance 6 cm

Running time ca. 20 min

Detection and result: The chromatogram was dried in a current of warm air and either immersed in reagent solution I for 1 s or placed for 15 min in a twin-trough chamber in whose second trough 5 ml of dipping solution I had been placed ca. 10 min previously. If the chromatogram was derivatized by dipping it had to be dried for ca. 1 min in a stream of hot air and allowed to cool to room temperature.

The chromatogram was then immersed for 1 s in reagent solution II to increase sensitivity and stabilize the fluorescence and then dried in a stream of hot air.

Thiamine (hR_f 40–45) yielded a yellow fluorescent zone on a dark background under long-wavelength UV light ($\lambda = 365$ nm).

Note: Vapor-phase derivatization was appreciably more homogeneous than that produced by dipping and simultaneously increased both the precision and the sensitivity of the method.

In situ quantitation: The fluorimetric analysis was performed in long-wavelength UV light ($\lambda_{exc} = 365$ nm, $\lambda_{fl} > 430$ nm). The detection limit for thiamine was less than 3 ng per chromatogram zone.

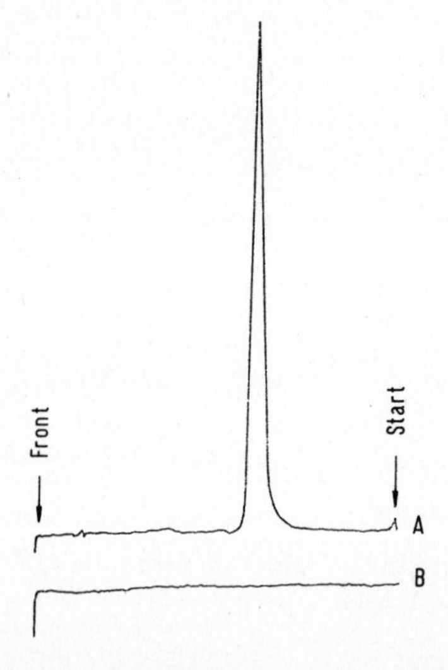

Fig. 1: Fluorescence scan of 30 ng thiamine after vapor-phase derivatization (A) and an accompanying blank track (B).

References

[1] Smith, B. G.: *J. Chromatogr.* **1973,** *82,* 95-100
[2] Mazur, R. H., Ellis, B. W., Cammarata, P. S.: *J. Biol. Chem.* **1962,** *237,* 1619-1621
[3] Derr, P.: Thesis, Fachhochschule Gießen, Fachbereich Technisches Gesundheitswesen, 1985
[4] Canstein, M. von: Thesis, Fachhochschule Gießen, Fachbereich Technisches Gesundheitswesen, 1984
[5] Jork, H.: Private communication, Universität des Saarlandes, Fachbereich 14, Saarbrücken 1986.

7-Chloro-4-nitrobenzo-2-oxa-1,3-diazole Reagent (NBD-Chloride Reagent)

Reagent for:

- Prim. and sec. aliphatic amines [1 – 3]
- Amino acids and peptides [1, 4, 5]
- Sulfonamides [6]
- Alkaloids [7]
- Phenols [7]

Cl

N O N

NO_2

$C_6H_2ClN_3O_3$

$M_r = 199.55$

Preparation of Reagent

Dipping solution I Dissolve 10 g sodium acetate in 20 ml water and dilute with 40 ml methanol.

Dipping solution II Dissolve 0.1 g NBD-chloride (7-chloro-4-nitrobenzofurazan, 7-chloro-4-nitro-2,1,3-benzoxadiazole) in 50 ml ethanol.

Spray solution I Dissolve 2 g sodium carbonate or 5 g sodium hydroxide in 100 ml water [6, 7].

Spray solution II Dissolve 0.1 – 0.5 g NBD-chloride in 100 ml ethanol [5], methanol [6] or acetonitrile [7].

Storage All solutions may be stored for an extended period of time.

Substances 7-Chloro-4-nitrobenzofurazan
Sodium acetate trihydrate
Sodium carbonate

Sodium hydroxide pellets
Methanol
Ethanol

Reaction

NBD-chloride reacts with nucleophilic compounds (amines, mercaptans etc.) to yield the corresponding 7-substituted 4-nitrobenzofurazan derivatives.

NBD-chloride + Prim. amine ($R{-}NH_2$) $\xrightarrow{-\,HCl}$ Reaction product

Method

The dried chromatograms are immersed in dipping solution I for 1 s, dried in a stream of warm air, dipped immediately after cooling in dipping solution II (1 s) and then heated to 100 °C for 2–3 min. Alternatively the chromatogram can be sprayed with the appropriate spray solutions.

Under long-wavelength UV light ($\lambda = 365$ nm) the chromatogram zones fluoresce yellow to yellow-green, they are sometimes detectable in daylight as colored zones, too.

Note: The reagent can be employed for qualitative and quantitative analysis on silica gel and RP layers. Ammonia, amine and acid-containing mobile phases should be completely removed beforehand. Amino phases cannot be employed. The NBD-chloride reagent is not as sensitive as the DOOB reagent (qv.) on RP phases.

Primary and secondary amines and phenols generally produce yellow to reddish-orange zones (serotonin violet) on a pale yellow background, under long-wavelength UV light ($\lambda = 365$ nm) the zones fluoresce yellowish green (serotonin appears as a dark zone). The plate background also fluoresces but appreciably less. Noradrenaline, 5-hydroxytryptophan and dopamine yield red to blue-violet zones on a yellow background, they do not fluoresce*. Sulfonamides and alkaloids produce greenish-yellow, olive brown or violet colors. Peptides and amino acids (except tryptophan) yield pale yellow fluorescent zones; tryptophan and proline produce orange-red zones. The detection limits are 100 – 800 ng substance per chromatogram zone, sometimes appreciably less (cf. "Procedure Tested" [8]).

Although the reagent itself is not fluorescent an excess of NBD-chloride can interfere in quantitative analysis. In such cases it should be checked whether prechromatographic derivatization produces better results [3, 4]. The reaction products can then be separated on polyamide layers.

Procedure Tested

Proline, Hydroxyproline [8]

Method Ascending, one-dimensional development in a trough chamber without chamber saturation.

Layer HPTLC plates Silica gel 60 (MERCK), which had been developed once with chloroform – methanol (50 + 50) and dried at 110 °C for 30 min before applying the samples.

Mobile phase Chloroform – methanol – ammonia solution (25%) (100 + 90 + 20).

Migration distance 7 cm

Running time 40 min

Detection and result: The chromatogram was freed from mobile phase in a stream of warm air, then immersed in dipping solution I for 1 s, dried for 90 s in a stream of warm air and after cooling immersed in dipping solution II for 1 s. After drying briefly in a stream of warm air it was heated to 110 °C for 2 min and treated with hydrochloric acid vapor for 15 min in a twin-trough chamber (37% hydrochloric acid in the vacant trough).

---

* Jork, H.: private communication, Universität des Saarlandes, Fachbereich 14 "Pharmazie und Biologische Chemie", D-6600 Saarbrücken 1988.

Hydroxyproline (hR_f 20–25) and proline (hR_f 25–30) yielded orange-colored chromatogram zones on a pale yellow background, under long-wavelength UV light ($\lambda = 365$ nm) they fluoresced yellow on a blue background. (Detection limits: ca. 5 ng substance per chromatogram zone).

The chromatogram was then immersed for 1 s in a mixture of liquid paraffin and *n*-hexane (1 + 2) to stabilize and enhance the fluorescence (by a factor of ca. 1.5). The detection limits were then ca. 1 ng substance per chromatogram zone.

In situ quantitation: The absorption photometric measurement in reflectance was performed at $\lambda = 490$ nm, the fluorimetric analysis at $\lambda_{exc} = 436$ nm and $\lambda_{fl} > 560$ nm (Fig. 1).

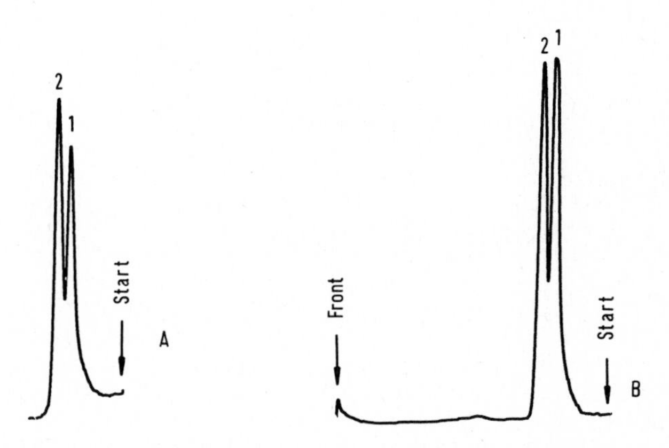

Fig. 1: Reflectance scan (A) and fluorescence scan (B) of a chromatogram track with 50 ng of each amino acid per chromatogram zone: hydroxyproline (1), proline (2).

References

[1] Ghosh, P. B., Whitehouse, M. W.: *Biochem. J.* **1968,** *108,* 155–156.
[2] Benjamin, D. M., McCormack, J. J., Gump, D. W.: *Anal. Chem.* **1973,** *45*, 1531–1534.
[3] Klimisch, H.-J., Stadler, L.: *J. Chromatogr.* **1974,** *90*, 141–148.
[4] Fager, R. S., Kutina, C. B., Abrahamson, E. W.: *Anal. Biochem.* **1973,** *53*, 290–294.
[5] Distler, W.: *Fresenius Z. Anal. Chem.* **1981,** *309*, 127–128.
[6] Reisch, J., Alfes, H., Kommert, H.-J.: *Fresenius Z. Anal. Chem.* **1969,** *245*, 390.
[7] Reisch, J., Kommert, H.-J., Clasing, D.: *Pharm. Ztg.* **1970,** *115*, 752–753.
[8] Netz, S., Funk, W.: Private communication, Fachhochschule Gießen, Fachbereich Technisches Gesundheitswesen, 1988.

Copper(II) Acetate – Phosphoric Acid Reagent

Reagent for:

- Prostaglandins [1–3]
- β-Sitosterol [4]
- Polar lipids [5]
- Phospholipids [5–17]
- Glycolipids [9]
- Esters of fatty acids [18, 19]

$(CH_3COO)_2Cu \cdot H_2O$	H_3PO_4
$M_r = 199.65$	$M_r = 98.00$
Copper(II) acetate monohydrate	*ortho*-Phosphoric acid

Preparation of the Reagent

Dipping solution Dissolve 3 g copper(II) acetate in 100 ml 8–15% aqueous phosphoric acid.

Storage The dipping solution can be stored for long periods of time.

Substances *ortho*-Phosphoric acid (85%)
Copper(II) acetate monohydrate

Reaction

The reaction mechanism has not yet been elucidated.

Method

The chromatograms are dried in a stream of warm air, immersed in the reagent solution for 2 s and then heated to 120–125 °C for 10–15 min. β-Sitosterol appears as a grey-blue zone on a colorless background [4] and prostaglandins yield variously colored zones [2], which absorb long-wavelength UV light ($\lambda = 365$ nm). The dipping solution can also be employed as a spray solution [11].

Note: Phospholipids and glycolipids appear after 6–25 min [5–13] and esters of fatty acids after heating for 2 h [18] at 180 °C (charring!) as grey-black zones on an almost colorless background. According to TOUCHSTONE [7] in the case of phospholipids it is only the unsaturated ones that react, while the saturated ones only react with the copper(II) sulfate – phosphoric acid reagent. The reagent may be applied to silica gel layers.

Procedure Tested

Prostaglandin E_1 [1]

Method Horizontal, one-dimensional development in a linear chamber (CAMAG).

Layer HPTLC plates Silica gel 60 (MERCK).
The layer was prewashed twice before application of the samples by developing first with cyclohexane and then, after drying for 30 min at 120 °C, with methanol; it was finally activated for 30 min at 120 °C.

Mobile phase Ethyl acetate – formic acid (80 + 1).

Migration distance ca. 5 cm

Running time 10–15 min

Detection and result: The chromatogram was dried in a stream of warm air, immersed in the reagent solution for 2 s and then heated to 120 °C for 15 min. Prostaglandin E_1 appeared as a yellow-brown zone (hR_f 25–30) on a colorless background.

In situ quantitation: Quantitation by reflectance had to be performed as soon as possible, since the color intensity of a zone decreased ca. 40% within a day. Detection wavelength: $\lambda = 345$ nm; detection limit: 2.5 ng per chromatogram zone (Fig. 1).

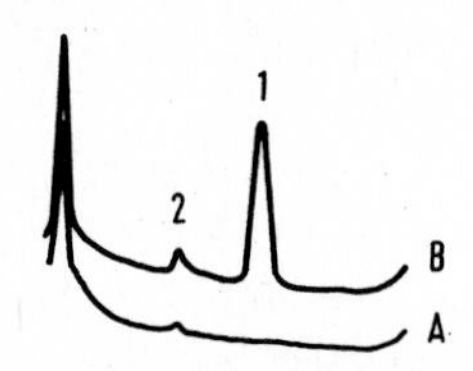

Fig. 1: Reflectance curve of a blank track (A) and a chromatogram (B) with 80 ng prostaglandin E_1 (1), β-front (2).

References

[1] Luitjens, K.-D., Funk, W., Rawer, P.: *J. High Resolut. Chromatogr. Chromatogr. Commun.* **1981,** *4,* 136–137.

[2] Ubatuba, F. B.: *J. Chromatogr.* **1978,** *161,* 165–177.

[3] Bruno, P., Caselli, M., Garappa, C., Traini, A.: *J. High Resolut. Chromatogr. Chromatogr. Commun.* **1984,** *7,* 593–595.

[4] Ronen, Z.: *J. Chromatogr.* **1982,** *236,* 249–253.

[5] Kolarovic, L., Traitler, H.: *J. High Resolut. Chromatogr. Chromatogr. Commun.* **1985,** *8,* 341–346.

[6] Touchstone, J. C., Chen, J. C., Beaver, K.: *Lipids* **1980,** , *15,* 61–62.

[7] Touchstone, J. C., Levin, S. S., Dobbins, M. F., Carter, P. J.: *J. Liq. Chromatogr.* **1983,** *6,* 179–192: *J. High Resolut. Chromatogr. Chromatogr. Commun.* **1981,** *4,* 423–424.

[8] Sherma, J., Bennett, S.: *J. Liq. Chromatogr.* **1983,** *6,* 1193–1211.

[9] Fewster, M. E., Burns, B. J., Mead, J. F.: *J. Chromatogr.* **1969,** *43,* 120–126.

[10] Kohl, H. H., Telander, D. H. jr., Roberts, E. C., Elliott, H. C. jr.: *Clin. Chem.* **1978,** *24,* 174–176.

[11] Mallikarjuneswara, V. R.: *Clin. Chem.* **1975,** *21,* 260–263.

[12] Pappas, A. A., Mullins, R. E., Gadsden, R. G.: *Clin. Chem.* **1982,** *28,* 209–211.

[13] Selvam, R., Radin, N. S.: *Anal. Biochem.* **1981,** *112,* 338–345.

[14] Sherma, J., Touchstone, J. C.: *J. High Resolut. Chromatogr. Chromatogr. Commun.* **1979,** *2,* 199–200.

[15] Watkins, T. R.: *J. High Resolut. Chromatogr. Chromatogr. Commun.* **1982,** *5,* 104–105.

[16] Spillman, T., Cotton, D., Lynn, S., Bretaudiere J.: *Clin. Chem.* **1983,** *29,*, 250–255.

[17] Seher, A., Spiegel, H., Könker, S., Oslage, H.: *Fette-Seifen-Anstrichm.* **1983,** *85,* 295–304

[18] Gosselin, L., Graeve, J. de: *J. Chromatogr.* **1975,** *110,* 117–124.

[19] Zeringue, H., J., Feuge, R. O.: *J. Assoc. Off. Anal. Chem.* **1976,** *53,* 567–571.

Copper(II) Nitrate Reagent

Reagent for:

- Stabilization of "ninhydrin" spots [1 – 3]

$Cu(NO_3)_2$
$M_r = 187.56$

Preparation of the Reagent

Dipping solution Mix 1 ml saturated aqueous copper(II) nitrate solution with 0.2 ml 10% nitric acid and make up to 100 ml with 95% ethanol [1].

Storage The reagent can be kept for several days.

Substances Copper(II) nitrate trihydrate
Nitric acid (65%)
Ethanol
Ammonia solution (25%)

Reaction

The blue-violet stain which forms on thin-layer chromatograms when amino acids are stained with ninhydrin is only stable for a short time. It rapidly begins to fade even on cellulose layers. The stability can be appreciably enhanced by complex formation with metal ions [3].

Method

The chromatograms stained with ninhydrin are immersed in the reagent solution for 1 s or sprayed evenly with it and then placed in the free half of a twin trough chamber containing 25% ammonia solution. Apart from proline and hydroxyproline, which yield yellow copper complexes, all the amino acids yield reddish-colored chromatogram zones [3].

Note: The copper in the reagent may be replaced by other metal ions. Table 1 [3] lists the colors obtainable with some cations.

Table 1: Colors of the metal cation complexes

Amino acid	Zn^{2+}	Cd^{2+}	Co^{2+}	Cu^{2+}
Proline	yellow	yellow	brown	yellow ochre
Hydroxyproline	grey	yellow	light brown	
Alanine	brick red	wine red	brown	salmon
Glycine	brick red	wine red	brown	salmon
Serine	brick red	wine red	brown	salmon
Threonine	brick red	wine red	brown	salmon
Phenylalanine	red	red ochre	brown	red ochre
Tryptophan	red	orange	brown	red ochre
Other amino acids	red	wine red	brown	salmon

The copper complex is very stable at neutral pH, but it fades very rapidly in the presence of hydrogen ions. Other complex formers such as tartaric acid or citric acid and thiourea interfere with the reaction and, therefore, should not be included in mobile phases used for the separation of amino acids [3].

When staining with ninhydrin the appearance of colors of various hues on TLC layers with and without fluorescence indicators is probably a result of complex formation between the "ninhydrin zones" and the cations of the inorganic fluorescence indicators.

Procedure Tested

Valine, Phenylalanine, Leucine, Isoleucine [1]

Method Ascending, one-dimensional development in a trough chamber without chamber saturation

Layer HPTLC plates Cellulose (MERCK) Before sample application the layers were prewashed by developing with chloroform – methanol (50 + 50) and then dried at 110 °C for 30 min.

Mobile phase 2-Propanol – 1-butanol – trisodium citrate solution ($c = 0.2$ mol/l in water) (50 + 30 + 20).

Migration distance 9 cm

Running time 120 min

Detection and result: After staining with ninhydrin the chromatogram was cooled to room temperature, immersed in the reagent solution for 1 s and after drying in the air it was then placed for 15 min in the empty half of a twin-trough chamber containing 25% ammonia solution.

Before treatment with ammonia valine (hR_f 45–50), isoleucine (hR_f 65–70) and leucine (hR_f 70–75) produced red chromatogram zones and phenylalanine (hR_f 55–60) violet chromatogram zones on a pale pink background. After treatment with ammonia all four amino acids appeared as violet chromatogram zones on a flesh-colored background; these zones were stable over an extended period.

In situ quantitation: The absorption-photometric quantitation was carried out in reflectance at $\lambda = 540$ nm. The detection limit was 5 ng substance per chromatogram zone.

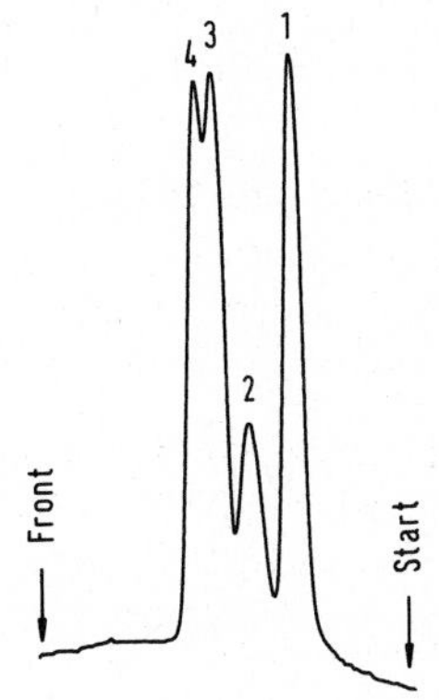

Fig. 1: Reflectance scan of a chromatogram track with ca. 100 ng amino acid per chromatogram zone. Valine (1), phenylalanine (2), isoleucine (3), leucine (4).

References

[1] Netz, S., Funk, W.: Private communication, Fachhochschule Gießen, Fachbereich Technisches Gesundheitswesen, Gießen 1987.

[2] Andermann, G., Andermann, C.: *J. High Resolut. Chromatogr. Chromatogr. Commun.* **1980,** *3,* 36–37.

[3] Kawerau, E., Wieland, T.: *Nature* (London) **1951,** *168,* 77–78.

Copper(II) Sulfate Reagent

Reagent for:

- Di-, tri- and polycarboxylic acids [1]
- Thioglycolic and dithioglycolic acids [2]
- Diuretics [3]
- Antithyroid pharmaceuticals [4]
  e.g. N-aryl-N′-benzenesulfonylthiocarbamides
- N-Arylthiosemicarbazides [5]
- Sterols and their esters [6]
  e.g. cholesterol and cholesteryl esters
- Metal chelates [7 – 9]

$Cu_2SO_4 \cdot 5H_2O$

$M_r = 249.68$

Preparation of Reagent

Dipping solution Dissolve 1.5 g copper(II) sulfate pentahydrate in a few milliliters of water and make up to 100 ml with methanol.

Storage The solution should always be freshly made up.

Substances Copper(II) sulfate pentahydrate
Methanol

Reaction

The substances listed above form colored copper complexes.

Method

The chromatograms are freed from mobile phase, immersed in the reagent solution for 10 s or homogeneously sprayed with it and then dried in a stream of warm air. Di- and tricarboxylic acids are reported to appear rapidly as blue zones on an evenly pale blue background [1]. On layers containing fluorescence indicators they can usually also be recognized under short-wavelength UV light ($\lambda = 254$ nm) as dark spots on a green fluorescent background [1]; thioglycolic and dithioglycolic acids yield green-colored zones [2] and N-aryl-N′-benzenesulfonylthiocarbamides yield yellow to violet-colored zones [4].

For the determination of sterols and their esters the chromatogram is immersed in a 10% aqueous copper(II) sulfate solution and then heated to 105 °C for 30 min. In this case green-yellow fluorescent chromatogram zones are visible in long-wavelength UV light ($\lambda = 365$ nm) [6].

Note: Silica gel, kieselguhr and polyamide layers can be used as stationary phases. Not all acids are stained on RP layers. Amino layers yield a pale blue background. The detection limits are in the µg range for carboxylic acids [1], thioglycolic and dithioglycolic acids [2] and for antithyroid pharmaceuticals [4]; they are about 5 ng per chromatogram zone for sterols and steryl esters [6].

Table 1: List of carboxylic acids that can be detected with copper(II) sulfate reagent [1]

Acid	Color	Fluorescence quenching ($\lambda = 254$ nm)	Sensitivity (µg/zone)	
			TLC	HPTLC
Malonic acid	pale blue	+ (weak)	20	
Succinic acid	blue	–	5	
Methyl-succinic acid	blue	–	10	
Adipic acid	blue	–	15	5
Pimelic acid	blue	–	5	
Suberic acid	blue	–	5	
Sebacic acid	blue	–	5	
Fumaric acid	pale blue	+	10	2.5
Maleic acid	blue	+	10	
Aconitic acid	blue	+	10	
Phthalic acid	pale blue	+	30	20
Terephthalic acid	blue	+	15	
α-Ketoglutaric acid	pale blue	+	25	

A mixture of 10% aqueous copper(II) sulfate solution and 25% ammonia solution (100 + 15) is recommended as spray solution for the detection of diuretics.

Procedure Tested

Organic Acids [1]

Method Ascending, one-dimensional development in a trough chamber with chamber saturation.

Layer TLC aluminium sheets, TLC plates. Silica gel 60 F_{254} (MERCK).

Mobile phase Diisopropyl ether – formic acid – water (90 + 7 + 3).

Migration distance 10 cm

Running time 40 min

Detection and result: The TLC plate was dried in the air for 30 min and heated to 110 °C for 10 min in order to remove the formic acid from the mobile phase, before immersing the chromatogram in the reagent solution for 10 s.

Tartaric acid (hR_f 5), malic acid (hR_f 15), maleic acid (hR_f 35), lactic acid (hR_f 40), adipic acid (hR_f 45) and fumaric acid (hR_f 55) yielded blue zones on a weakly bluish background; these were still to be recognized unchanged after several weeks.

In situ quantitation: The quantitation (Fig. 1, succinic acid) was performed in reflectance with visible light ($\lambda = 690$ nm).

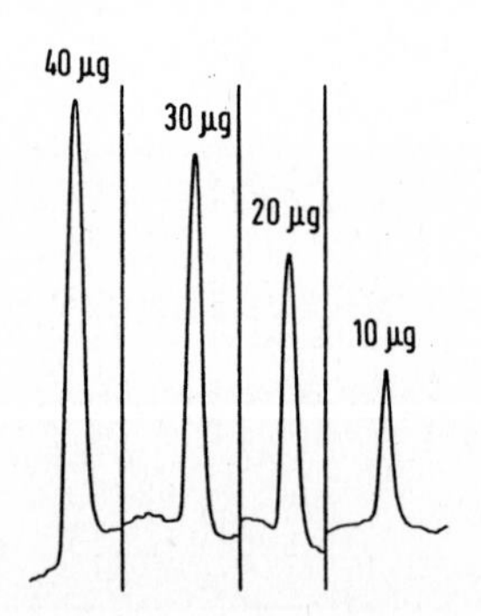

Fig. 1: Absorption scans of a range of dilutions of succinic acid [1].

References

[1] Gaberc-Porekar, V., Socic, H.: *J. Chromatogr.* **1979,** *178,* 307–310.
[2] Klaus, R.: *Chromatographia* **1985,** *20,* 235–238.
[3] Guven, K. C., Cobanlar, S.: *Eczacilik. Bul.* **1967,** *9,* 98–103.
[4] Upadhyaya, J. S., Upadhyaya, S. K.: *Fresenius Z. Anal. Chem.* **1979,** *294,* 407.
[5] Srivastava, S. P., Dua, V. K., Upadhyaya, J. S.: *Fresenius Z. Anal. Chem.* **1977,** *286,* 249.
[6] Tiefenbacher, K., Woidich, H. in: Proceedings of the International Symposium on Instrumental HPTLC, Würzburg. Bad-Dürkheim: IfC-Verlag, 1985.
[7] König, K., Becker, J., Henke, W., Stenshorn, J., Werner, H., Ballschmiter, K.: *Z. Anal. Chem.* **1972,** *259,* 11–16.
[8] Malissa, H., Kellner, R., Prokopowski, P.: *Analyt. chim. Acta* **1973,** *63,* 225–229.
[9] Ballschmiter, K.: *Z. Anal. Chem.* **1971,** *254,* 348–353.

2,6-Dibromoquinone-4-chloroimide Reagent
(Gibbs' Reagent)

Reagent for:

- Phenols [1 – 3, 5]
  e.g. vitamin B_6 [1, 3]
- Cumarins [15]
- Thiol and thione compounds [4, 5]
- Antioxidants [6]
- Prim. and sec. aliphatic amines [7]
- Prim., sec. and tert. aromatic amines [7]
- Carbazoles and aromatic hydrocarbons [7]
- Indoles and other N-containing heterocyclics [7]
- 2,4-Pentanedione [7]
- Pesticides [8, 9, 13]
- Antiepileptics [10, 11]
- Barbiturates [10]

Br Br O NCl

$C_6H_2Br_2ClNO$

$M_r = 299.36$

Preparation of the Reagent

Dipping solution Dissolve 0.1 g 2,6-dibromoquinone-4-chloroimide in 10 ml dimethyl sulfoxide saturated with sodium hydrogen carbonate. Then make up to 100 ml with chloroform.

Spray solution Dissolve 0.4 to 1 g 2,6-dibromoquinone-4-chloroimide in 100 ml methanol or ethanol and filter if necessary.

Storage The solutions may be stored in the refrigerator for ca. 2 weeks [3].

Care: 2,6-dibromoquinone-4-chloroimide can decompose explosively [9]; so only small amounts should be stored in the refrigerator!

Substances 2,6-Dibromoquinone-4-chloroimide
Sodium hydrogen carbonate
Dimethyl sulfoxide (DMSO)
Chloroform
Methanol
Ethanol

Reaction

2,6-Dibromoquinone-4-chloroimide, in whose stead 2,6-dichloroquinone-4-chloroimide (q.v.) may also be employed, reacts primarily with phenols and anilines which are not substituted in the *p*-position.

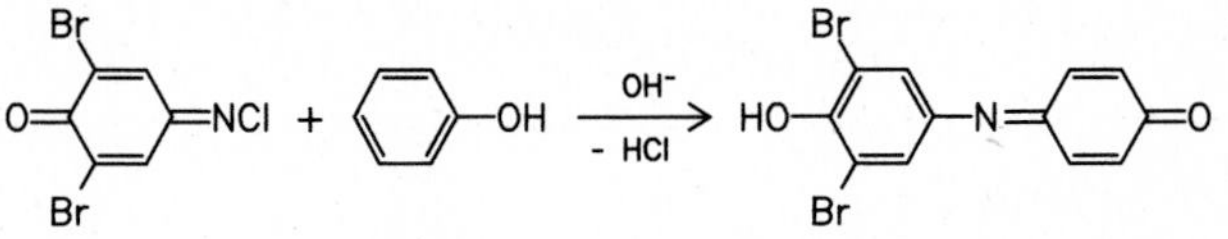

Method

The chromatogram is dried in a stream of cold air, heated to 110 °C for 10 min, cooled and immersed in the dipping solution for 5 s or sprayed with spray solution and finally heated to 110 °C for 2 min.

Note: If the spray solution or a nonbasic dipping solution [2] is employed for detection then it is advisable to spray again with a 10% aqueous sodium carbonate solution. The necessary basicity can also often be achieved by placing the treated chromatogram in a twin-trough chamber whose second trough contains 5 ml ammonia solution (25%).

In general no warming is necessary to produce the colored zones: pyridoxine is colored intense blue, pyridoxamine violet and pyridoxal dark blue [1, 3]. Phenols that are substituted in the *p*-position exhibit characteristic color differences. Thiol compounds and substances containing sulfhydryl groups yield yellow zones, while thiones are colored red. Thiourea appears as a brown zone on a light background [4].

In some cases, e.g. the detection of antioxidants [6], the plate is heated to 105 °C for 5 min after being sprayed and the still hot plate placed immediately in an ammonia-vapor chamber. The blue color of the tryptamine derivatives is also stabilized by spraying afterwards with a 5% methanolic ammonia solution [12].

The sensitivity of detection is usually between 0.2 and 0.5 µg per chromatogram zone. This is also true for pesticides based on organophosphorus acids [9].

The reagent can be employed on silica gel, cellulose and polyamide layers.

Procedure Tested

Antiepileptics, Barbiturates [10]

Method Ascending, one-dimensional development in a trough chamber. After sample application the HPTLC plates were equilibrated in a conditioning chamber at 42% relative humidity for 30 min and then developed immediately.

Layer HPTLC plates Silica gel 60 (MERCK) which were prewashed by developing three times in chloroform – methanol (50 + 50) and drying at 110 °C for 30 min.

Mobile phase Chloroform – acetone (80 + 15).

Migration distance 7 cm

Running time 20 min

Detection and result: The chromatogram was freed from mobile phase (first dried for 5 min in a stream of cold air, then heated for 10 min at 110 °C and allowed to cool), immersed for 5 s in the reagent solution and finally heated for 2 min at 110 °C. Grey to grey-blue zones were formed on a light background. The following hR_f values were obtained: primidone (hR_f 10 – 15); carbamazepine (hR_f 40 – 45); phenytoin (hR_f 50 – 55); phenobarbital (hR_f 60); ethosuximide (hR_f 75); hexobarbital (hR_f 90 – 95).

In situ quantitation: Reflectance measurements were carried out at a wavelength of $\lambda = 429$ nm. The detection limit lay at 50 to 200 ng per chromatogram zone (Fig. 1).

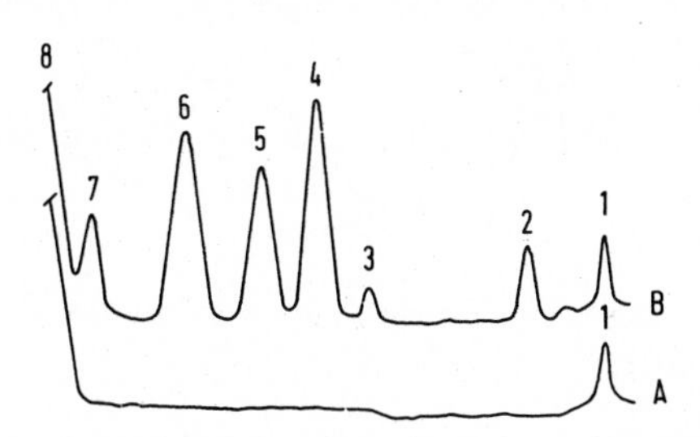

Fig. 1: Reflectance scan of a blank (A) and of a mixture of antiepileptics with 500 ng substance per chromatogram zone (B). Start (1), primidone (2), carbamazepine (3), phenytoin (4), phenobarbital (5), ethosuximide (6), hexobarbital (7) and solvent front (8).

References

[1] Nürnberg, E.: *Dtsch. Apoth. Ztg.* **1961,** *101,* 268–269.
[2] Linnenbrink, N., Kraus, L.: *GIT Fachz. Lab.* **1979,** *23,* 666–667.
[3] Reio, L.: *J. Chromatogr.* **1958,** *1,* 338–373: **1960,** *4,* 458–476.
[4] Stenerson, J.: *J. Chromatogr.* **1971,** *54,* 77–81.
[5] Laub, E., Geisen, M.: *Lebensmittelchem. gerichtl. Chemie* **1976,** *30,* 129–132.
[6] Peteghem, C. H. van, Dekeyser, D. A.: *J. Assoc. Off. Anal. Chem.* **1981,** *64,* 1331–1335.
[7] Ross, J. H.: *Anal. Chem.* **1968,** *40,* 2138–2143.
[8] Chmil, V. D., Grahl, K., Stottmeister, E.: *J. Anal. Chem.* (USSR) **1978,** *33,* 1862–1866
[9] Kömives, T., Katona, A., Marton, A. F.: *Elelmiszervizsgalati Közl.* **1978,** *23,* 244–248.
[10] Canstein, M. von: Thesis, Fachhochschule Gießen, Fachbereich Technisches Gesundheitswesen, 1984.
[11] Funk, W., Canstein, M. von, Couturier, Th. Heiligenthal, M., Kiefer, U., Schlierbach, S. Sommer, D. in: Proceedings of the 3rd International Symposium on Instrumental HPTLC, Würzburg; Bad Dürkheim: IfC-Verlag 1985, p. 281–311.
[12] Studer, A., Traitler, H.: *J. High Resolut. Chromatogr. Chromatogr. Commun.* **1982,** *5,* 581–582.
[13] Stenerson, J., Gilman, A., Vardanis, A: *J. agric. Food Chem.* **1973,** *21,* 166–171.
[14] Kadoum, A. M.: *J. agric. Food Chem.* **1970,** *18,* 542–543.
[15] Daenens, P., Boven, M. van: *J. Chromatogr.* **1971,** *57,* 319–321.

2,6-Dichlorophenolindophenol Reagent
(Tillmans' Reagent)

Reagent for:

- Organic acids [1, 2]
- Reductones
  e.g. ascorbic acid (vitamin C) [2, 3]

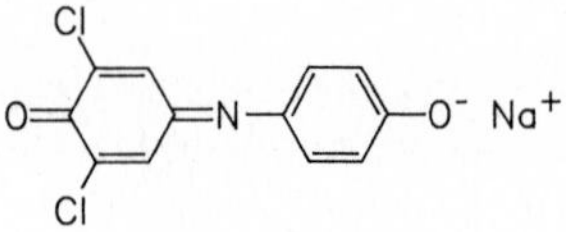

$C_{12}H_6Cl_2NNaO_2$
$M_r = 290.09$

Preparation of the Reagent

Dipping solution Dissolve 40 mg 2,6-dichlorophenolindophenol sodium in 100 ml ethanol.

Spray solution Dissolve 100 mg 2,6-dichlorophenolindophenol sodium in 100 ml ethanol [1, 2].

Storage The reagent solutions may be stored in the refrigerator for several weeks.

Substances 2,6-Dichlorophenolindophenol sodium salt
Ethanol

Reaction

Organic acids convert the blue mesomerically stabilized phenolate anion to the red undissociated acid. Reductones (e.g. ascorbic acid) reduce the reagent to a colorless salt.

Organic acids:

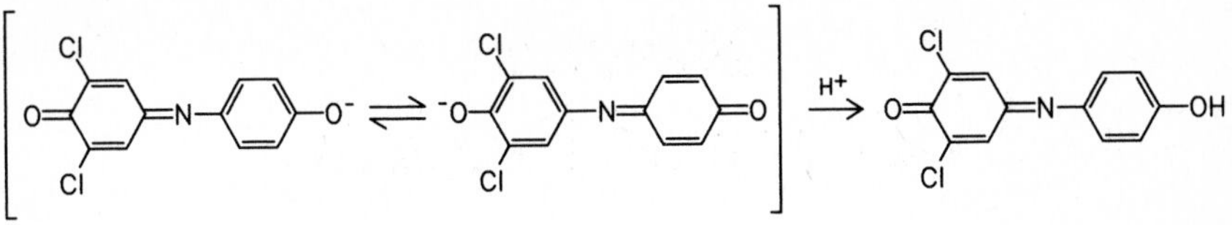

H^+

blue red

Reductones:

Reduction

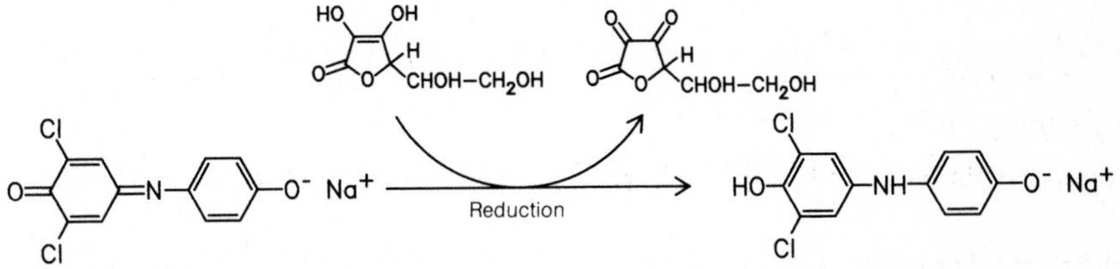

blue colorless

Method

The chromatograms are freed from mobile phase, immersed in the dipping solution for 1 s or sprayed evenly with the spray solution and heated to 100 °C for 5 min. Carboxylic acids yield red-orange zones and reductones colorless zones on a blue-violet background.

Note: Acids contained in the mobile phase must be completely removed otherwise pale violet zones will be produced instead of the red-orange zones on a violet background. If heating is prolonged the pink color of keto acids changes to white [1]. The colored zones of the carboxylic acids appear sharper if the chromatograms are exposed to an atmosphere of ammonia for a few seconds after heating [1]. Ascorbic and dehydroascorbic acids can also be detected fluorimetrically if the chromatogram is immersed after heating in liquid paraffin – *n*-hexane (1 + 2). In long-wavelength UV light ($\lambda = 365$ nm) they appear as green fluorescent zones on a dark background [3].

The reagent can be employed on silica gel, kieselguhr, Si 50 000, cellulose and polyamide layers.

Procedure Tested

Organic Acids [4]

Method Ascending, one-dimensional development in a trough chamber with chamber saturation.

Layer TLC plates Silica gel 60 (MERCK).

Mobile phase Diisopropyl ether – formic acid – water (90 + 7 + 3).

Migration distance 10 cm

Running time 30 min

Detection and result: The chromatogram was freed from mobile phase, immersed for 1 s in the dipping reagent and then heated on a hotplate at 100 °C for 5 min.

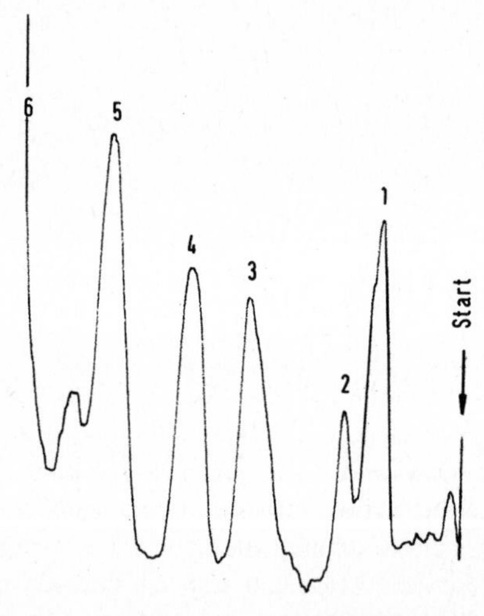

Fig. 1: Absorption scan of a chromatogram with 10 μg (!) per chromatogram zone of the carboxylic acids tartaric acid (1), malic acid (2), lactic acid (3), succinic acid (4), fumaric acid (5), stearic acid + front (6).

The carboxylic acids tartaric acid (hR_f 15–20), malic acid (hR_f 25–30), lactic acid (hR_f 45–50), succinic acid (hR_f 60–65), fumaric acid (hR_f 80) and stearic acid (hR_f 95–100) yielded orange-red zones on a violet background.

In situ quantitation: The photometric evaluation was carried out at $\lambda = 600$ nm. The detection limits of the acids were at 200–500 ng per chromatogram zone (Fig. 1).

References

[1] Passera, C., Pedrotti, A., Ferrari, G.: *J. Chromatogr.* **1964,** *14,* 289–291.
[2] Chan, H. T., Chang, T. S., Chenchin, E.: *J. Agric. Food Chem.* **1972,** *20,* 110–113.
[3] Schnekenburger, G.: Thesis, Fachhochschule Gießen, Fachbereich Technisches Gesundheitswesen, 1987.
[4] Jork, H., Kany, E., Klein, I.: GDCh-training course Nr. 300 „Einführung in die Dünnschicht-Chromatographie" Universität des Saarlandes, Saarbrücken, 1987.

2,6-Dichloroquinone-4-chloroimide Reagent

Reagent for:

- Phenols [1 – 3, 6]
  e.g. vitamin B_6
- Antioxidants [2 – 4]
- Prim. and sec. aliphatic amines [5]
- Prim., sec. and tert. aromatic amines [5]
- Carbazoles and aromatic hydrocarbons [5]
- Indoles and other N-containing heterocyclics [5]
- 2,4-Pentandione [5]
- Pharmaceuticals [7]
  e.g. barbiturates, amphetamines, diuretics, antihistamines, narcotics etc. [7]
- Phenoxyacetic acid herbicides [8, 9]

Cl
O= =NCl
Cl

$C_6H_2Cl_3NO$
$M_r = 210.45$

Preparation of Reagent

Dipping solution Dissolve 0.1 g 2,6-dichloroquinone-4-chloroimide in 10 ml dimethyl sulfoxide saturated with sodium hydrogen carbonate. Finally make up to 100 ml with chloroform.

Spray solution Dissolve 1 to 2 g 2,6-dichloroquinone-4-chloroimide in 100 ml ethanol [5, 6, 11] or toluene [8, 9] and filter, if necessary.

Storage The solutions should always be freshly made up.
Caution: 2,6-dichloroquinone-4-chloroimide can decompose exothermically (5); it should, therefore, only be stored in small quantities in the refrigerator!

Substances 2,6-Dichloroquinone-4-chloroimide
Sodium hydrogen carbonate
Dimethyl sulfoxide (DMSO)
Chloroform
Ethanol

Reaction

2,6-Dichloroquinone-4-chloroimide, which can be replaced by 2,6-dibromoquinone-4-chloroimide (qv.) reacts preferentially with phenols and anilines which are not substituted in the *p*-position [2].

Phenol + 2,6-Dichloroquinone-4-chloroimide $\xrightarrow[-\ HCl]{OH^-}$ Reaction product

Uric acid couples with 2,6-dichloroquinone-4-chloroimide in position N-7 to yield a yellow dyestuff.

Uric acid + 2,6-Dichloroquinone-4-chloroimide $\longrightarrow$ Yellow reaction product

Method

The chromatogram is dried in a stream of cold air and heated to 110 °C for 10 min, after cooling it is immersed in the reagent solution for 5 s or sprayed with it and then heated to 110 °C for 2 min.

Note: If the spray solution or a nonbasic dipping solution is employed for detection then it is advisable to spray afterwards with a 10% aqueous solution of sodium carbonate or a 2% solution of borax in ethanol – water (1 + 1). It is often possible to achieve the required basicity by placing the chromatogram in a twin-trough chamber one of whose troughs contains 5 ml 25% ammonia. This is not suitable for the Chiralplate® (MACHEREY-NAGEL) because in this case the plate background acquires a dark violet coloration.

In general chromatogram zones of differing colors are formed even without warming. Thus, gallates produce brown zones, butylhydroxytoluene orange ones and butylhydroxyanisole blue ones on a light background [11]. The detection sensitivity is usually between 0.2 and 0.5 μg substance per chromatogram zone; *o*-phenylphenol and cephaeline can be detected 10 times more sensitively. Despite its free amino group serotonin does not produce a coloration [1].

The reagent can be employed on silica gel, kieselguhr, Si 50 000, NH_2, cellulose and polyamide layers.

Procedure Tested

Isoquinoline Alkaloids [12]

Method	Ascending, one-dimensional double development in a trough chamber with chamber saturation.
Layer	HPTLC plates Silica gel 60 (MERCK).
Mobile phase	Chloroform – methanol (15 + 5).
Migration distance	2 × 5 cm
Running time	2 × 12 min

Detection and result: The chromatogram was dried for 5 min in a stream of warm air, immersed in the dipping solution for 5 s and then heated to 110 °C for 2 min.

Cephaeline (hR_f 40–45) produces a blue color spontaneously immediately on immersion, while emetine (hR_f 60–65) only does so on heating; on storage this color slowly changes to brown (background light brown). The detection limits for both substances are ca. 10 ng per chromatogram zone.

In situ quantitation: The absorption photometric analysis was made at $\lambda = 550$ nm (Fig. 1).

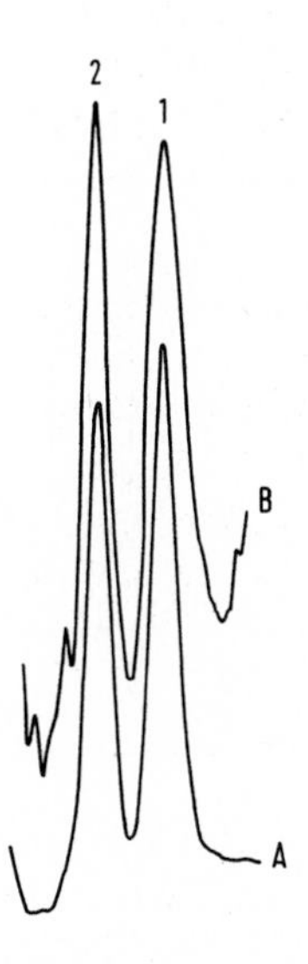

Fig. 1: Absorption scans of the pure substances (ca. 500 ng of each) cephaeline and emetine (A) and an extract of Ipecacuanhae Radix (B): cephaeline (1), emetine (2).

References

[1] Studer, A., Traitler, H.: *J. High Resolut. Chromatogr. Chromatogr. Commun.* **1982,** *5,* 581–582; *Proc. Int. Symp. Instrum. High Perform. Thin-Layer Chromatogr. 2nd,* Interlaken 1982, p. 62–73.
[2] Josephy, P. D., Lenkinski, R. E.: *J. Chromatogr.* **1984,** *294,* 375–379.
[3] Van der Heide, R. F.: *J. Chromatogr.* **1966,** *24,* 239–243.
[4] Seher, A.: Fette, Seifen, Anstrichm. **1959,** *61,* 345–351.
[5] Ross, J. H.: *Anal. Chem.* **1968,** *40,* 2138–2143.
[6] Raghuveer, K. G., Govindarajan, V. S.: *J. Assoc. Off. Anal. Chem.* **1979,** *62,* 1333–1337.

[7] Vinson, J. A., Hooyman, J. E.: *J. Chromatogr.* **1975**, *105*, 415-417.
[8] Sattar, M. A., Paasivirta, J.: *J. Chromatogr.* **1980**, *189*, 73-78.
[9] Sattar, M. A.: *J. Chromatogr.* **1981**, *209*, 329-333.
[10] Kovar, K.-A., Teutsch, M.: *Arch. Pharm.* (Weinheim) **1986**, *319*, 81-83.
[11] Groebel, W., Wessels, A.: *Dtsch. Lebensm. Rundsch.* **1973**, *69*, 453-459.
[12] Kany, E., Jork, H.: GDCh-training course Nr. 300 „Einführung in die Dünnschicht-Chromatographie“, Universität des Saarlandes, Saarbrücken 1987.

2,5-Dimethoxytetrahydrofuran – 4-Dimethylaminobenzaldehyde Reagent

Reagent for:

- Primary amines [1]
- Amino acids [1]
- Benzodiazepines [1]
- Panthenol [1]

$C_6H_{12}O_3$
$M_r = 132.16$
2,5-Dimethoxy-tetrahydrofuran

$C_9H_{11}NO$
$M_r = 149.19$
4-Dimethylamino-benzaldehyde

Preparation of the Reagent

Dipping solution I Mix 1 ml 2,5-dimethoxytetrahydrofuran with 99 ml glacial acetic acid [1] or ethyl acetate – glacial acetic acid (95 + 5).

Dipping solution II Dissolve 2 g 4-dimethylaminobenzaldehyde in 100 ml of a mixture of glacial acetic acid and 32% hydrochloric acid (85 + 15).

Storage Dipping solution I may be kept for several days and dipping solution II for several weeks at room temperature.

Substances 2,5-Dimethoxytetrahydrofuran
4-Dimethylaminobenzaldehyde
Acetic acid (glacial acetic acid)
Hydrochloric acid (32%)
Ethyl acetate

Reaction

Dimethoxytetrahydrofuran forms pyrrole derivatives with primary amines, these derivatives then condense with 4-dimethylaminobenzaldehyde in acid milieu to yield colored products [1]:

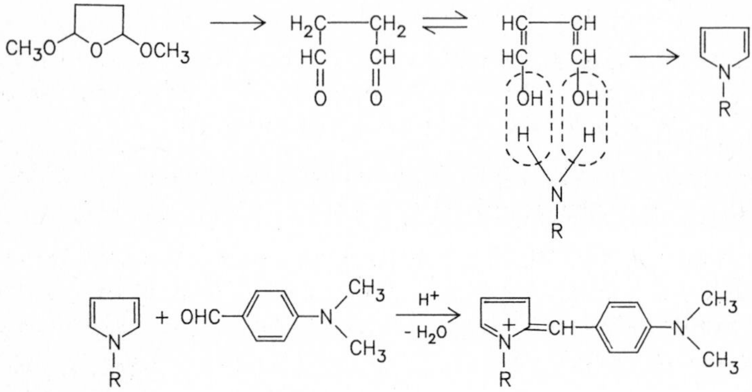

Method

The chromatogram is freed from mobile phase in a stream of warm air, dipped in dipping solution I for 3 s, then heated to 120 °C for 5 – 10 min, cooled to room temperature and then immersed in dipping solution II for 3 s. The final drying of the chromatogram should take place in a stream of cold air in order to avoid strong background coloration.

Red-violet chromatogram zones on a weakly ochre-colored background are yielded within a few seconds (< 30 s) on silica gel layers.

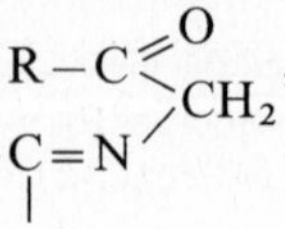

Note: Traces of ammonia left by the mobile phase should be completely removed from the chromatograms before the reagent is applied in order to avoid strong background coloration. The dipping solutions may also be applied as spray solutions. Secondary amines, amides, pyrimidines and purines do not react with the reagent [1]. In the case of benzodiazepines only those substances react which contain the structural element R–C(=O)–CH₂–N=C– , such as, for example, diazepam or nitrazepam; other benzodiazepines, such as, for example, chlorodiazepoxide or medazepam do not react [1]. The detection limits are at 5 – 50 ng for amino acids and benzodiazepines and at 500 ng substance per chromatogram zone for panthenol [1].

The reagent can be applied to silica gel, kieselguhr, Si 50 000, RP and CN layers. The detection sensitivity is reduced by a factor of 100 on cellulose layers; NH_2 layers are not suitable, as would be expected.

Occasionally, instead of dipping solution I it may be advisable to employ the following alternative dipping solution Ia mixed with citrate buffer in order to increase the detection sensitivity (e.g. with nitrazepam) [1].

Dipping solution Ia: Mix 1 ml 2,5-dimethoxytetrahydrofuran with 99 ml glacial acetic acid – citrate buffer, pH = 6.6 (1 + 2).

Citrate buffer solution: Dissolve 210 g citric acid in 400 ml caustic soda solution ($c = 5$ mol/l) and make up to 1 l with water. Mix 530 ml of this solution with 470 ml caustic soda solution ($c = 1$ mol/l) and adjust to pH 6.6 with caustic soda solution or citric acid [1].

Procedure Tested

Amino acids [2]

Method Ascending, one-dimensional development in a trough chamber with chamber saturation.

Layer HPTLC plates Silica gel 60 F_{254} (MERCK).

Mobile phase 2-Propanol – glacial acetic acid – water (16 + 3 + 3).

Migration distance 5 cm

Running time 55 min

Detection and result: The chromatogram was freed from mobile phase in a stream of warm air, immersed in reagent solution I for 3 s, heated to 120 °C for 5 – 10 min, cooled to room temperature and finally immersed in reagent solution II for 3 s. It was then dried in a stream of cold air.

Glycine (hR_f 20 – 25), DL-alanine (hR_f 30 – 35), DL-valine (hR_f 45 – 50) and L-leucine (hR_f 55 – 60) yielded red-violet chromatogram zones on a pale yellow background. The detection limits for these amino acids were 5 ng substance per chromatogram zone. With the exception of glycine the detection sensitivity was greater than with ninhydrin reagent.

In situ quantitation: Direct quantitative analysis was performed in reflectance at $\lambda = 580$ nm (Fig. 1).

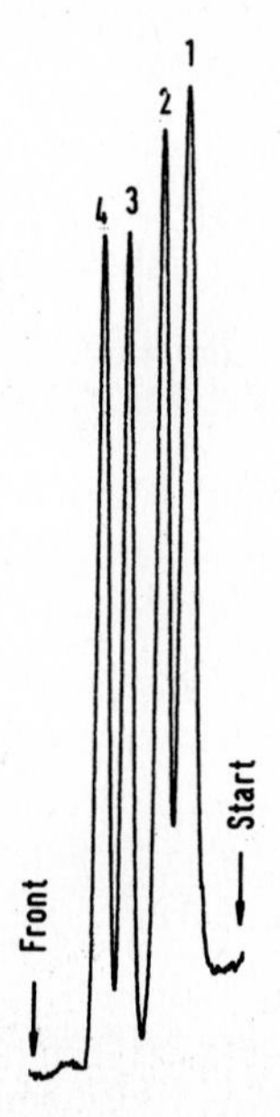

Fig. 1: Scanning curve of a chromatogram track with 100 ng per chromatogram zone of the amino acids glycine (1), alanine (2), valine (3), leucine (4).

References

[1] Haeffelfinger, P.: *J. Chromatogr.* **1970**, *48*, 184–190.
[2] Jork, H., Klingenberg, R.: GDCh-training course Nr. 300 „Einführung in die Dünnschicht-Chromatographie“, Universität des Saarlandes, Saarbrücken 1987.

4-Dimethylaminocinnamaldehyde – Hydrochloric Acid Reagent

Reagent for:

- Primary amines [1, 2, 9]
- Hydrazines [3]
- Indoles [4, 5, 13–15]
- Organoarsenic compounds [6]
- Urea [7, 8] and thiourea derivatives [9]
- Biotin [7]
- Sulfonamides [10]
- Pyrroles

$(CH_3)_2N-C_6H_4-CH=CH-CHO$

$C_{11}H_{13}NO$ HCl

$M_r = 175.23$ $M_r = 36.46$

4-Dimethylamino-cinnamaldehyde

Preparation of the Reagent

Dipping solution Dissolve 0.5 g 4-dimethylaminocinnamaldehyde in 50 ml hydrochloric acid (c = 5 mol/l) and make up to 100 ml with ethanol.

Spray solution Dilute the dipping solution 1 + 1 with ethanol for spraying.

Storage Both solutions may be stored for several days in the refrigerator.

Substances 4-Dimethylaminocinnamaldehyde
Ethanol
Hydrochloric acid 5 mol/l Combi-Titrisol

Reaction [11]

4-Dimethylaminocinnamaldehyde reacts with primary amines to form colored or fluorescent SCHIFF's bases (I). Pyrroles react with the reagent to form colored or fluorescent condensation products (II):

I: $(CH_3)_2N-C_6H_4-CH{=}CH-\overset{H}{C}{=}O + H_2N-R \xrightarrow{-H_2O} (CH_3)_2N-C_6H_4-CH{=}CH-\overset{H}{C}{=}N-R$

(1) (2)

II: $(1) + \text{pyrrole} \xrightarrow{H^+} (CH_3)_2N-C_6H_4-CH{=}CH-CH{=}\text{pyrrolium } (N^+H)$

(3)

Method

The chromatogram is freed from mobile phase and dipped in the dipping solution for 1 s or uniformly sprayed with spray solution and then heated to 100 °C for ca. 10 min. Colored zones are produced on a colorless to yellow background, after being left in the air for ca. 5 min they fluoresce greenish-yellow under long-wavelength UV light ($\lambda = 365$ nm).

Note: Sulfuric acid (4%) can also be employed in place of hydrochloric acid [3]. If ammoniacal mobile phases are employed the ammonia should be removed completely (e.g. heat to 105 °C for 10 min) before dipping or spraying; otherwise background discoloration can occur. The addition of titanium(III) chloride to the reagent allows also the staining of aromatic nitro compounds [6].

Aminoglycoside antibiotics and β-substituted indoles are stained red. Pyrrole derivatives with free β-positions react at room temperature to yield blue-colored zones [11]. Exposure to the vapors of *aqua regia* deepens the colors. This reaction sometimes produces fluorescence [3]. The detection limit for monomethylhydrazine is 200 pg per chromatogram zone [3].

Silica gel, cellulose [3, 13] and ion exchanger layers are amongst the stationary phases that can be employed.

Procedure Tested

Gentamycin Complex [1, 12]

Method Ascending, one-dimensional development at 10–12 °C in a twin-trough chamber with 5 ml conc. ammonia solution in the trough containing no mobile phase (chamber saturation 15 min).

Layer HPTLC plates Silica gel 60 (MERCK), prewashed by triple development with chloroform – methanol (50 + 50) and drying for 30 min at 110 °C.

Mobile phase Chloroform – ethanol – ammonia solution (25%) (10 + 9 + 10); the *lower organic phase* was employed.

Migration distance 5 cm

Running time 20 min

Detection and result: The chromatogram was freed from mobile phase (NH_3!) in a stream of cold air for 45 min. It was then immersed in the dipping solution for 1 s and heated to 100 °C for ca. 20 min. The chromatograms could be further treated with ammonia vapor if a colored background was found to be troublesome. The pale red zones then became bright red spots and the background frequently

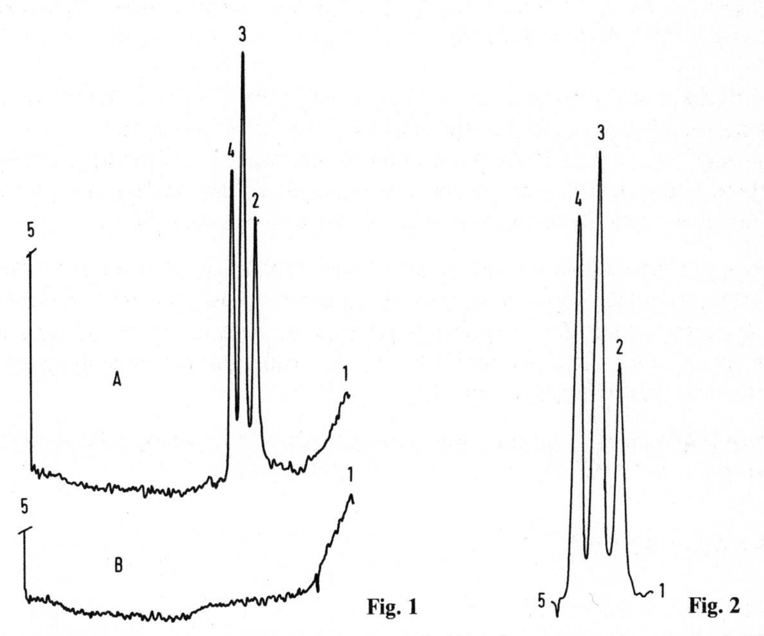

Fig. 1: Absorption scan of a chromatogram track (A) of a gentamycin standard (600 ng gentamycin C complex) and of an accompanying blank (B). Start (1), gentamycin C_1 (2), gentamycin C_2 and C_{2a} (3), gentamycin C_{1a} (4), solvent front (5).

Fig. 2: Fluorescence scan of a gentamycin C complex. Peak order and amount applied as in Figure 1.

acquired a strong yellow coloration. Greenish-yellow fluorescent zones appeared after allowing to lie in the air for ca. 5 min.

In situ quantitation: Gentamycins could be both absorption photometrically and fluorimetrically quantitated.

The absorption scans were made at a wavelength of $\lambda = 505$ nm (Fig. 1). The limit of detection was 100 ng gentamycin C complex. The best conditions for fluorimetric determination (Fig. 2) were excitation at $\lambda_{exc} = 313$ nm and detection at $\lambda_{fl} > 390$ nm.

References

[1] Kiefer, U.: Thesis, Fachhochschule Gießen, Fachbereich Technisches Gesundheitswesen, 1984.
[2] Ritter, W.: *J. Chromatogr.* **1977,** *142,* 431–440. Otherwise in: *Proceedings of the 2nd. International Symposium on HPTLC,* Interlaken, Bad Dürkheim: IfC-Verlag 1982, p. 100–113.
[3] Andary, C., Privat, G., Bourrier, M. J.: *J. Chromatogr.* **1984,** *287,* 419–424.
[4] Harley-Mason, J., Archer, A. A. P. G.: *Biochem. J.* **1958,** *69,* 60p.
[5] Reio, L.: *J. Chromatogr.* **1960,** *4,* 458–476.
[6] Morrison, J. L.: *J. Agric. Food Chem.* **1968,** *16,* 704–705.
[7] Bikfalvi, B., Szep, E., Berndorfer-Kraszner, E.: *Szeszipar,* **1978,** *26,* 35–39.
[8] Maulding, H. V., Nazareno, J., Polesuk, J., Michaelis, A. F.: *J. Pharm. Sci.* **1972,** *61,* 1389–1393
[9] Marshall, W. D., Singh, J.: *J. Agric. Food Chem.* **1977,** *25,* 1316–1320.
[10] Pauncz, J. K.: *J. High Resolut. Chromatogr. Chromatogr. Commun.* **1981,** *4,* 287–291.
[11] Strell, M., Kalojanoff, A.: *Chem. Ber.* **1954,** *87,* 1025–1032.
[12] Funk, W., Canstein, M. von, Couturier, Th., Heiligenthal, M., Kiefer, U., Schlierbach, S., Sommer, D. in: *Proceedings of the 3rd International Symposium on HPTLC,* Würzburg. Bad Dürkheim: IfC-Verlag 1985, p. 281–311.
[13] Baumann, P. Matussek, N.: *Z. Klin. Chem. Klin. Biochem.* **1972,** *10,* 176.
[14] Baumann, P., Scherer, B., Krämer, W., Matussek, N. : *J. Chromatogr.* **1971,** *59,* 463–466.
[15] Narasimhachari, N., Plaut, J. M., Leiner, K. Y.:: *Biochem. Med.* **1971,** *5,* 304–310.

2,4-Dinitrophenylhydrazine Reagent

Reagent for:

- Free aldehyde and keto groups [1 – 3]
  e.g. in aldoses and ketoses
- *o*- and *m*-Dihydroxybenzenes, dinitrophenols [1]
- Dehydroascorbic acid [4]
- Alkaloids [5, 6]
- Zearalenone [7, 8]
- Flavonoids
  e.g. silymarin, silydianine [9]
- Phospholipids [10]
- Prostaglandins [11, 12]
- Valepotriates [13]

O_2N–C₆H₃(NO₂)–NH–NH₂

$C_6H_6N_4O_4$

$M_r = 198.14$

Preparation of the Reagent

Dipping solution Dissolve 75 mg 2,4-dinitrophenylhydrazine in a mixture of 25 ml ethanol and 25 ml *ortho*-phosphoric acid (85%).

Spray solution Dissolve 100 mg 2,4-dinitrophenylhydrazine in a mixture of 90 ml ethanol and 10 ml conc. hydrochloric acid.

Storage Both solutions may be stored for several days in the refrigerator.

Substances
2,4′-Dinitrophenylhydrazine
Hydrochloric acid (32%)
Hydrochloric acid (0.5 mol/l)
Ethanol
ortho-Phosphoric acid (85%)

Reaction

2,4-Dinitrophenylhydrazine reacts with carbonyl groups with the elimination of water to yield hydrazones (I) and with aldoses or ketoses to yield colored osazones (II).

I: $O_2N-C_6H_3(NO_2)-NH-NH_2 + O{=}C(R^1)(R^2) \xrightarrow{-H_2O} O_2N-C_6H_3(NO_2)-NH-N{=}C(R^1)(R^2)$

II: $H-C(=O)-C(H)(OH)-R + 3\ H_2N-NH-C_6H_3(NO_2)-NO_2 \xrightarrow[-\ H_2N-C_6H_3(NO_2)-NO_2]{-\ 2\ H_2O,\ -\ NH_3} H-C(=N-NH-C_6H_3(NO_2)-NO_2)-C(=N-HN-C_6H_3(NO_2)-NO_2)-R$

Method

The chromatogram is freed from the mobile phase in a current of warm air (2 to 10 min), immersed in the dipping solution for 2 s or sprayed evenly with the spray solution and then dried in a stream of warm air (or 10 – 20 min at 110 °C).

Substances containing aldehyde or keto groups yield yellow to orange-yellow chromatogram zones on an almost colorless background [1, 11]. Silymarin appears red-blue and silydianine ochre-colored [9].

Note: The spray reagent can be made up with sulfuric acid instead of hydrochloric acid [9]. In contrast to other prostaglandins containing carbonyl groups which yield yellow-orange colored chromatogram zones, some of them without heating, 6-oxo-PGF_1 does not react even when heated [11].

The reagent can be employed on silica gel and cellulose layers.

Procedure Tested

Dehydroascorbic Acid [4]

Method Ascending, one-dimensional development in a trough chamber with chamber saturation.

Layer HPTLC plates Silica gel 60 (MERCK).

Mobile phase Acetone – toluene – formic acid (60 + 30 + 10).

Migration distance 6 cm

Running time 11 min

Detection and result: The chromatogram was freed from mobile phase in a current of warm air (2 min), immersed in the reagent solution for 5 s and dried in a current of warm air or at 110 °C for 10 min.

Dehydroascorbic acid (hR_f 45–50) appears as a yellow zone on a colorless background.

The detection limit for dehydroascorbic acid is 10 ng per chromatogram zone.

In situ quantitation: The quantitative analysis is performed by measuring the absorption of the chromatogram zone in reflectance at $\lambda = 440$ nm (Fig. 1).

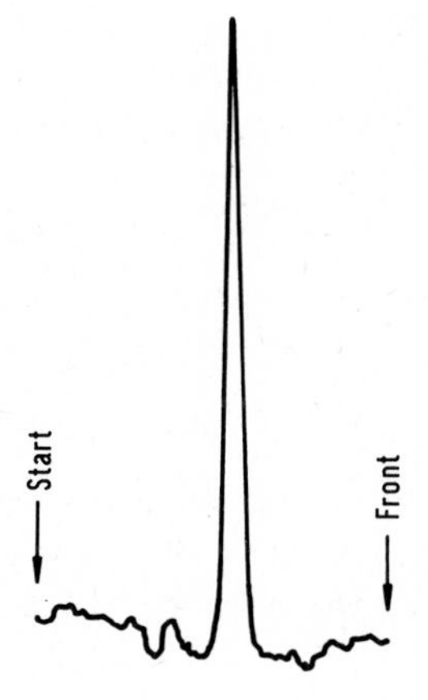

Fig. 1: Absorption scan of a chromatogram containing 200 ng dehydroascorbic acid per chromatogram zone.

References

[1] Reio, L.: *J. Chromatogr.* **1958,** *1,* 338–373; **1960,** *4,* 458–476.
[2] Lemberkovics, E., Verzar-Petri, G., Nagy, E.: *3rd Danube Symposium on Chromatography, Siofok* **1981,** 166–172.

[3] Peczely, P.: *Acta Biol. Hung.* **1985,** *36,*, 45–70.
[4] Schnekenburger, G.: Thesis, Fachhochschule Gießen, Fachbereich Technisches Gesundheitswesen, 1987.
[5] Minh Hoang, P. T., Verzar-Petri, G., Fuggerth, E.: *Acta Pharm. Hung.* **1979,** *49,* 105–113.
[6] Verzar-Petri, G., Minh Hoang, P. T., Oroszlan, P., Meszaros, S.: *3rd Danube Symposium on Chromatography, Siofok* **1981,** 269–272.
[7] Harrach, B., Palyusik, M.: *Acta Vet. Hung.* **1979,** *27,* 77–82.
[8] Kamimura, H., Nishijima, M., Yasuda, K., Saito, K., Ibei, A., Nagayama, T., Ushiyama, H., Naoi, Y.: *J. Assoc. Off. Anal. Chem.* **1981,** *64,* 1067–1073.
[9] Halbach, G., Winkler, W.: *Z. Naturforsch.* **1971,** *26B,* 971–972.
[10] Aman, M. B., Alekaev, N. S., Bekhova, E. A.: *Izv. Vyssh. Uchebn. Zaved. Pishch. Tekhnol.* **1970,** *6,* 22–24.
[11] Ubatuba, F. B.: *J. Chromatogr.* **1978,** *161,* 165–177.
[12] Bygdeman, M., Svanborg, K., Samuelson, B.: *Clin. Chim. Acta* **1969,** *26,* 373 –379.
[13] Braun, R., Dittmar, W., Machut, M., Wendland, S.: *Dtsch. Apoth. Ztg.* **1983,** *123,* 2474–2477.

Diphenylboric acid-2-aminoethyl Ester Reagent
(Flavone Reagent According to NEU)

Reagent for:

- Flavonoids [1 – 16, 21]
- Penicillic acid [17, 18]
- Carbohydrates [19]
  e.g. glucose, fructose, lactose
- Anthocyanidines [20]
- Hydroxy- and methoxycinnamic acid [20]

$(C_6H_5)_2B-O-CH_2-CH_2-NH_2$

$C_{14}H_{16}BNO$
$M_r = 225.10$

Preparation of the Reagent

Dipping solution Dissolve 1 g diphenylboric acid-2-aminoethyl ester (diphenylboric acid β-aminoethyl ester, diphenylboryloxyethylamine, "Naturstoffreagenz A", Flavognost®) in 100 ml methanol.

Storage The reagent solution may be kept for several days in the refrigerator. However, it should always be freshly prepared for quantitative analyses.

Substances Naturstoffreagenz A
Methanol

Reaction

Diphenylboric acid-2-aminoethyl ester reacts to form complexes with 3-hydroxyflavones* with bathochromic shift of their absorption maximum.

* Hohaus, E.: Private communication, Universität-GH Siegen, Analytische Chemie, D-5900 Siegen, 1987.

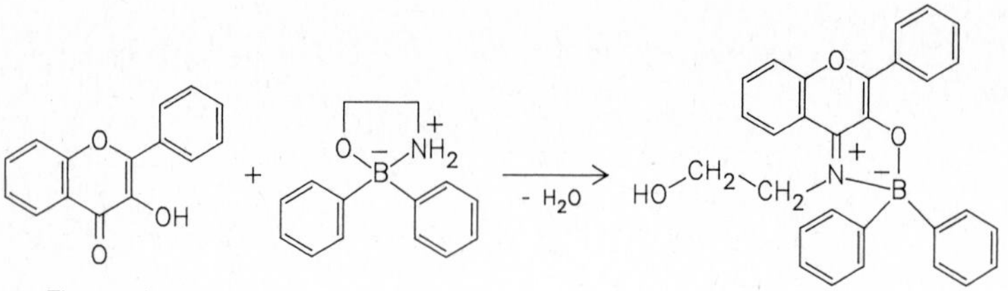

Flavonol

Method

The chromatograms are heated to 80 °C for 10 min and allowed to cool in a desiccator for 10 min, they are then immersed in the dipping solution for 1 s or sprayed with it, dried in a stream of warm air for ca. 1 min and stored in the desiccator for 15 min. Finally, they are immersed for 1 s in a solution of liquid paraffin – *n*-hexane (1 + 2) to stabilize the fluorescence and – after drying for 1 min in a stream of warm air – irradiated for ca. 2 min with intense long-wavelength UV light ($\lambda = 365$ nm).

Substances are produced with a characteristic fluorescence in long-wavelength UV light ($\lambda = 365$ nm).

Note: A 5% solution of polyethylene glycol 4000 in ethanol can be sprayed onto the chromatogram [2, 4] for the purpose of increasing and stabilizing the fluorescence instead of dipping it in liquid paraffin – *n*-hexane (1 + 2). If this alternative is chosen the plate should not be analyzed for a further 30 min since it is only then that the full intensity of the fluorescence develops [6].

In order to detect penicillic acid (detection limit: ca. 5 ng) the plate is heated to 110 °C for 15 min after it has been sprayed with reagent; this causes penicillic acid to produce pale blue fluorescent zones [17, 18].

In long-wavelength UV light ($\lambda = 365$ nm) carbohydrates, e.g. glucose, fructose and lactose, yield pale blue fluorescent derivatives on a weakly fluorescent background. In situ quantitation can be performed at $\lambda_{exc} = 365$ nm and $\lambda_{fl} = 546$ nm (monochromatic filter M 546) [19]. Further differentiation can be achieved by spraying afterwards with *p*-anisidine-phosphoric acid reagent [8].

The reagent can be employed on silica gel, kieselguhr, cellulose, RP, NH_2 and polyamide-11 layers.

Procedure Tested

Flavonoids [21]

Method Ascending, one-dimensional development in a trough chamber with chamber saturation. After application of the samples but before development the layer was conditioned for 30 min over water.

Layer HPTLC plates Silica gel 60 (MERCK). The layer was prewashed by developing once in methanol and then drying at 110 °C for 30 min.

Mobile phase Ethyl acetate – formic acid – water (85 + 10 + 15).

Migration distance 6 cm

Running time 30 min

Detection and result: The developed chromatogram was heated to 80 °C for 15 min; the warm plate was sprayed first with "Flavone Reagent" (3% in methanol)

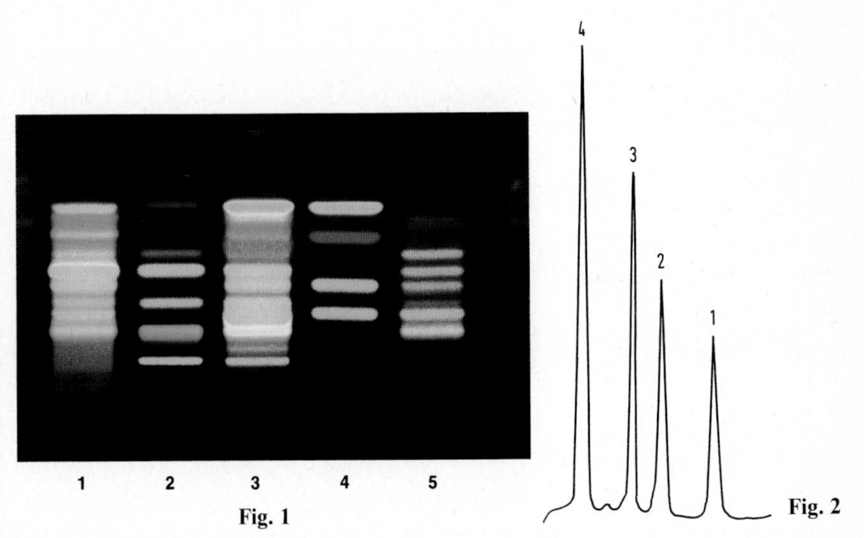

Fig. 1: Chromatography of flavonoids. 1. Extract. Solidaginis; 2. Rutin – chlorogenic acid – isoquercitrin – quercitrin; 3. Extract. Hyperici; 4. Hyperosid – quercetin-3-arabinosid – hypericin – quercetin; 5. Extract. Betulae

Fig. 2: Fluorescence scan of a chromatogram track with 30 ng substance per chromatogram zone of rutin (1), hyperoside (2), quercitrin (3) and quercetin (4).

and then immediately with polyethylene glycol 6000 (5% in ethanol). After drying at room temperature for 30 min the chromatogram was inspected under long-wavelength UV light (λ = 365 nm).

Colored substance zones were obtained which could be analyzed quantitatively. The (visual) detection limits were: hypericin 1 ng, rutin and chlorogenic acid 5 ng, hyperoside – quercetin 10 ng per mm chromatogram zone. The detection limits for densitometric analysis are between 20 and 50% of those for visual detection.

The following hR_f values were obtained: quercetin 90–95, hypericin 75–80, quercitrin 60–65, quercetin-3-arabinoside 55–60, isoquercitrin 45–50, hyperoside 40–45, chlorogenic acid 30–35, rutin 20–25 (Fig. 1).

In situ quantitation: The in situ fluorescence determination was carried out by excitation at λ_{exc} = 436 nm and detection at λ_{fl} = 546 nm (monochromate filter M 546) (Fig. 2).

References

[1] Neu, R.: *Naturwissenschaften* **1957,** *44,* 181-182.
[2] Wagner, H., Diesel, P., Seitz, M.: *Arzneim.-Forsch.* **1974,** *24,* 466-471; Wagner, H. Tittel, G. Bladt, S.: *Dtsch. Apoth. Ztg.* **1983,** *123* 515-521.
[3] Pachaly, P.: *Dtsch. Apoth. Ztg.* **1984,** *124,* 2153-2161.
[4] Hiermann, A., Karting, Th.: *J. Chromatogr.* **1977,** *140,* 322-326.
[5] Becker, H., Exner, J., Bingler, T.: *J. Chromatogr.* **1979,** *172,* 420-423.
[6] Stahl. E., Juell, S.: *Dtsch. Apoth. Ztg.* **1982,** *122,* 1951-1957.
[7] Wildanger, W., Herrmann, K.: *J. Chromatogr.* **1973,** *76,* 433-440.
[8] Niemann, G. J.: *J. Chromatogr.* **1979,** *170,* 227.
[9] Gilles, F.: Thesis, Univ. Gießen, Fachbereich Agrarwissenschaften, 1987.
[10] Chirikdjian, J. J.: *Pharmazie* **1974,** *29,* 292-293.
[11] Henning, W., Herrmann, K.: *Phytochemistry* **1980,** *19,* 2727-2729.
[12] Ulubelen, A., Kerr, K. M., Mabry, T. J.: *Phytochemistry* **1980,** *19,* 1761-1766; **1982,** *21,* 1145-1147.
[13] Kunde, R., Issac, O.: *Planta Med.* **1979,** *37,* 124-130.
[14] Wollenweber, E., Seigler, D. S.: *Phytochemistry* **1982,** *21,* 1063-1066.
[15] Rauwald, H., Miething, H.: *Dtsch. Apoth. Ztg.* **1985,** *125,* 101-105.
[16] Brasseur, T., Augenot, L.: *J. Chromatogr.* **1986,** *351,* 351-355.
[17] Johann, H., Dose, K.: *Fresenius Z. Anal. Chem.* **1983,** *314,* 139-142.
[18] Ehnert, M., Popken, A. M., Dose, K.: *Z. Lebensm. Unters. Forsch.* **1981,** *172,* 110-114.
[19] Funk, W.: Private communication, Fachhochschule Gießen, Fachbereich: Technisches Gesundheitswesen 1988.
[20] Somaroo, B. H., Thakur, M. L., Grant, W. F.: *J. Chromatogr.* **1973,** *87,* 290-293.
[21] Hahn-Deinstrop, E.: Private communication, Fa. Heumann, Abt. Entwicklungsanalytik, D-8500 Nürnberg 1.

Diphenylboric Anhydride Reagent (DBA Reagent)

Reagent for:

- 2-(2-Hydroxyphenyl)benztriazole derivatives (UV absorber in plastics)*

B–O–B

$C_{24}H_{20}B_2O$
$M_r = 346.04$

Preparation of the Reagent

Dipping solution Dissolve 350 mg diphenylboric anhydride (DBA) in 100 ml methanol.

Storage DBA should be stored dry, cool and protected from light. The solution should be freshly made up every day.

Substances Diphenylboric anhydride
Methanol

Reaction

The reaction takes place with the formation of fluorescent chelates according to the following scheme:

* Hohaus E.: Private communication, Universität-GH Siegen, Analytische Chemie, D-5900 Siegen, 1985.

2-(2-Hydroxy-5-methyl-phenyl)benztriazole (Tinuvin P, CIBA-GEIGY) + DBA $\xrightarrow{-H_2O}$ 2 Fluorescent reaction product

Method

The chromatogram is freed from mobile phase in a stream of warm air, immersed for 1 s in the reagent solution or sprayed evenly with it and dried in a stream of cold air.

Yellow-green fluorescent zones are formed on a dark background in long-wavelength UV light ($\lambda = 365$ nm).

Note: Since neither the DBA reagent nor 2-(2-hydroxyphenyl)benzotriazole are intrinsically fluorescent the chromatogram is not affected by interfering signals.

The reagent can be employed, for example, on silica gel, kieselguhr and Si 50000 layers.

Procedure Tested

2-(2-Hydroxy-5-methylphenyl)benztriazole *

Method Ascending, one-dimensional development in a twin-trough chamber with chamber saturation.

Layer TLC plates Silica gel 60 (MERCK).

* Hohaus, E., Monien, H., Overhoff-Pelkner, A.-P.: private communication, Universität-GH Siegen, Analytische Chemie, D-5900 Siegen 1985.

Mobile phase Carbon tetrachloride.

Migration distance 10 cm

Running time ca. 12 min

Detection and result: The chromatogram was freed from mobile phase in a stream of warm air, immersed in the reagent solution for 1 s and dried in a stream of warm air.

Under long-wavelength UV light ($\lambda = 365$ nm) the 2-(2-hydroxyphenyl)-benztriazoles yielded yellow-green fluorescent chromatogram zones, which were, in the cases of Tinuvin P (hR_f 20 – 25) and Tinuvin 343 (2-[2-hydroxy-3-(1-methylpropyl)-5-*tert*-butylphenyl]benztriazole; hR_f 45 – 50), suitable for quantitation.

In situ quantitation: Quantitative determination is performed fluorimetrically ($\lambda_{exc} = 365$ nm, $\lambda_{fl} = 535$ nm), the detection limits are 250 ng substance per chromatogram zone.

Diphenylboric Anhydride – Salicylaldehyde Reagent (DOOB Reagent)

Reagent for:

- Primary (!) amines
  e.g. alkyl amines [1 – 3]
  lipid amines [4]
  α, ω-diamines [5, 6]
  polyamines [6]
  alkanol amines [7]
  subst. anilines [8]
  aminoglycoside antibiotics [9, 10]
  biogenic amines [11]
  hydrazines

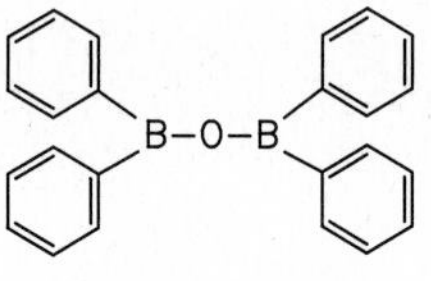

$C_{24}H_{20}B_2O$

$M_r = 346.04$

Diphenylboric anhydride

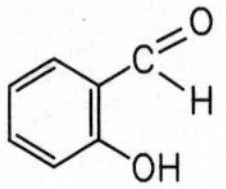

$C_7H_6O_2$

$M_r = 122.12$

Salicylaldehyde

Preparation of the Reagent

Dipping solution Dissolve 35 mg diphenylboric anhydride (DBA) and 25 mg salicylaldehyde (= 2-hydroxybenzaldehyde) in 100 ml chloroform. This results in the formation of 2,2-diphenyl-1-oxa-3-oxonia-2-boratanaphthalene (DOOB).

Storage DBA solutions should only be stored for a short time even in the refrigerator. On the other hand, DOOB reagent solution in bottles with ground-glass stoppers may be stored in the refrigerator for at least 2 weeks.

Substances Diphenylboric anhydride
2-Hydroxybenzaldehyde
Chloroform

Reaction

The DOOB reagent, which is formed by reaction of diphenylboric anhydride with salicylaldehyde, yields fluorescent reaction products with primary amines [1].

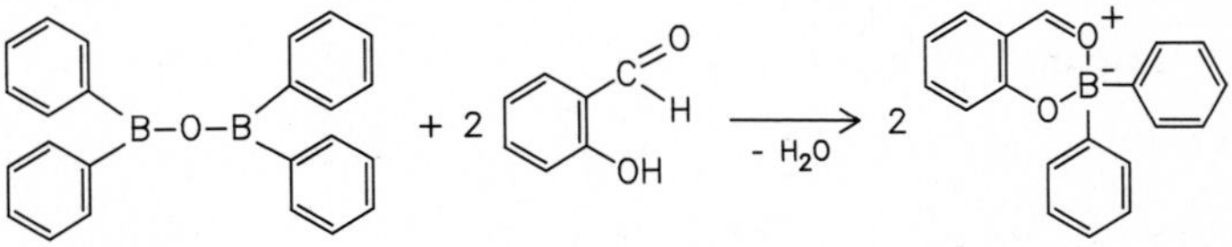

Diphenylboric anhydride + 2 2-Hydroxybenzaldehyde $\xrightarrow{-H_2O}$ 2 DOOB reagent

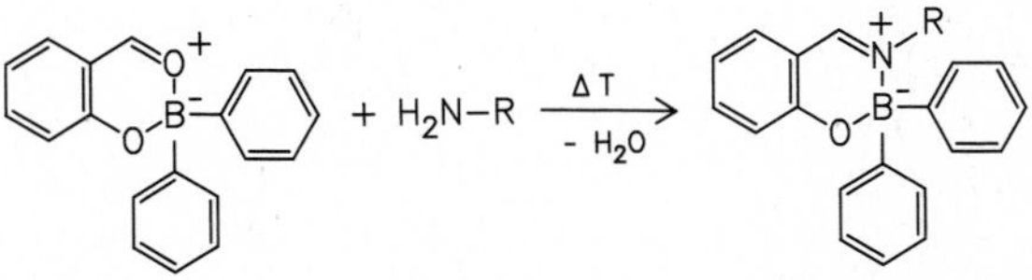

DOOB reagent + $H_2N{-}R$ (Amine) $\xrightarrow[-H_2O]{\Delta T}$ Fluorescent reaction product

Method

The chromatogram is freed from mobile phase (stream of warm air, 15 min), immersed for 2 s in the reagent solution after cooling to room temperature and heated to 110–120 °C for 10–20 min. The chromatogram is then briefly immersed in liquid paraffin – *n*-hexane (1 + 6) in order to enhance and stabilize the fluorescence.

After evaporation of the hexane blue fluorescent chromatogram zones are visible on a dark background under long-wavelength UV light (λ = 365 nm).

Note: The dipping solution may also be used as spray solution [1]. The reagent may be applied to RP layers; it is not suitable for amino phases.

Procedure Tested

Netilmicin and Gentamycin-C Complex [10]

Method Ascending, one-dimensional development in a HPTLC-trough chamber with chamber saturation.

Layer KC8F or KC18F plates (WHATMAN).

Mobile phase 0.1 mol/l LiCl in 32% ammonia solution – methanol (25 + 5).

Migration distance 10 cm

Running time 30 min

Detection and result: The chromatogram was dried in a stream of warm air for 15 min, cooled and immersed in the reagent solution for 2 s. It was then heated to 110–120 °C for 10–20 min, allowed to cool to room temperature and immersed for 2 s in liquid paraffin – *n*-hexane (1 + 6) to enhance and stabilize the fluores-

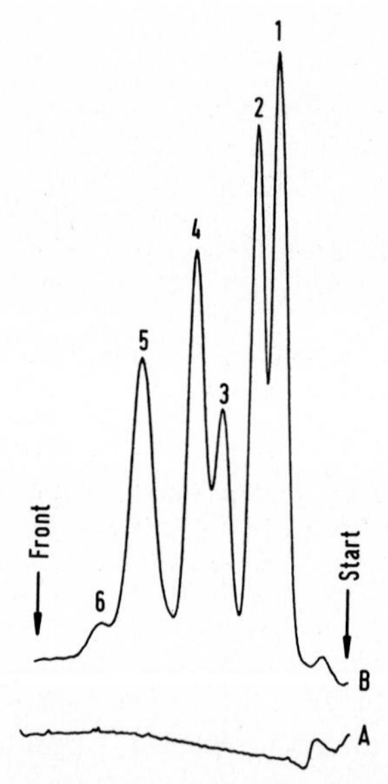

Fig. 1: Fluorescence scan of a blank track (A) and of a chromatogram track (B) with 500 ng netilmicin and 1 µg gentamycin C complex as sample. Netilmicin (1), gentamycin C_1 (2), gentamycin C_{2a} (3), gentamycin C_2 (4), gentamycin C_{1a} (5), impurity in netilmicin (6).

cence. After the evaporation of the *n*-hexane the netilmicin and the gentamycin components appeared as blue fluorescent chromatogram zones on a dark background under long-wavelength UV light ($\lambda = 365$ nm). The same applied to neomycin. The detection limits were 10 ng per chromatogram zone for netilmicin and 15–20 ng per chromatogram zone for the various gentamycin components.

Note: If netilmicin is to be chromatographed alone it is recommended that the methanol content of the mobile phase be increased (e.g. to 23 + 7), in order to increase the value of the hR_f. The detection limit for the substances in the application tested was more sensitive using DOOB reagent on RP layers than when NBD chloride, fluorescamine or *o*-phthalaldehyde were employed. The derivatives so formed were stable and still fluoresced after several weeks if they were stored in the dark.

In situ quantitation: Quantitative analysis was performed fluorimetrically ($\lambda_{exc} = 365$ nm, $\lambda_{fl} > 430$ nm) (Fig. 1).

References

[1] Hohaus, E.: *Fresenius Z. Anal. Chem.* **1982,** *310*, 70–76.
[2] Hohaus, E.: *Bunseki Kagaku* **1984,** *33*, E55–E60.
[3] Hohaus, E.: *Fresenius Z. Anal. Chem.* **1984,** *319*, 533–539.
[4] Claas, K. E., Hohaus, E., Monien, H.: *Fresenius Z. Anal. Chem.* **1986,** *325*, 15–19.
[5] Claas, K. E., Hohaus, E., Monien, H.: *Fresenius Z. Anal. Chem.* **1983,** *316*, 781–784.
[6] Winkler, E., Hohaus, E., Felber, E.: *J. Chromatogr.* **1988,** *436*, 447–454; BASF-Analysenmethoden Bd. V, Nr. 1615a and b.
[7] Claas, K. E., Hohaus, E.: *Fresenius Z. Anal. Chem.* **1985,** *322*, 343–347.
[8] Hohaus, E.: Private communication, Universität-GH-Siegen, Analytische Chemie, Siegen 1986–1989.
[9] Kunz, F. R., Jork, H.: *Proceedings of the 4th International Symposium on Instrumental TLC (Planar Chromatography), Selvino.* Bad Dürkheim: IfC-Verlag 1987, S. 437–451.
[10] Kunz, F. R., Jork, H.: *Fresenius Z. Anal. Chem.* **1988,** *329*, 773–777.
[11] Stark, G.: Thesis, Universität des Saarlandes, Fachbereich 14, Saarbrücken 1988.

Fast Blue Salt B Reagent

Reagent for:

- Phenols
  e.g. *n*-alkylresorcinol homologues [1]
  thymol derivatives [2]
  hydoxyanthraquinones [3]
  ratanhia phenols [4]
  cardol, anacardol [5]
  phloroglucinol derivatives [6, 7]
- Tannins [8, 9]
- α- and γ-Pyrone derivatives
  e.g. cumarins [10]
  flavonols [11]
- Phenolcarboxylic acids [12]
- Cannabinoids [13 – 18]
- Amines capable of coupling
  e.g. carbamate pesticides [19]
  pharmaceutical metabolites [20 – 22]

$Cl^- \left[N{\equiv}\overset{+}{N}-C_6H_3(OCH_3)-C_6H_3(OCH_3)-\overset{+}{N}{\equiv}N \right]^{2+} Cl^- \cdot ZnCl_2$

$C_{14}H_{12}Cl_2N_4O_2 \cdot ZnCl_2$

$M_r = 339.18$

Preparation of Reagent

Dipping solution Dissolve 140 mg fast blue salt B in 10 ml water and make up to 40 ml with methanol. Add 20 ml of this solution with stirring to a mixture of 55 ml methanol and 25 ml dichloromethane.

Spray solution I Dissolve 0.1 to 5 g fast blue salt B in water [7, 8], 70% ethanol [17] or acetone-water (9 + 1) [14].

Spray solution II Caustic soda solution ($c = 0.1$ mol/l)

Storage The reagent solutions should always be freshly made up and fast blue salt B should always be stored in the refrigerator.

Substances Fast blue salt B
Methanol
Dichloromethane
Sodium hydroxide solution (0.1 mol/l)

Reaction

Fast blue salt B couples best with phenols in alkaline medium, e.g. with 11-nor-Δ^9-THC-9-carboxylic acid to yield the following red-colored product:

11-Nor-Δ^9-THC-9-carboxylic acid + Cl^- $[N{\equiv}N]^+$–(H$_3$C–O)…(O–CH$_3$)–$[N{\equiv}N]^+$ $Cl^- \cdot ZnCl_2$ + 11-Nor-Δ^9-THC-9-carboxylic acid

Fast blue salt B

red-colored azo dye

Method

The chromatograms are dried in a stream of warm air, then placed for 10 – 15 min in the empty half of a twin-trough chamber whose other trough contains 25 ml conc. ammonia solution (equilibrate for 60 min!) and then immediately immersed in the dipping reagent for 5 s [18] and dried for 5 min in a stream of warm air.

Alternatively the dried chromatograms can be homogeneously sprayed with spray solutions I and II consecutively.

Chromatogram zones of various colors are produced (yellow, red, brown and violet) on an almost colorless background.

Note: The detection is not affected if the dipping solution exhibits a slight opalescent turbidity. Fast blue salt BB [18] or fast blue salt RR [18, 19] can be employed in the reagent in place of fast blue salt B. It is occasionally preferable not to apply spray solutions I and II separately but to work directly with a 0.1% solution of fast blue salt B in caustic soda solution ($c = 1 - 2$ mol/l) [13, 15] or in 0.5% methanolic caustic potash [3].

A dipping solution consisting of 0.2% fast blue salt B in hydrochloric acid ($c = 0.5$ mol/l, immersion time: 30 s) has been reported for the detection of resorcinol homologues [1].

Detection limits of 250 ng per chromatogram zone have been reported for THC-11-carboxylic acid [15] and of 500 ng per chromatogram zone for carbamate pesticides [19].

The reagent can be employed on cellulose, silica gel and polyamide layers [11]; kieselguhr, RP and Si 50 000 layers are also suitable.

Caution: Fast blue salt B may be carcinogenic [23].

Procedure Tested

THC Metabolites in Urine after Consumption of Hashish [18]

Method Ascending, one-dimensional multiple development method (stepwise technique, drying between each run) in two mobile phase systems in a twin-trough chamber without chamber saturation (equilibration: 30 min at 20 – 22 °C) at a relative humidity of 60 – 70%.

Layer HPTLC plates Silica gel 60 F_{254s} (MERCK) or HPTLC plates Silica gel 60 WRF_{254s} (MERCK).

Mobile phase 1 Toluene – ethyl acetate – formic acid [16 + 4 + 0.5 (0.4*)].

Mobile phase 2 Toluene – ethyl acetate – formic acid [17+3+0.25 (0.2*)].

Migration distances Eluent 1: 7.5 mm, 15.0 mm and 22.5 mm
Eluent 2: 30.0 mm, 37.5 mm and 45 mm

Detection and result: The chromatogram was dried in a stream of warm air for 5 min (it is essential to remove all traces of formic acid) and then placed for 10–15 min in the empty half of a twin-trough chamber whose other trough contained 25 ml conc. ammonia solution (equilibrated for 60 min!). It was immersed immediately afterwards in the dipping reagent for 5 s and dried in a stream of warm air for 5 min.

In visible light the chromatogram zones were red on an almost colorless background (Fig. 1), they slowly turned brownish on exposure to air and appeared

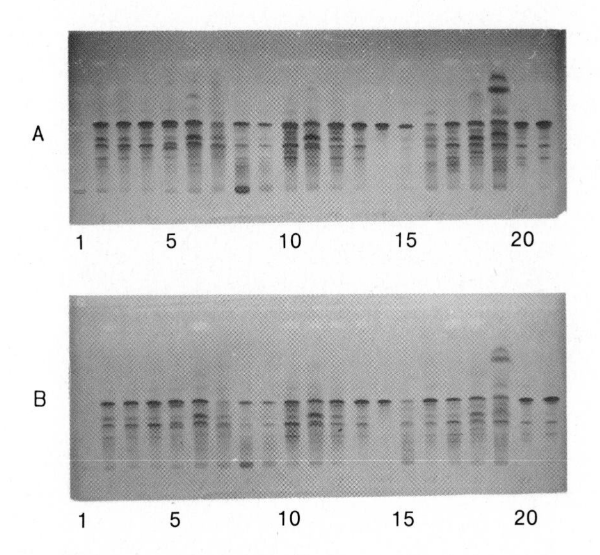

Fig. 1: Chromatograms of urine samples containing THC metabolites; detection with fast blue salt RR (A) and fast blue salt B (B). Track A_1 and B_1: metabolite-free urines; tracks A_{16} and B_{15} represent ca. 60 ng total cannabinoids per ml urine (determined by RIA).

* These quantities apply to the WRF_{254s} layer.

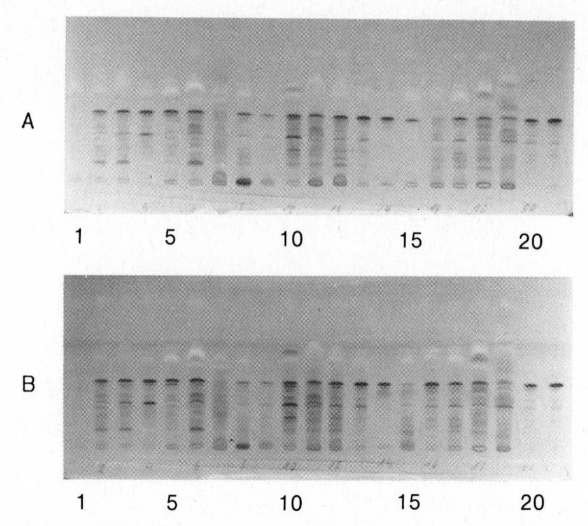

Fig. 2: Chromatogram from Fig. 1 observed under long-wavelength UV light ($\lambda = 365$ nm).

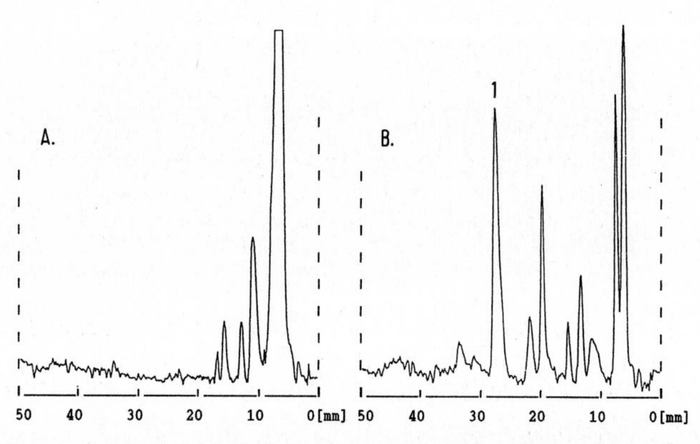

Fig. 3 Reflectance scan of a urine free from THC metabolites (A) and one containing THC metabolites (B): Peak 1 corresponds to ca. 100 ng 11-nor-Δ^9-THC-9-carboxylic acid per ml urine.

under UV light on the indicator-containing layer as dark blue zones on a pale blue fluorescent background.

Additional dipping in a 0.001% solution of dansyl semipiperazide or bis-dansyl piperazide in dichloromethane – liquid paraffin (75 + 25) stabilized the color of the chromatogram for a period of months.

Note: The alternative fast blue salt BB produced the most intensely colored chromatogram zones for visual analysis in daylight, while fluorescence quenching in UV light ($\lambda = 254$ nm) was greater with fast blue salt B and fast blue salt RR (Figs. 1 and 2).

In situ quantitation: The absorption photometric analysis was performed at $\lambda = 546$ nm (Fig. 3).

References

[1] Kozubek, A.: *J. Chromatogr.* **1984,** *295*, 304-307.
[2] Willuhn, G.: *Planta Med.* **1972,** *22*, 1-33.
[3] Van den Berg, A. J. J., Labadie, R. P.: *Planta Med.* **1981,** *41*, 169-173.
[4] Stahl, E., Ittel, I.: *Planta Med.* **1981,** *42*, 144-154.
[5] Stahl, E., Keller, K., Blinn, C.: *Planta Med.* **1983,** *48*, 5-9.
[6] Widen, C. J., Vida, G., von Euw, J., Reichstein, T.: *Helv. Chim. Acta* **1971,** *54*, 2824-2850.
[7] Von Euw, J., Reichstein, T., Widen, C. J.: *Helv. Chim. Acta* **1985,** *68*, 1251-1274.
[8] Byung-Zun Ahn, Gstirner, F.: *Arch. Pharm.* (Weinheim) **1973,** *306*, 6-17.
[9] Gstirner, F., Flach, G.: *Arch. Pharm.* (Weinheim) **1970,** *303*, 339-345.
[10] Reichling, J., Beiderbeck, R., Becker, H.: *Planta Med.* **1979,** *36*, 322-332.
[11] Hösel, W., Barz, W.: *Biochim. Biophys. Acta* **1972,** *261*, 294-303.
[12] Hartley, R. D.: *J. Chromatogr.* **1971,** *54*, 335-344.
[13] Scherrmann, J. M., Hoellinger, H., Sonnier, M., Hoffelt, J., Nguyen-Hoang-Nam: *J. Chromatogr.* **1980,** *196*, 342-346.
[14] Neuninger, H.: *Arch. Kriminol.* **1981,** *167*, 99-109.
[15] Kanter, S. L., Hollister, L. E.: *J. Chromatogr.* **1978,** *151*, 225-227; **1982,** *234*, 201-208.
[16] Nakamura, G. R., Stall, W. J., Folen, V. A., Masters, R. G.: *J. Chromatogr.* **1983,** *264*, 336-338.
[17] Verzar-Petri, G., Ladocsy, T., Oroszlan, P.: *Acta Botanica* **1982,** *28*, 279-290.
[18] Hänsel, W., Strömmer, R.: *GIT Fachz. Lab.* **1988,** *32*, 156-166.
[19] Tewari, S. N., Singh, R.: *Fresenius Z. Anal. Chem.* **1979,** *294*, 287; *J. Chromatogr.* **1979,** *172*, 528-530.
[20] Kröger, H., Bohn, G., Rücker, G.: *Dtsch. Apoth. Ztg.* **1977,** *117*, 1787-1789.
[21] Goenechea, S., Eckhardt, G., Goebel, K. J.: *J. Clin. Chem. Clin. Biochem.* **1977,** *15*, 489-498.
[22] Goenechea, S., Eckhardt, G., Fahr, W.: *Arzneim.-Forsch.* **1980,** *30*, 1580-1584.
[23] Hughes, R. B., Kessler, R. R.: *J. Forens. Sci.* **1979,** *24*, 842-846.

Fluorescamine Reagent

Reagent for:

- Primary amines, amino acids [1 – 5]
  e.g. histamine [4], sympathomimetics [6],
  catecholamines, indolamines [5],
  arylamines [7], gentamicins [8]
- Peptides [2, 5]
- Secondary amines [9]
- Sulfonamides [10 – 14]

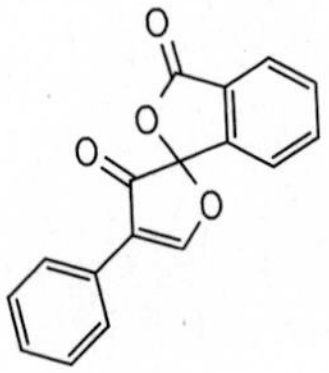

$C_{17}H_{10}O_4$

$M_r = 278.27$

Preparation of the Reagent

Dipping solution I Dissolve 10 – 50 mg fluorescamine (Fluram®) in 100 ml acetone [2, 9, 10].

Dipping solution II Make 10 ml triethylamine up to 100 ml with dichloromethane [2].

Spray solution I Dissolve 1.9 g *di*-sodium tetraborate decahydrate in 100 ml water and adjust the pH to 10.5 with caustic soda solution.

Spray solution II Dissolve 3.6 g *di*-sodium hydrogen phosphate dihydrate in 100 ml water and adjust the pH to 7.5 with phosphoric acid. Dissolve 2.5 g taurine in this.

Storage The dipping solutions and the spray solutions may be stored for at least one week in the refrigerator [2].

Substances Fluorescamine
Triethylamine
Acetone
Dichloromethane

di-Sodium tetraborate decahydrate
Taurine
di-Sodium hydrogen phosphate dihydrate

Reaction

Fluorescamine reacts directly with primary amines to form fluorescent products. Secondary amines yield nonfluorescent derivatives which can be transformed into fluorescent products by a further reaction with primary amines.

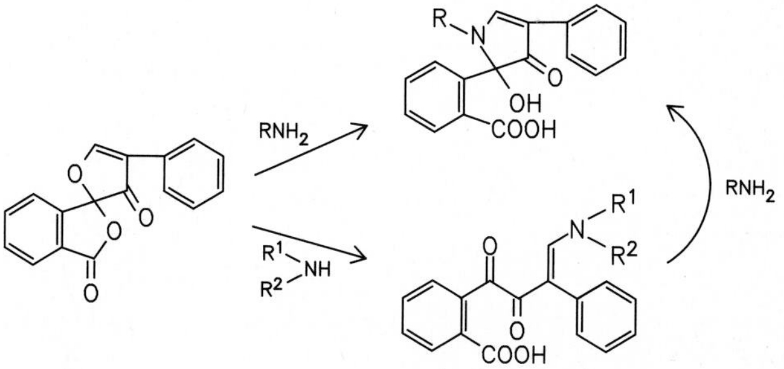

Fluorescamine

Fluorescent reaction product

Method

Primary amines: The chromatograms are dried for 10 min at 110 °C, cooled to room temperature and then immersed in dipping solution I or sprayed evenly with it and then dried for a few seconds in the air.

Then they are immersed in dipping solution II for 1 s or sprayed with it.

Fluorescent blue-green chromatogram zones appear on a dark background in long-wavelength UV light ($\lambda = 365$ nm).

Secondary amines: The chromatograms are first freed from mobile phase and then sprayed evenly with spray solution I and heated to 110 °C for 15 min.
Then they are immersed in dipping solution I or sprayed with it, stored in the dark for 15 min, then sprayed with spray solution II and heated to 60 °C for 5 min.

Sulfonamides: The chromatograms are freed from mobile phase and immersed in dipping solution I for 1 s or sprayed evenly with it [10, 11].
After ca. 15 min green-yellow fluorescent chromatogram zones appear on a dark background in long-wavelength UV light ($\lambda = 365$ nm).

Note: The pre- and post-treatment of the chromatograms with the basic triethylamine solution, which can be replaced by an alcoholic solution of sodium hydroxide [1, 4] or a phosphate buffer solution pH = 8.0 ($c = 0.2$ mol/l) [5], serves to stabilize the fluorescence of the amino derivatives [2]. A final spraying with methanolic hydrochloric acid ($c_{HCl} = 5$ mol/l) or 70% perchloric acid renders the detection reaction highly specific for histamine [4] and for catecholamines and indolamines [5].

The reagent can also be employed for prechromatographic derivatization by overspotting [6] or dipping [5].

The detection limit for amines and sulfonamides lies in the low nanogram range.

The layers with which the reagent can be employed include silica gel, kieselguhr, Si 50000 and cellulose.

Procedure Tested

Primary Amines [15]

Method	Ascending, one-dimensional development in a trough chamber with chamber saturation.
Layer	HPTLC plates Cellulose (MERCK).
Mobile phase	1-Butanol – glacial acetic acid – water (15 + 5 + 5).
Migration distance	5 cm
Running time	25 min

Detection and result: The chromatogram was freed from mobile phase, immersed in dipping solution I for 1 s, dried briefly in a stream of cold air and then immersed for 1 s in dipping solution II. The layer was then dipped in a mixture of liquid paraffin – *n*-hexane (1 + 5) in order to enhance the sensitivity of the fluorescence by 25–40% and to stabilize it.

Glycine (hR_f 20–25), alanine (hR_f 35–40), valine (hR_f 55–60) and leucine (hR_f 65–70) appeared as blue-green fluorescent chromatogram zones in long-wavelength UV light ($\lambda = 365$ nm); these zones could be quantitatively determined after 15 min (Fig. 1).

In situ quantitation: The fluorimetric determination was performed at λ_{exc} = 365 nm and $\lambda_{fl} > 460$ nm. The detection limits were 5–20 ng substance per chromatogram zone.

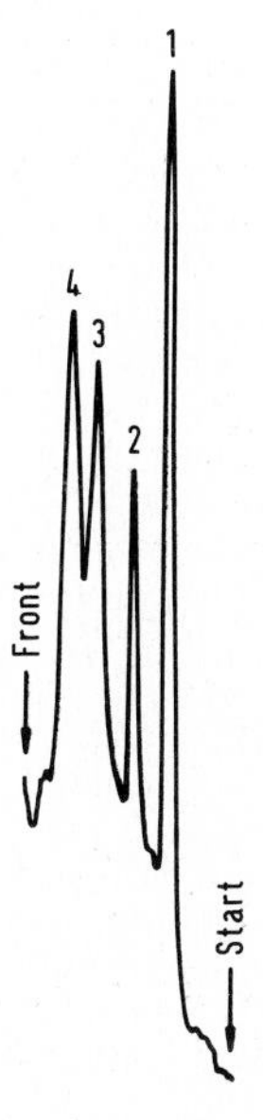

Fig. 1: Fluorescence scan of a chromatogram track with 200 ng of each substance per chromatogram zone. Glycine (1), alanine (2), valine (3), leucine (4).

References

[1] Sarhan, S., Seiler, N., Grove, J. Bink, G.: *J. Chromatogr.* **1979,** *162,* 561–572.
[2] Felix, A. M., Jimenez, M. H.: *J. Chromatogr.* **1974,** *89,* 361–364.
[3] Furlan, M., Beck, E. A.: *J. Chromatogr.* **1974,** *101,* 244–246.
[4] Lieber, E. R., Taylor, S. L.: *J. Chromatogr.* **1978,** *160,* 227–237.
[5] Nakamura, H., Pisano, J. J.: *J. Chromatogr.* **1976,** *121,* 79–81; **1978,** *152,* 153–165; **1978,** *154,* 51–59.
[6] Wintersteiger, R., Gübitz, G., Hartinger, A.: *Chromatographia* **1980,** *13,* 291–294.
[7] Ha, Y.-D., Bergner, K. G.: *Dtsch. Lebensm. Rundsch.* **1980,** *76,* 390–394: **1981,** *77,* 102–106.

[8] Funk, W., Canstein, M. von, Couturier, T., Heiligenthal, M., Kiefer U., Schlierbach, S., Sommer, D.: Proceedings of the 3rd International Symposium on Instrumental HPTLC, Würzburg. Bad Dürkheim: IfC-Verlag 1985, p. 281–311.
[9] Nakamura, H., Tsuzuki, S., Tamura, Z. Yoda, R., Yamamoto, Y.: *J. Chromatogr.* **1980,** *200,* 324–329.
[10] Thomas, M. H., Epstein, R. L., Ashworth, R. B., Marks, H.: *J. Assoc. Off. Anal. Chem.* **1983,** *66,* 884–892.
[11] Knupp, G., Pollmann, H., Jonas, D.: *Chromatographia* **1986,** *22,* 21–24.
[12] Sherma, J., Duncan, M.: *J. Liq. Chromatogr.* **1986,** *9,* 1861–1868.
[13] Schlatterer, B.: *Z. Lebensm. Unters. Forsch.* **1982,** *175* 392–398; **1983,** *176,* 20–26.
[14] Thomas, M. H., Soroka, K. E., Simpson, R. M., Epstein, R. L.: *J. Agric. Food Chem.* **1981,** *29,* 621–624.
[15] Kany, E., Jork, H.: GDCh-training course Nr. 300: Einführung in die DC, Universität des Saarlandes, 66 Saarbrücken 1987.

Formaldehyde – Sulfuric Acid Reagent
(Marquis' Reagent)

Reagent for:

- Aromatic hydrocarbons
  and heterocyclics [1]
- *β*-Blockers
  e.g. pindolol, alprenolol, propranolol,
  oxprenolol, nadolol etc. [2, 3]
- Alkaloids
  e.g. morphine, codeine, heroin
  6-monoacetylmorphine [4]
- Amphetamines
  e.g. 2,5-dimethoxy-4-bromoamphetamine [5]
- Methyl esters of fatty acids [6]
- Phenothiazines [7]
- Tannins [8]
- Guaifenesin [9]

H–C(=O)–H

CH_2O	H_2SO_4
$M_r = 30.03$	$M_r = 98.08$
Formaldehyde	Sulfuric acid

Preparation of Reagent

Dipping solution Add 10 ml sulfuric acid (95 – 97%) carefully to 90 ml methanol. To this add 2 ml formaldehyde solution (37%).

Spray solution Add 0.2 – 1 ml formaldehyde solution (37%) carefully to 10 ml conc. sulfuric acid [1 – 3].

Storage The dipping solution may be stored for ca. 4 weeks in the refrigerator.

Substances Formaldehyde solution (≥ 37%)
Sulfuric acid (95–97%)
Methanol

Reaction

Morphine (1) reacts with formaldehyde in acidic solution to yield a cyclic ketoalcohol (2) which is transformed into the colored oxonium (3) or carbenium ion (4) in acidic conditions [10].

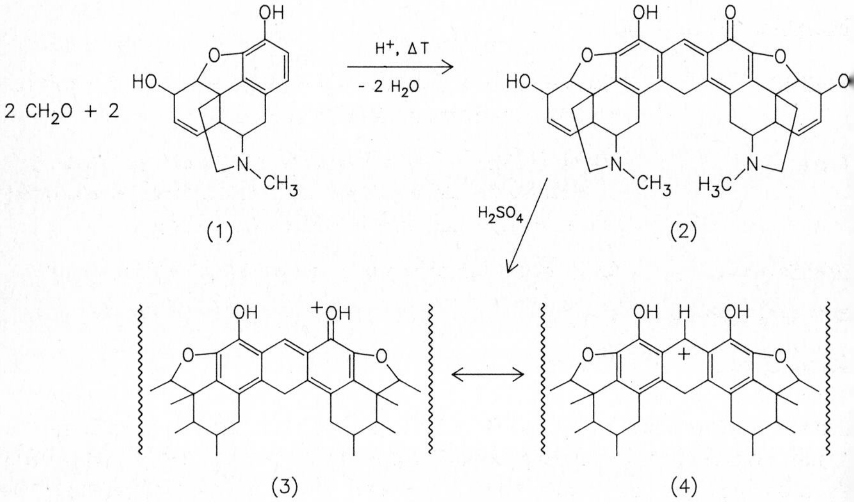

Method

The chromatograms are freed from mobile phase (5 min in a stream of warm air), immersed in the dipping solution for 4 s or sprayed evenly with it and then heated to 110 °C for 20 min (the methyl esters of fatty acids are heated to 140 °C for 10 min [6]).

Variously colored chromatogram zones are formed on a pale pink background, some of them before heating. These zones frequently fluoresce in long-wavelength UV light ($\lambda = 365$ nm).

Note: The dipping reagent is to be preferred because of the strongly irritating effects of formaldehyde on the respiratory tract. Detection limits of ca. 10—40 ng have been reported for alkaloids [4] and 50 ng—1 µg for β-blockers [2, 3].

The reagent can be employed on silica gel, kieselguhr, Si 50 000 and aluminium oxide layers.

Procedure Tested

Morphine Alkaloids [4]

Method Ascending, one-dimensional development in a trough chamber without chamber saturation.

Layer HPTLC plates Silica gel 60 (MERCK) which had been prewashed by developing once with chloroform — methanol (50 + 50) and then dried at 110 °C for 20 min.

Mobile phase Methanol — chloroform — water (12 + 8 + 2).

Migration distance 6 cm

Running time 20 min

Detection and result: The chromatogram was dried in a stream of warm air for 5 min, immersed in the dipping solution for 6 s and heated to 110 °C for 20 min. After drying in a stream of cold air morphine (hR_f 25—30), 6-monoacetylmorphine (hR_f 40—45) and heroin (hR_f 50—55) yielded reddish chromatogram zones and codeine (hR_f 30—35) yielded blue chromatogram zones on a pale pink background.

If a quantitative fluorimetric analysis was to follow the chromatogram was then exposed to ammonia vapor for 20 min (twin-trough chamber with 25% ammonia in the vacant trough) and then immediately immersed for 2 s in a 20% solution of dioctyl sulfosuccinate in chloroform. After drying in a stream of cold air the morphine alkaloids investigated appeared under long-wavelength UV light (λ = 365 nm) as pink to red fluorescent zones on a blue fluorescent background.

In situ quantitation: The absorption-photometric determination in a reflectance mode was performed at $\lambda = 330$ nm (detection limit ca. 40 ng per chromatogram zone). The fluorimetric analysis was carried out at $\lambda_{exc} = 313$ nm and $\lambda_{fl} > 560$ nm (detection limits: ca. 10 ng per chromatogram zone) (Fig. 1).

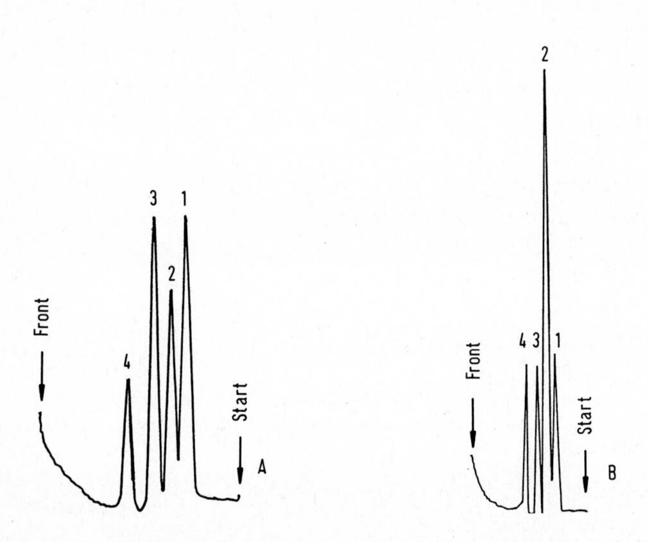

Fig. 1: Reflectance scan (A) and fluorescence scan (B) of a mixture of alkaloids with 725 ng (A) and 100 ng (B) substance per chromatogram zone. Morphine (1), codeine (2), 6-monoacetylmorphine (3), heroin (4).

References

[1] Kucharczyk, N., Fohl, J., Vymetal, J.: *J. Chromatogr.* **1963,** *11*, 55–61.
[2] Jack, D. B., Dean, S., Kendall, M. J.: *J. Chromatogr.* **1980,** *187*, 277–280.
[3] Jack, D. B., Dean, S., Kendall, M. J., Laugher, S.: *J. Chromatogr.* **1980,** *196*, 189–192.
[4] Patzsch, K., Funk, W., Schütz, H.: *GIT Fachz. Lab., Supplement „Chromatographie"*, **1988,** *32*, 83–91.
[5] Gielsdorf, W., Klug, E.: *Dtsch. Apoth. Ztg.* **1981,** *121*, 1003–1005.
[6] Acher, A. J., Kanfer, J. N.: *J. Lipid Res.* **1972,** *13*, 139–142.
[7] Kraus, L., Richter, R.: *Dtsch. Apoth. Ztg.* **1980,** *120*, 2349–2350.
[8] Botha, J. J., Viviers, P. M., Ferreira, D., Roux D. G.: *Phytochemistry* **1982,** *21*, 1289–1294.
[9] Kauert, G., von Meyer, L., Drasch, G.: *Dtsch. Apoth. Ztg.* **1979,** *119*, 986–988.
[10] Pindur, U.: *Pharm. Unserer Zeit* **1982,** *11*, 74–82.

Hydrochloric Acid Vapor Reagent

Reagent for:

- Antiepileptics
  e.g. carbamazepine [1, 2]
  primidone, phenytoin, phenylethylmalonamide, phenobarbital [3]
- Chalcones [4]
- pH indicators
  e.g. 4-aminoazobenzene derivatives [5]
- Digitalis glycosides [6 – 10]
  e.g. acetyldigoxin [6], digoxin [7, 9]
  digitoxin [8]
- Carbohydrates, diazepam, testosterone [11]
- Alkaloids
  e.g. papaverrubines [12]
- Anabolics
  e.g. trenbolone [13 – 15]
- Chloroplast pigments [16]
- Bitter principles [17]
- Penicillic acid [18]

HCl
$M_r = 36.46$

Preparation of Reagent

Reagent solution Place 10 ml conc. sulfuric acid in a twin-trough chamber and add 2 ml conc. hydrochloric acid dropwise and with care.

Storage The reagent should be made up fresh daily.

Substances Hydrochloric acid (32%)
Sulfuric acid (95–97%)

Reaction

The reaction mechanism has not yet been elucidated.

Method

Free the chromatogram from mobile phase (first in a stream of cold air for a few minutes, then at 100°C for 5 min), place in the free trough of the prepared twin-trough chamber for 5 min and then evaluate immediately in the case of chalcones or 4-aminoazobenzene derivatives. Digitalis glycosides and carbohydrates as well as diazepam and testosterone are first viewed after reheating (to 160–165°C for 15 min) [6, 7, 11] or after irradiation with unfiltered UV light from a mercury lamp [8]. In the case of antiepileptics the chromatogram is irradiated for 15 min with unfiltered UV light from a mercury lamp immediately after exposure to the HCl vapors and then inspected [1, 2].

Chalcones yield orange-red to brown-colored zones [4] as do 4-aminoazobenzene derivatives, but their colors begin to change after 10 min and slowly fade [5]. Penicillic acid is visible as a greyish-black zone [18].

Antiepileptics [3], carbohydrates, diazepam and testosterone [11] as well as digitalis glycosides are visible as light blue fluorescent zones on a dark background in long-wavelength UV light ($\lambda = 365$ nm); the blue fluorescence of the digitalis glycosides turns yellow after irradiation with UV light [7, 8]. Trenbolone fluoresces green; its detection limit in meat is 5 ppb [14].

Note: The reagent can be employed on silica gel, kieselguhr, Si 50 000 and RP layers [6]. The fluorescence intensities of the chromatogram zones can be increased by dipping in a solution of liquid paraffin in hexane or chloroform [8, 11].

Procedure Tested

Digitalis Glycosides [9]

Method Ascending, one-dimensional double development (stepwise technique) in a trough chamber with chamber saturation.

Layer HPTLC plates Silica gel 60 with concentrating zone (MERCK).

Application technique The samples were applied as bands (ca. 5 mm long) at right angles to the lower edge of the plate (that is parallel to the future direction of development), e.g. with the Linomat III (CAMAG).

Mobile phase 1. Chloroform;
2. Acetone – chloroform (65 + 35).

Migration distance Mobile phase 1: 7 cm
Mobile phase 2: 4 cm

Detection and result: The chromatogram was freed from mobile phase in a stream of warm air and exposed to hydrochloric acid gas (LINDE) in a Teflon autoclave (diameter 20 cm, height 8 cm). For this purpose the HPTLC plate was laid on a

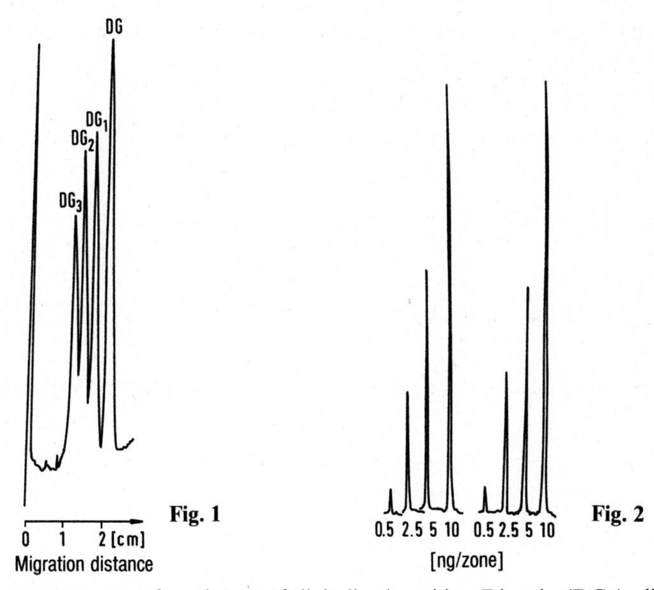

Fig. 1: Fluorescence scan of a mixture of digitalis glycosides. Digoxin (DG_3), digoxigenin bis-digitoxoside (DG_2), digoxigenin mono-digitoxoside (DG_1), digoxigenin (DG).

Fig. 2: Fluorimetric analysis of various quantities of digoxin for evaluation of the detection limit.

Teflon support with its glass side facing the gas inlet; the autoclave was evacuated (10 min, < 1 mbar), carefully filled with HCl gas (1 bar) and heated to 110 °C for 15 min. After opening the autoclave the HCl vapors were blown out with a stream of cold air (5 min) and the HPTLC plate was cooled to room temperature. In order to intensify and stabilize the fluorescence the plate was dipped in a solution of liquid paraffin in chloroform (30 + 70) for 15 s.

The digitalis glycosides yielded blue fluorescent zones in long- wavelength UV light ($\lambda = 365$ nm) (Fig. 1).

In situ quantitation: The fluorimetric analysis was carried out in UV light (λ_{exc} = 365 nm, $\lambda_{fl} > 430$ nm). The detection limit for digoxin was 300 pg per zone (Fig. 2).

References

[1] Hundt, H. K. L., Clark, E. C.: *J. Chromatogr.* **1975,** *107,* 149-154.
[2] Funk, W., Canstein, M. v., Couturier, T., Heiligenthal, M. Kiefer, U., Schlierbach, S., Sommer, D.: Proceedings of the 3rd International Symposium on Instrumental HPTLC, Würzburg. Bad Dürkheim: IfC-Verlag, 1985, p. 281-311.
[3] Canstein, M. v.: Thesis, Fachhochschule Gießen, Fachbereich Technisches Gesundheitswesen, 1984.
[4] Stanley, W. L.: *J. Ass. Off. Agric. Chem.* **1961,** *44,* 546-548.
[5] Topham, J. C. Westrop, J. W.: *J. Chromatogr.* **1964,** *16,* 233-234.
[6] Winsauer, K., Buchberger, W.: *Chromatographia* **1981,** *14,* 623-625.
[7] Bottler, R.: *Kontakte (Merck)* **1978,** *2,* 36-39.
[8] Faber, D. B., Kok, A. de, Brinkmann, U. A. Th.: *J. Chromatogr.* **1977,** *143,* 95-103.
[9] Reh, E., Jork, H.: *Fresenius Z. Anal. Chem.* **1984,** *318,* 264-266.
[10] Lugt, C. B.: *Pharm. Weekbl. Ned.* **1976,** *111,* 405-417.
[11] Zhou, L., Shanfield, H., Wang, F. S., Zlatkis, A.: *J. Chromatogr.* **1981,** *217,* 341-348.
[12] Slavik, J.: *Collect. Czech. Chem. Commun.* **1980,** *45,* 2706-2709.
[13] Schopper, D., Hoffmann, B.: *Arch. Lebensm. Hyg.* **1981,** *32,* 141-144.
[14] Hohls, F. W., Stan, H. J.: *Z. Lebensm. Unters. Forsch.* **1978,** *167,* 252-255.
[15] Oehrle, K.-L., Vogt, K., Hoffmann, B.: *J. Chromatogr.* **1975,** *114,* 244-246.
[16] Petrovic, S. M., Kolarov, L. A.: *J. Chromatogr.* **1979,** *171,* 522-526.
[17] Schneider, G., Mielke, B.: *Dtsch. Apoth. Ztg.* **1978,** *118,* 469-472.
[18] Vesely, D., Vesela, D.: *Chem. Listy* (ČSSR) **1980,** *74,* 289-290.

Hydrogen Peroxide Reagent

Reagent for:

- Aromatic acids [1]
- Thiabendazole [2]

H_2O_2

$M_r = 34.01$

Preparation of Reagent

Dipping solution *Acids:* Make up 1 ml 30% hydrogen peroxide (Perhydrol®) to 100 ml with water [1].

Spray solution *Thiabendazole:* Make up 3 ml 30% hydrogen peroxide to 100 ml with 10% acetic acid [2].

Storage The reagent solutions should always be freshly made up.

Substances Perhydrol® 30% H_2O_2
Acetic acid

Reaction

Many carboxylic acids are converted into fluorescent derivatives by oxidation and UV irradiation. The reaction mechanism has not been elucidated.

Method

The chromatograms are freed from mobile phase, immersed for 3 s in the reagent solution or homogeneously sprayed with it and then subjected to intense UV radiation ($\lambda = 365$ nm) for 30 s to 3 min while still moist [3]. Generally, after a few minutes, aromatic acids produce blue fluorescent chromatogram zones under long-wavelength UV light ($\lambda = 365$ nm) [3] which are not detectable in visible light. Within a few minutes thiabendazole yields bright fluorescent zones on a dark background under long-wavelength UV light ($\lambda = 356$ nm) [2].

Note: The dipping solution can also be sprayed on. The detection of the aromatic acids is best performed on cellulose layers, if ammonia-containing mobile phases have been employed. The reagent can also be employed on silica gel, aluminium oxide, RP 18 and polyamide layers.

Procedure Tested

Organic Acids [1]

Method	Ascending, one-dimensional development in a trough chamber with chamber saturation.
Layer	SIL G-25 UV_{254} plates (MACHEREY-NAGEL).
Mobile phase	Diisopropyl ether – formic acid – water (90 + 7 + 3).
Migration distance	10 cm
Running time	30 min

Detection and result: The chromatogram was dried in a stream of warm air for 10 min, immersed in the reagent solution for 3 s and then subjected to intense UV radiation (high pressure lamp, $\lambda = 365$ nm) for up to 10 min. Terephthalic (hR_f 0–5), pimelic (hR_f 55), suberic (hR_f 60), sebacic (hR_f 65–70) and benzoic acids (hR_f 70–75) together with sorbic, malic, adipic, citric, tartaric, lactic and fumaric acids only exhibited a reaction on silica gel layers at higher concentrations. 4-Hydroxybenzoic, salicylic and acetylsalicylic acids fluoresced light blue after irradiation. The detection limit per chromatogram zone was 0.5 µg for salicylic acid and more than 5 µg for benzoic acid.

In situ quantitation: The reagent had no advantages for the direct determination of the acids investigated; the determination of the intrinsic absorption or intrinsic fluorescences was to be preferred.

References

[1] Jork, H., Klein, I.: GDCh-training course Nr. 300 „Einführung in die Dünnschicht-Chromatographie“, Universität des Saarlandes, Saarbrücken 1987.
[2] Laub, E., Geisen, M.: *Lebensmittelchem. gerichtl. Chem.* **1976,** *30,* 129–132.
[3] Grant, D. W.: *J. Chromatogr.* **1963,** *10,* 511–512.

8-Hydroxyquinoline Reagent

Reagent for

- Cations [1 – 10]
- 1,4-Benzodiazepines [11]

C_9H_7NO
$M_r = 145.16$

Preparation of the Reagent

Dipping solution Dissolve 0.5 g 8-hydroxyquinoline in 100 ml ethyl acetate.

Storage The solution may be kept for several days.

Substances 8-Hydroxyquinoline
Ethyl acetate
Ethanol
Ammonia solution (25%)

Reaction

8-Hydroxyquinoline forms colored and fluorescent complexes with numerous metal cations.

$$2\ C_9H_7NO + M^{2+} \xrightarrow{-\ 2\ H^+} M(C_9H_6NO)_2$$

Method

The chromatograms are dried in a stream of warm air, immersed in the reagent solution for 5 s, dried in the air and then placed for 5 min in a twin-trough chamber whose second trough contains 5 ml 25% ammonia solution. Directly after it is removed the plate is viewed under UV light ($\lambda = 254$ nm or 365 nm). The zones are mainly yellow in color; the colors of their fluorescence are listed in Table 1.

The dipping solution can also be employed as spray reagent, in this event 60% [6], 80% [5] or 95% ethanol [9] have been recommended in the literature. The reagent can be employed on silica gel, Dowex 50X4 (Na^+), cellulose and starch layers. The detection limit for 1,4-benzodiazapines is 100 ng per chromatogram zone [11].

Table 1: Colors of the 8-hydroxyquinoline complexes

Ion	Color		Detection limit [ng]
	Daylight	Fluorescence ($\lambda_{exc} = 365$ nm)	
Be^{2+}	green-yellow	yellow	200
Mg^{2+}	green-yellow	yellow	100
Ca^{2+}	yellow	yellow	500
Sr^{2+}	yellow	green-blue	200
Ba^{2+}	yellow	blue	500
$Sn^{2+/4+}$	yellow	*	*
$Cr^{3+/6+}$	black	*	*
$Fe^{2+/3+}$	black	purple	*
Al^{3+}	yellow	yellow	*
Ni^{3+}	yellow	red	*
Co^{2+}	yellow	red	*
Cu^{2+}	yellow	red	*
Bi^{2+}	yellow	red	*
Zn^{2+}	yellow	yellow	*
Cd^{2+}	yellow	yellow	*
Hg^{2+}	yellow	red	*

* = not reported

Procedure Tested

Alkaline Earths [4, 12]

Method Ascending, one-dimensional development in a HPTLC chamber with chamber saturation.

Layer HPTLC plates Cellulose (MERCK).

Mobile phase Methanol – hydrochloric acid (25%) (80 + 20).

Migration distance 5 cm

Running time 20 min

Detection and result: The chromatogram was dried in a stream of warm air for 15 min, immersed in the reagent solution for 5 s, then dried in the air and finally placed in an ammonia chamber for 5 min. The plate was inspected under long-wavelength UV light ($\lambda = 365$ nm) immediately after removal from the chamber. The separation (Fig. 1) corresponded to the order of the group in the periodic table: Be^{2+} (hR_f 95–98), Mg^{2+} (hR_f 75–80), Ca^{2+} (hR_f 50–55), Sr^{2+} (hR_f 30–35, slight tailing) Ba^{2+} (hR_f 15–20).

The proportion of hydrochloric acid in the mobile phase was not to exceed 20%, so that complex formation did not occur and zone structure was not adversely affected. An excess of accompanying alkaline earth metal ions did not interfere with the separation but alkali metal cations did. The lithium cation fluoresced blue and lay at the same height as the magnesium cation, ammonium ions interfered with the calcium zone.

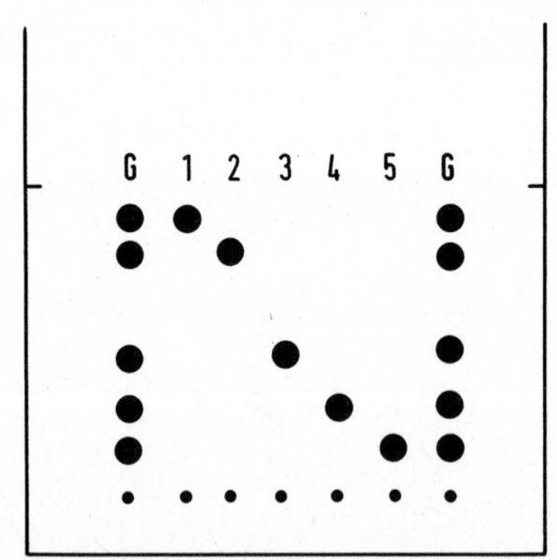

Fig. 1: Chromatographic separation of the alkaline earth cations. Mixture (G), Be^{2+} (1), Mg^{2+} (2), Ca^{2+} (3), Sr^{2+} (4), Ba^{2+} (5).

References

[1] Reeves, W. A., Crumler, Th. B.: *Anal. Chem.* **1952,** *23,* 1576-1578.
[2] Arden, T. V., Burstall, F. H., Davies, G. R., Lewis, J. A., Linstead, R. P.: *Nature* (London) **1948,** *162,* 691-692.
[3] Perisic-Janjic, N., Canic, V., Radosavljevic, S.: *Chromatographia,* **1983,** *17,* 454-455.
[4] Gagliardi, E., Likussar, W.: *Mikrochim. Acta* (Vienna) **1965,** 765-769.
[5] Vries, G. de, Schütz, G. P., Dalen, E. van: *J. Chromatogr.* **1964,** *13,* 119-127.
[6] Ryabchikov, D. J., Volynets, M. P., Kopneva, L. A.: *J. Anal. Chem.* (USSR) **1969,** *24,* 65-67.
[7] Rai, J., Kukreja, V. P.: VI. Intern. chromatogr. Symp., Brussels **1970,** 453-461, *Chromatographia* **1970,** *3,* 499-500.
[8] Oguma, K., Kuroda, R.: *J. Chromatogr.* **1971,** *61,* 307-316.
[9] Lepri, L., Desideri, P. G., Mascherini, R: *J. Chromatogr.* **1972,** *70,* 212-215.
[10] Lederer, M., Rinalduzzi, B.: *J. Chromatogr.* **1972,** *68,,* 237-244.
[11] Kosa, Z.: *Gyogyszereszet* **1977,** *31,* 289-290.
[12] Kany, E., Jork, H.: GDCh-training course Nr. 301 „Dünnschicht-Chromatographie für Fortgeschrittene", Universität des Saarlandes, Saarbrücken 1985.

Iron(III) Chloride – Perchloric Acid Reagent (FCPA Reagent)

Reagent for:

- Indole alkaloids [1 – 10]
  e.g. from *rauwolfia* [1]
  tabernaemontana [2, 7, 9]
  mitragyna [3, 4]
  strychnos [5]
  synclisia [6]
  cinchona [8]
- Indoles

$FeCl_3 \cdot 6H_2O$
$M_r = 270.30$
Iron(III) chloride hexahydrate

$HClO_4$
$M_r = 100.46$
Perchloric acid

Preparation of the Reagent

Solution I Mix 2 ml perchloric acid (70%) carefully into 100 ml ethanol.

Solution II Dissolve 1.35 g iron(III) chloride hexahydrate in 100 ml ethanol.

Dipping solution Mix 100 ml solution I with 2 ml solution II [10].

Spray solution Dissolve 5 g iron(III) chloride hexahydrate in a mixture of 50 ml water and 50 ml perchloric acid (70%).

Storage The dipping solution may be stored for at least a week.

Substances Iron(III) chloride hexahydrate
Perchloric acid (70%)
Ethanol

Reaction

The reagent mechanism has not yet been elucidated.

Method

The chromatogram is freed from mobile phase in a stream of warm air, immersed for 4 s in the dipping solution or sprayed evenly with the spray solution until just transparent, dried briefly in the air and then heated to 110–120 °C for 20–60 min.

Variously colored chromatogram zones are produced on a colorless background. Some of the zones appear before heating; their hues, which are to a great extent structure-specific, change during the heating process and they not infrequently fluoresce under short or long-wavelength UV light ($\lambda = 254$ or 365 nm).

Note: Indoles, that are substituted with oxygen in position 2 or 3, do not react [11]. The reagent can be employed on silica gel, kieselguhr and Si 50 000 layers. Aluminium oxide layers are not suitable [3].

Danger warning: Mists of perchloric acid can condense in the exhausts of fume cupboards and lead to uncontrolled explosions! So dipping is to be preferred.

Procedure Tested

Strychnine, Brucine [10]

Method Ascending, one-dimensional development in a trough chamber without chamber saturation.

Layer HPTLC plates Silica gel 60 (MERCK). The layers were prewashed once with chloroform – methanol (1 + 1) and dried at 110 °C for 10 min before application of samples.

Mobile phase Acetone – toluene – 25% ammonia solution (40 + 15 + 5).

Migration distance 5 cm

Running time 10 min

Detection and result: The chromatogram was freed from mobile phase in a stream of warm air (45 min), immersed in the dipping solution for 4 s, dried briefly in the air and heated to 110 °C for 20 min.

Strychnine (hR_f 50–55) appeared as a red and brucine (hR_f 35) as a yellow chromatogram zone on a colorless background. The detection limit for both substances was 10 ng per chromatogram zone.

In situ quantitation: The light absorption in reflectance was measured at a wavelength $\lambda = 450$ nm (Fig. 1).

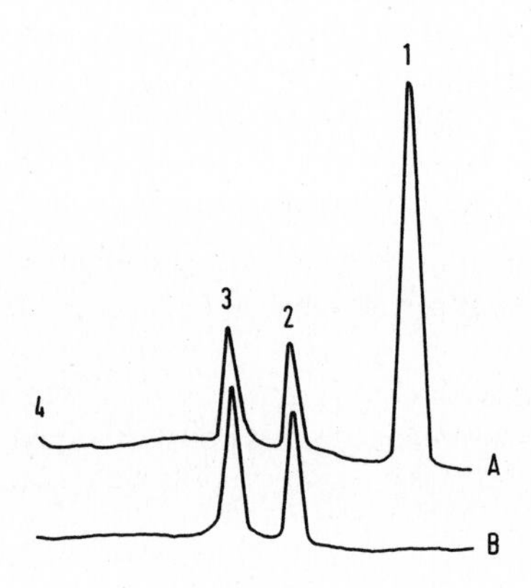

Fig. 1: Reflectance scan of a *nux vomica* extract (A) and a reference chromatogram containing 1 µg each of strychnine and brucine (B). Start (1), brucine (2), strychnine (3), front (4).

References

[1] Court, W. E., IWU, M. M.: *J. Chromatogr.* **1980,** *187,* 199–207.
[2] Beek, T. A. van, Verpoorte, R., Baerheim Svendsen, A.: *J. Chromatogr.* **1984,** *298,* 289–207; *J. Nat. Prod.* **1985,** *48,* 400–423.
[3] Shellard, E. J., Alam, M. Z.: *J. Chromatogr.* **1968,** *33,* 347–369.
[4] Shellard, E. J., Lala, P. K.: *Planta Med.* **1978,** *33,* 63–69.
[5] Verpoorte, R., Joosse, F. T., Groenink, H., Baerheim Svendsen, A.: *Planta Med.* **1981,** *42,* 32–36.
[6] Ohiri, F. C., Verpoorte, R., Baerheim Svendsen, A.: *Planta Med.* **1983,** *47,* 87–89.

[7] Beek, T. A. van, Kuijlaars, F. L. C., Thomassen, P. H. A. M., Braga, R. M., Leitão Filko, H. F., Reis, F. De A. M.: *Phytochemistry* **1984,** *23,* 1771-1778.
[8] Mulder-Krieger, T., Verpoorte, R., De Water, A., Gressel, M. van, Oeveren, B. C. J. A., Baerheim Svendsen, A.: *Planta Med.* **1982,** *46,* 19-24.
[9] Beek, T. A. van, Verpoorte, R., Baerheim Svendsen, A.: *Planta Med.* **1983,** *47,* 83-86.
[10] Müller, J.: Thesis, Fachhochschule Gießen, Fachbereich Technisches Gesundheitswesen, 1987.
[11] Clotten, R., Clotten, A.: *Hochspannungs-Elektrophorese.* Stuttgart, G. Thieme 1962.

Isonicotinic Acid Hydrazide Reagent
(INH Reagent)

Reagent for:

- Δ^4-3-Ketosteroids [1 – 6]
  e.g. corticosteroids, androgens, gestagens

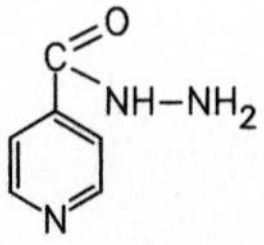

$C_6H_7N_3O$
$M_r = 137.14$

Preparation of the Reagent

Dipping solution Dissolve 1 g isonicotinic acid hydrazide (4-pyridinecarboxylic acid hydrazide, isoniazide) in 100 ml ethanol and add 500 µl trifluoroacetic acid [1] or 1 ml glacial acetic acid.

Spray solution Dissolve 0.8 g isonicotinic acid hydrazide in 200 ml methanol and add 1 ml hydrochloric acid (25%) [6] or glacial acetic acid [2].

Storage Both solutions may be kept in the refrigerator for one week at 4°C.

Substances 4-Pyridinecarboxylic acid hydrazide
Hydrochloric acid (25%)
Trifluoroacetic acid
Ethanol
Methanol

Reaction

Isonicotinic acid hydrazide forms fluorescent hydrazones with ketosteroids.

Testosterone + Isonicotinic acid hydrazide $\xrightarrow{-H_2O}$ Testosterone-isonicotinic acid hydrazone

Method

The chromatograms are dried in a stream of cold air, immersed in the dipping solution for 20 s or sprayed evenly with spraying solution, then dried in the air and left for a few minutes at room temperature.

The layer can then be dipped into liquid paraffin – *n*-hexane (1 + 2) to intensify the fluorescence [1].

Colored hydrazones are formed which fluoresce on a dark background in long-wavelength UV light ($\lambda = 365$ nm).

Note: Δ^4-, $\Delta^{4,6}$-, $\Delta^{1,4}$- and Δ^1-3-Ketosteroids react at different rates – as a function amongst other things of the acid strength of the reagent – so they can be differentiated [3, 5, 6].

The dipping solution may also be used as a spray solution. If this is done the reaction occurs more rapidly than reported in the literature for alcoholic spray solutions.

Silica gel, kieselguhr, Si 50000 and cellulose layers, amongst others, can be used as stationary phases.

Procedure Tested

Testosterone [1]

Method Ascending, one-dimensional development in a trough chamber with chamber saturation.

Layer HPTLC plates Silica gel 60 (MERCK).
The layer was prewashed by developing once in chloroform and then twice in toluene – 2-propanol (10 + 1), with drying at 110 °C for 30 min after each step.

Mobile phase Toluene – 2-propanol (10 + 1).

Migration distance 5 cm

Running time 10 min

Detection and result: The chromatograms were freed from mobile phase (stream of cold air), immersed in the reagent solution for 20 s, then dried in the air and finally kept at room temperature for 20 min. Testosterone (hR_f 35 – 40) fluoresced pale blue in long-wavelength UV light ($\lambda = 365$ nm, Fig. 1).

In situ quantitation: The chromatogram was then dipped into liquid paraffin – *n*-hexane (10 + 20) to increase the intensity of the fluorescence by a factor of ten and to stabilize it. The detection limit for testosterone is less than 500 pg per chromatogram zone ($\lambda_{exc} = 365$ nm; $\lambda_{fl} > 430$ nm).

The separation of a mixture of Δ^4-3-ketosteroids is illustrated in Figure 2.

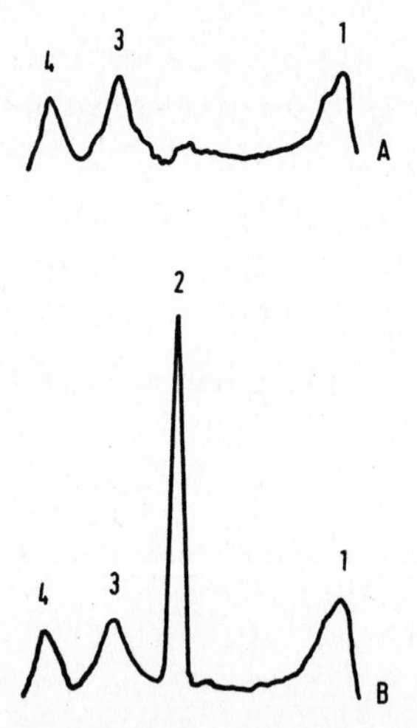

Fig. 1: Fluorescence scan of a blank track (A) and of the testosterone derivative (B, 2 µg). Start (1), testosterone-INH derivative (hR_f 39) (2), unknown substances (3, 4).

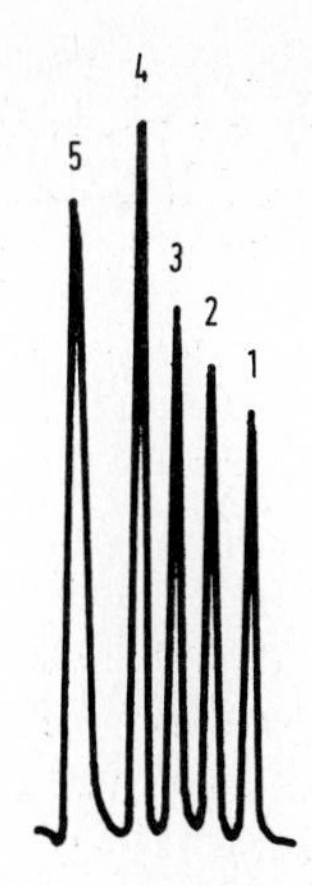

Fig. 2: Fluorescence scan of a Δ^4-3-ketosteroid mixture after INH treatment. Aldosterone (1), corticosterone (2), cortexolone (3), testosterone (4), progesterone (5).

References

[1] Funk, W.: *Fresenius Z. Anal. Chem.* **1984,** *318,* 206–219.

[2] Lisboa, B. P.: *Acta Endocrinol.* **1963,** *43,* 47–66

[3] Lisboa, B. P.: *J. Chromatogr.* **1964,** *16,* 136–151: **1965,** *19,* 81–104.

[4] Vaedtke, J., Gajewska, A.: *J. Chromatogr.* **1962,** *9,* 345–347.

[5] Weichselbaum, T. E., Margraf, H. W.: *J. Clin. Endocrinol. Metab.* (Copenhagen) **1957,** *17,* 959–965.

[6] Smith, L. L., Foell, T.: *Anal. Chem.* **1959,** *31,* 102–105.

Lead(II) Acetate Basic Reagent

Reagent for:

- Flavonoids [1–3]
- Aryl-substituted thiourea derivatives (thiocarbamide derivatives) [4]
- Oligogalactoside uronic acids [5]

$Pb(CH_3COO)_2 \cdot n\,Pb(OH)_2$

Preparation of Reagent

Dipping solution Basic lead(II) acetate solution (lead content: 17.5–19%).

Storage The reagent solution may be stored over a longer period of time, providing it is tightly stoppered.

Substances Lead acetate (17.5–19% Pb)
(lead(II) acetate solution, basic)

Reaction

Lead(II) acetate yields colored lead salts with flavonoids and thiourea derivatives.

$$>N-C(=S)-SH + Pb^{2+} \xrightarrow[-H^+]{} \left[>N-C(=S)-S\right]^- \; Pb^{2+}/2$$

Thiourea derivatives Lead salt

Method

The chromatogram is freed from mobile phase in a stream of warm air, immersed in the reagent solution for 1 s or homogeneously sprayed with it until the layer starts to be transparent and heated to 50 – 80 °C for 5 – 20 min.

Chromatogram zones of various colors sometimes appear even before heating, these fluoresce under long-wavelength UV light ($\lambda = 365$ nm) in the case of some substances.

Note: The detection limits per chromatogram zone are ca. 1 µg substance in the case of aryl-substituted thioureas [4], but even at 50 µg per zone diallate and triallate did not produce any reaction [6]. The reagent should be employed undiluted (cf. "Procedure Tested", Fig. 2).

The reagent can be employed on silica gel, kieselguhr, Si 50 000, cellulose and polyamide layers.

Procedure Tested

Flavone Glucosides [7]

Method	Ascending, one-dimensional development in a HPTLC trough chamber with chamber saturation.
Layer	HPTLC plates Silica gel 60 (MERCK).
Mobile phase	Ethyl acetate – dichloromethane – formic acid – water (35 + 15 + 5 + 3).
Migration distance	5 cm
Running time	13 min

Detection and result: The chromatogram was freed from mobile phase, immersed in the reagent solution for 1 s and then heated to 80 °C for 10 min. Rutin (hR_f 5 – 10), kaempferol glucoside (hR_f 10 – 15), hyperoside (hR_f 20 – 25), isoquercitrin (hR_f 25 – 30), quercitrin (hR_f 40 – 45), luteolin (hR_f 75 – 80), quercetin (hR_f 80 – 85) and isorhamnetin (hR_f 80 – 85) yielded yellow to brown chromatogram zones on a pale background; under long-wavelength UV light ($\lambda = 365$ nm) these produced yellow to orange fluorescence. The detection limits were 50 – 100 ng substance per chromatogram zone (Fig. 1). The influence of the reagent concentration on the sensitivity of detection is illustrated in Figure 2.

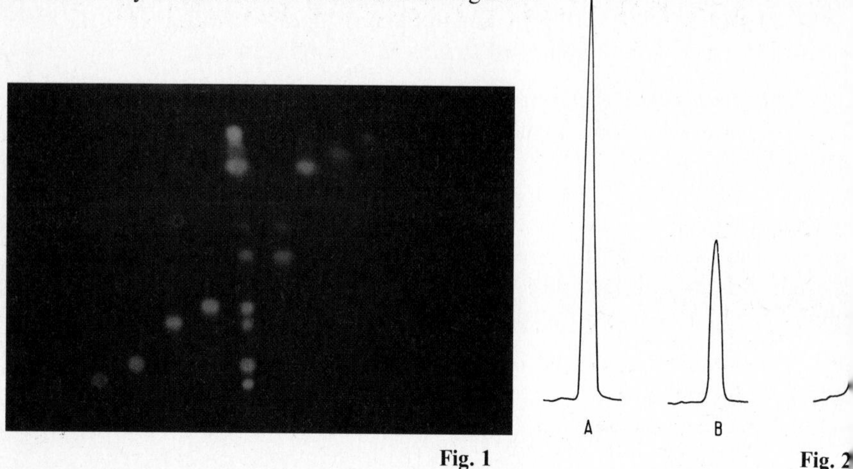

Fig. 1: Chromatographic separation of flavone glucosides

Fig. 2: The influence of the reagent concentration on the sensitivity of detection; detection of equal amounts of luteolin with basic lead acetate solution that was (A) undiluted, (B) diluted 1 + 4 and (C) diluted 1 + 50.

References

[1] Hörhammer, L., Wagner, H., Hein, K.: *J. Chromatogr.* **1964,** *13*, 235–237.
[2] Poethke, W., Schwarz, C., Gerlach, H.: *Planta Med.* **1971,** *19*, 177–188.
[3] Willuhn, G., Röttger, P.-M.: *Dtsch. Apoth. Ztg.* **1980,** *120*, 1039–1042.
[4] Upadhyaya, J. S., Upadhyaya, S. K.: *Fresenius Z. Anal. Chem.* **1979,** *294*, 407; **1980,** *304*, 144.
[5] Markovic, O., Slezarik, A.: *J. Chromatogr.* **1984,** *312*, 492–496.
[6] Jork, H.: Private communication, Universität des Saarlandes, Fachbereich 14 „Pharmazie und Biologische Chemie", Saarbrücken 1988.
[7] Kany, E., Jork, H.: GDCh-training course Nr. 301 „Dünnschicht-Chromatographie für Fortgeschrittene", Universität des Saarlandes, Saarbrücken 1988.

Lead(IV) Acetate – Dichlorofluorescein Reagent

Reagent for:

- Vicinal diols [1 – 5]
  e.g. sugar acids [1, 2]
  monosaccharides [2 – 4]
  oligosaccharides [2, 3]
  sugar alcohols [2 – 4]
  cyclitols [5]
  glycerol [2, 3]
- Glycosides
  e.g. menthyl glucoside [6]
  arbutin [7]
- Phenols [8]

$(CH_3COO)_4Pb$

$C_8H_{12}O_8Pb$
$M_r = 443.37$

Lead(IV) acetate

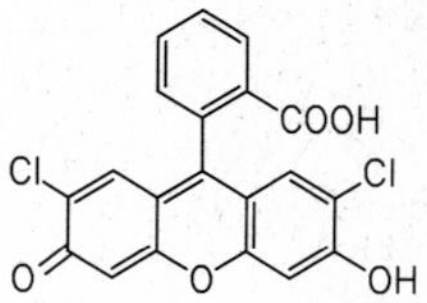

COOH
Cl
Cl
O
O
OH

$C_{20}H_{10}Cl_2O_5$
$M_r = 401.21$

Dichloro-
fluorescein

Preparation of the Reagent

Solution I Two percent to saturated solution of lead(IV) acetate in glacial acetic acid.

Solution II Dissolve 0.2 to 1 g 2′,7′-dichlorofluorescein in 100 ml ethanol.

Dipping solution Mix 5 ml solution I and 5 ml solution II and make up to 200 ml with toluene immediately before use.

Storage Solutions I and II on their own are stable for several days, the dipping solution should always be freshly made up before use.

Substances Lead(IV) acetate
2′,7′-Dichlorofluorescein
Acetic acid (glacial acetic acid) 100%
Ethanol
Toluene

Reaction

The reaction is based, on the one hand, on the oxidative cleavage of vicinal diols by lead(IV) acetate and, on the other hand, on the reaction of dichlorofluorescein with lead(IV) acetate to yield a nonfluorescent oxidation product. The dichlorofluorescein only maintains its fluorescence in the chromatogram zones where the lead(IV) acetate has been consumed by the glycol cleavage reaction [1].

Method

The chromatogram is freed from mobile phase, immersed in the reagent solution for 8 – 10 s, dried briefly in a stream of warm air and heated in the drying cupboard to 100 °C for 3 – 30 min.

Chromatogram zones are formed that exhibit yellow fluorescence in shortwave UV light ($\lambda = 254$ nm); they are sometimes also recognizable as orange-red spots in visible light.

Note: The full fluorescence intensity usually only develops about 30 min after the dipping process; it then remains stable for several days if the chromatograms are stored in the dark [1, 5]. Fluorescein sodium can be employed in the reagent in place of 2′,7′-dichlorofluorescein [5]. The detection limits lie in the lower nanogram to picogram range [1, 5].

The reagent can be employed on silica gel, Si 50 000 and kieselguhr layers.

Remark: All substances with vicinal diol groups (sugars, sugar alcohols, glycosides etc.) yield a yellow-green fluorescence with this reagent. In order to determine which zones are produced by sugars the plate can be sprayed later with naphthalene-1,3-diol – sulfuric acid reagent which colors the sugars but not the sugar alcohols [4, 9].

Procedure Tested

Mono- and Sesquiterpene Glucosides [6], Arbutin and Methylarbutin [7]

Method	Ascending, one-dimensional development in a trough chamber with chamber saturation.
Layer	HPTLC plates Silica gel 60 (MERCK).
Mobile phase	*Terpene glucosides:* chloroform – methanol (80 + 20). *Arbutin:* ethyl acetate – methanol – water (100 + 17 + 14).
Migration distance	5 cm
Running time	15 min

Detection and result: The chromatogram was freed from mobile phase, immersed in the reagent solution for 1 s, dried briefly in a stream of warm air and heated to 100 °C for 5 min.

The glucosides of menthol, citronellol, nerol, geraniol, *cis*-myrtenol, L-borneol, linalool and α-terpineol yielded yellow-green fluorescent chromatogram zones in long-wavelength UV light ($\lambda = 365$ nm). The same applied to arbutin (hR_f 45–50).

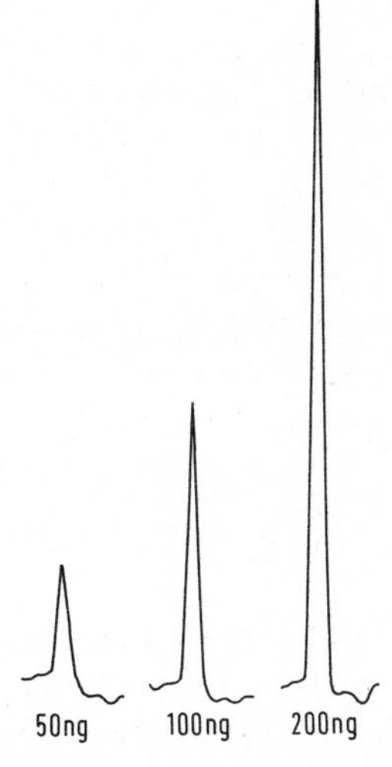

Fig. 1: Fluorescence scan of chromatogram tracks with 50, 100 and 200 ng arbutin per chromatogram zone.

In situ quantitation: The fluorimetric analysis of monoterpene glucosides could be performed with advantage at $\lambda_{exc} = 313$ nm and $\lambda_{fl} > 390$ nm. The detection limits of arbutin and L-menthylglucoside were 1 – 5 ng and 15 ng substance per chromatogram respectively (Fig. 1).

References

[1] Gübitz, G., Frei, R. W., Bethke, H.: *J. Chromatogr.* **1976,** *117,* 337-343.
[2] Klaus, R., Ripphahn, J.: *J. Chromatogr.* **1982,** *244,* 99-124.
[3] Klaus, R., Fischer, W.: *Chromatographia* **1987,** *23,* 137-140.
[4] Schmoldt, A., Machut, M.: *Dtsch. Apoth. Ztg.* **1981,** *121,* 1006-1009.
[5] Stepanek, J.: *J. Chromatogr.* **1983,** *257,* 405-410.
[6] Ishag, K. E. A.: Dissertation, Universität des Saarlandes, Saarbrücken 1984.
[7] Jork, H., Kany, E.: GDCh-training course Nr. 301 „Dünnschicht-Chromatographie für Fortgeschrittene", Universität des Saarlandes, Saarbrücken 1986.
[8] Klaus, R., Fischer, W., Bayer, H.: *J. Chromatogr.* **1987,** *398,* 300-308.
[9] Wimmer, T.: Dissertation, Universität Munich, 1985.

Lead(IV) Acetate – Fuchsin Reagent

Reagent for:

- α-Diol groupings
  e.g. sugar alcohols [1]
  sugars [4]

$(CH_3COO)_4Pb$

$C_8H_{12}O_8Pb$
$M_r = 443.37$
Lead(IV) acetate

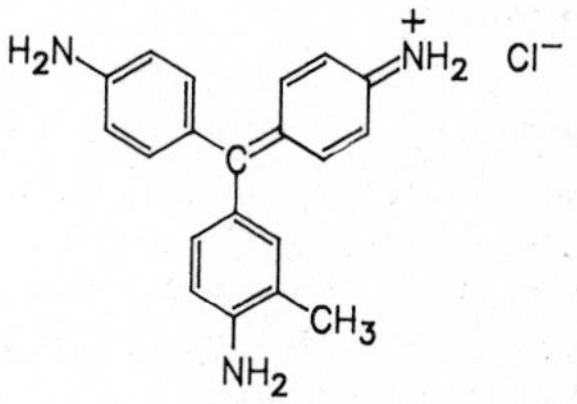

$C_{20}H_{20}ClN_3$
$M_r = 337.85$
Fuchsin

Preparation of Reagent

Dipping solution I Dissolve 1 g lead(IV) acetate (lead tetraacetate) in 100 ml ethanol.

Dipping solution II Dissolve 10 mg fuchsin in 100 ml methanol.

Spray solution I Dissolve 1 g lead(IV) acetate in 100 ml toluene.

Spray solution II Dissolve 50 mg fuchsin in 100 ml methanol.

Storage Solution I should always be freshly prepared, solution II may be stored for a longer period of time.

Substances Lead(IV) acetate
Fuchsin
Toluene
Methanol
Ethanol

Reaction

The reaction depends, on the one hand, on the fact that fuchsin is decolorized by oxidizing agents (e.g. lead(IV) acetate) and, on the other hand, on the fact that lead(IV) acetate is reduced by compounds containing α-diol groups. It is, therefore, no longer available to decolorize the fuchsin. The fuchsin undergoes a SCHIFF reaction with the aldehydes that are formed [2].

$$R{-}CH(OH){-}CH(OH){-}R \xrightarrow[-\,2\,AcOH]{Pb(OAc)_4} 2\,R{-}CHO + Pb(OAc)_2$$

Diol　　　Aldehyde　Lead(II) acetate

$$2\,R{-}CHO + \text{Fuchsin} \xrightarrow{-\,2\,H_2O} \text{SCHIFF's base}$$

Aldehyde　Fuchsin　SCHIFF's base

Method

The chromatogram is freed from mobile phase, immersed for 3 s in dipping solution I or sprayed homogeneously with spray solution I and then dried for ca. 10 min in a stream of cold air. It is then immersed for 1 s in dipping solution II or sprayed evenly with spray solution II.

Reaction usually occurs immediately or occasionally after heating briefly to 140 °C, to yield red-violet chromatogram zones on a pale yellow-beige background.

Note: The color of the zones persists for a long period, but changes to blue-violet [1]. Rosaniline [1, 2] can be employed instead of fuchsin. With sugar alcohols lead(IV) acetate alone yields white zones on a brown background (detection limit 1 – 2 µg per chromatogram zone) [3].

The reagent can be employed on silica gel, kieselguhr and Si 50 000 layers.

Procedure Tested

Mono-, Di- and Trisaccharides [4]

Method	Ascending, one-dimensional double development in a trough chamber at 80 °C without chamber saturation.
Layer	HPTLC plates Si 50000 (MERCK), which had been prewashed before sample application by developing once with chloroform – methanol and then dried at 110 °C for 30 min.
Mobile phase	Acetonitrile – water (85 + 15).
Migration distance	2 × 6 cm
Running time	2 × 8 min (5 min intermediate drying in stream of warm air)

Detection and result: The chromatogram was dried for 15 min in a stream of cold air, immersed in dipping solution I for 3 s and dried for 10 min in a stream of cold air. It was then immersed for 1 s in dipping solution II, dried briefly in a stream of warm air and heated to 140 °C for 1 min (hot plate).

Raffinose (hR_f 5 – 10), lactose (hR_f 15 – 20), sucrose (hR_f 30 – 35), glucose (hR_f 45 – 50) and fructose (hR_f 60 – 65) yielded violet chromatogram zones with a weak,

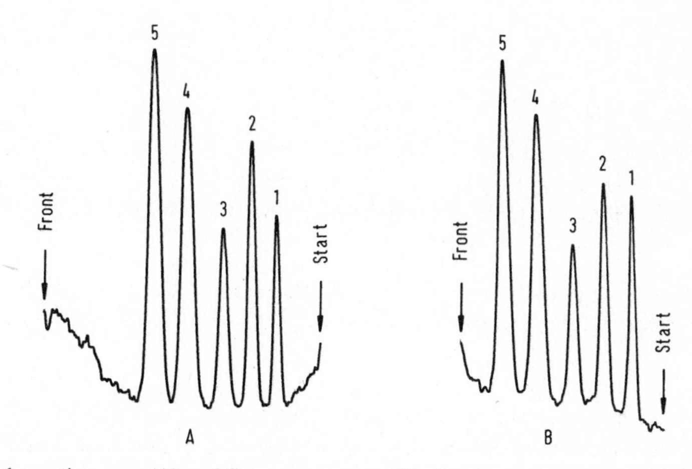

Fig. 1: Absorption scan (A) and fluorescence scan (B) of a chromatogram track with 200 ng sugar per chromatogram zone: raffinose (1), lactose (2) sucrose (3), glucose (4) and fructose (5).

pale blue fluorescence on a pale beige background. The detection limits were 20 ng substance per chromatogram zone. When the chromatogram was subsequently treated with hydrochloric acid vapor for 15 min (twin-trough chamber, 32% hydrochloric acid in the unoccupied half) the zones became more intensely violet-colored against a pale violet background; under long-wavelength UV light ($\lambda = 365$ nm) the zones fluoresced red against a dark background. Here too the detection limits were 20 ng substance per chromatogram zone.

In situ quantitation: The absorption photometric determination in reflectance was performed at $\lambda = 560$ nm, the fluorimetric analysis was performed at $\lambda_{exc} = 436$ nm and $\lambda_{fl} > 560$ nm (Fig. 1).

References

[1] Illner, E.: *Pharmazie* **1984,** *39*, 689–690.
[2] Sampson, K., Schild, F., Wicker, R. J.: *Chem. Ind.* (London) **1961,** 82.
[3] Wright, J.: *Chem. Ind.* (London) **1963,** 1125–1126.
[4] Netz, S., Funk, W.: Private communication, Fachhochschule Gießen, Fachbereich Technisches Gesundheitswesen, 1988.

Manganese(II) Chloride – Sulfuric Acid Reagent

Reagent for:

- Cholesterol, cholesteryl esters [1–4]
- Phospholipids, lipids [1, 5]
- Triglycerides, free fatty acids [1]
- Bile acids [2]
- Ketosteroids [3]
- Estrogens [6]

$MnCl_2 \cdot 4H_2O$	H_2SO_4
$M_r = 197.91$	$M_r = 98.08$
Manganese(II) chloride tetrahydrate	Sulfuric acid

Preparation of Reagent

Dipping solution Dissolve 0.2 g manganese chloride tetrahydrate in 30 ml water and add 30 ml methanol. Mix well and then carefully add 2 ml conc. sulfuric acid dropwise.

Storage The solution should always be freshly made up for quantitative evaluations.

Substances Manganese(II) chloride tetrahydrate
Methanol
Sulfuric acid (95–97%)

Reaction

The reaction mechanism has not yet been elucidated.

Method

The chromatograms are freed from mobile phase in a stream of cold air, immersed in the reagent solution for 1 s and then heated to 100–120 °C for 10–15 min. Colored zones appear on a colorless background [2] (Table 1) and these fluoresce pale blue in long-wavelength UV light ($\lambda = 365$ nm) [1].

Table 1: Colors of the chromatogram zones

Substance	Color
Cholesterol	pink
Cholic acid	deep yellow
Chenodesoxycholic acid	grey-green
Desoxycholic acid	yellow
Hyodesoxycholic acid	brown
Lithocholic acid	pale pink

The pink color of the cholesterol begins to fade after 5 min while the color of the bile acids deepens [2]. The visual detection limit in visible light is 1 µg for cholesterol and 2 µg per chromatogram zone for bile acids [2]. Fluorimetric detection is more sensitive by a factor of 1000!

Note: The reagent, which may also be used as spray solution, can be applied to silica gel, RP-2, RP-8, RP-18 and CN layers.

Procedure Tested

Cholesterol, Coprostanone, Coprostanol, 4-Cholesten-3-one, 5α-Cholestan-3-one [3]

Method Ascending, one-dimensional development in a trough chamber. The HPTLC plates were preconditioned for 30 min at 0% rel. humidity (over conc. sulfuric acid) after sample application and then developed immediately.

Layer HPTLC plates Silica gel 60 F_{254} (MERCK), which had been prewashed by developing once up to the edge of the plate with chloroform – methanol (1 + 1) and then activated to 110 °C for 30 min.

Mobile phase Cyclohexane – diethyl ether (1 + 1).

Migration distance 6 cm

Running time ca. 15 min

Detection and result: The developed chromatogram was dried in a stream of cold air, immersed in the reagent solution for 1 s and then heated to 120 °C for 15 min.

The result was that light blue fluorescent zones were visible under long-wavelength UV light (λ = 365 nm). Before fluorimetric analysis the chromatogram was dipped for 1 s into liquid paraffin – *n*-hexane (1 + 2) to enhance (by a factor of 2 to 8) and stabilize the intensity of the fluorescence and then dried for 1 min in a stream of cold air. The quantitation (λ_{exc} = 365 nm; λ_{fl} > 430 nm) was carried out after 1 h since it was only then that the fluorescence intensity had stabilized.

Note: Coprostanone, 4-cholesten-3-one and 5α-cholestan-3-one could be detected more sensitively, if the dried chromatogram was irradiated with intense long-wavelength UV light (λ = 365 nm) for 2 min before being immersed in the reagent solution. Figure 1 illustrates a separation of these substances. The detection limits

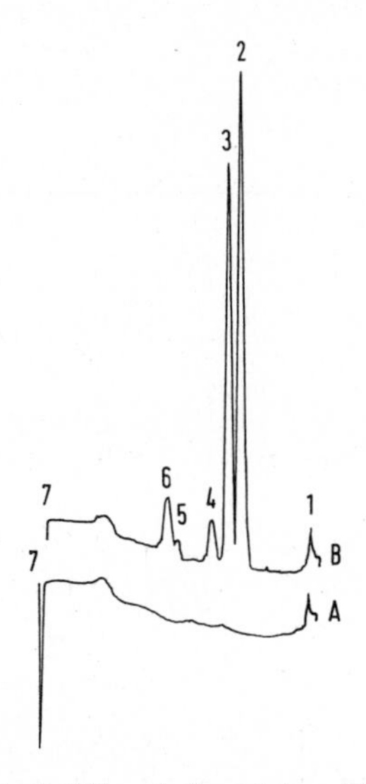

Fig. 1: Fluorescence scans of a blank (A) and of a mixture (B) of cholesterol (2), coprostanol (3), 4-cholesten-3-one (4), 5α-cholestan-3-one (5) and coprostanone (6), start (1), solvent front (7).

were 1 ng per chromatogram zone for cholesterol (hR_f 20–25) and coprostanol (hR_f 25–30) and 50 ng per chromatogram zone for 4-cholesten-3-one (hR_f 40–45), 5α-cholestan-3-one (hR_f 60–65) and coprostanone (hR_f 70–75).

References

[1] Halpaap, H.: *Kontakte* (MERCK) **1978**, 32–34.
[2] Goswami, S. K., Frey, C. F.: *J. Chromatogr.* **1970**, *53*, 389–390.
[3] Schade, M.: Thesis, Fachhochschule Gießen, Fachbereich Technisches Gesundheitswesen, 1986.
[4] Hauck, H. E., Jost, W.: *GIT Fachz. Lab. Suppl. 3 „Chromatographie“* **1983**, 3–7.
[5] Jork, H., Wimmer, H.: *Quantitative Auswertung von Dünnschicht-Chromatogrammen.* Darmstadt: GIT-Verlag, p. III./3–82.
[6] Jost, W., Hauck, H. E. in: Proceedings of the 3rd International Symposium on Instrumental HPTLC, Würzburg. Bad Dürkheim: IfC-Verlag, 1985, p. 83–91.

Mercury(I) Nitrate Reagent

Reagent for:

- Barbiturates and barbiturate metabolites [1 – 4]
- Various pharmaceuticals
  e.g. pentenamide [1]
  gluthethimide, primidone, phenytoin [2]
- Organophosphorus insecticides [5]
- Succinimides [6]
- Thiourea [7]

$Hg_2(NO_3)_2 \cdot 2H_2O$

$M_r = 561.22$

Preparation of Reagent

Spray solution Grind 5 g mercury(I) nitrate dihydrate with 100 ml water in a mortar and tranfer it to a storage vessel along with the sediment; the supernatant is fit for use as long as the sediment remains white [1].

Storage The reagent may be stored for an extended period; new reagent must be made up when the sediment becomes grey or yellow in color [1].

Substances Mercury(I) nitrate dihydrate

Reaction

The reaction mechanism has not yet been elucidated.

Method

The chromatograms are dried in a stream of warm air or at 105 °C for 10 min, cooled to room temperature and sprayed homogeneously with the spray reagent until they start to become transparent.

Grey-black chromatogram zones are produced on a white background, usually appearing immediately but sometimes only after a few minutes. In the case of insecticides the chromatograms are heated after spraying [5].

Note: Traces of ammonia from the mobile phase should be removed from the plate completely to avoid background discoloration (grey veil) [1]. If the layer is sprayed too heavily the initially grey-black chromatogram zones can fade again [2]. The reagent which is usually employed as a 1 to 2% solution [2, 3, 6, 7] can be treated with a few drops of nitric acid to clarify the solution [2].

Barbiturate metabolites are more heavily colored by the mercury(I) nitrate reagent (exception: allobarbital), while unaltered barbiturates react more sensitively to the mercury(II)-diphenylcarbazone reagent (q.v.) [1].

The reagent can be employed on silica gel, kieselguhr, aluminium oxide and cellulose layers.

Procedure Tested

Barbiturates and Metabolites [1]

Method Ascending, one-dimensional development in a trough chamber with chamber saturation.

Layer TLC plates Silica gel 60 F_{254} (MERCK).

Mobile phase Chloroform – acetone (80 + 20).

Migration distance ca. 6 cm

Running time 7 min

Detection and result: The developed chromatogram was dried for 5 min in a stream of hot air (120 – 150 °C) then treated with a stream of cold air for 5 min. It was then sprayed with reagent solution until the layer began to be transparent. After 5 min barbiturates and their metabolites produced grey to black zones on a colorless background (Fig. 1).

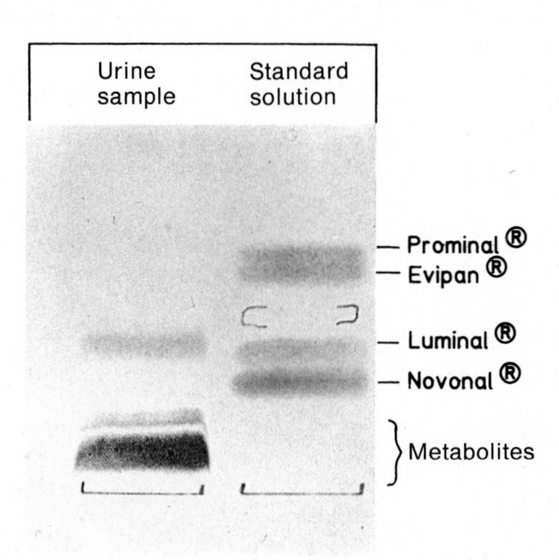

Fig. 1: Chromatograms of a urine sample and a standard solution.

References

[1] Interschick, E., Wüst, H., Wimmer, H.: *GIT Fachz. Lab.* **1981,** *4*, 412–440.
[2] Berry, D. J., Grove, J.: *J. Chromatogr.* **1973,** *80*, 205–219.
[3] Srivastava, S. P., Reena: *J. Liq. Chromatogr.* **1985,** *8*, 1265–1278.
[4] Cserhati, T., Bojarski, J., Fenyvesi, E.: *J. Chromatogr.* **1986,** *351*, 356–362.
[5] Mirashi, S. V., Kurhekar, M. P., D'Souza, F. C.: *J. Chromatogr.* **1983,** *268*, 352–354.
[6] Nuhn, P., Woitkowitz, P.: *Pharmazie* **1978,** *33*, 202–205.
[7] Hashmi, M. H., Chughtai, N. A., Ahmad, I.: *Mikrochim. Acta (Vienna)* **1970,** 254–257.

Mercury(II) Salt – Diphenylcarbazone Reagent

Reagent for:

- Barbiturates, barbiturate metabolites [1–6]
- Sulfonamides [2]
- Various pharmaceuticals
  e.g. glutethimide [2–4]
  lysergic acid [2]
  primidone, phenytoin [3]
- Organophosphorus insecticides [7]

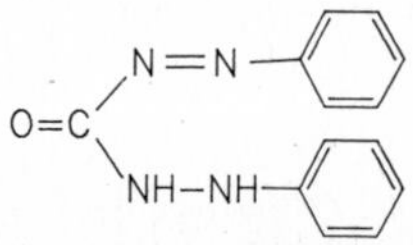

$HgCl_2$	$Hg(NO_3)_2 \cdot H_2O$	$C_{13}H_{12}N_4O$
$M_r = 271.50$	$M_r = 342.62$	$M_r = 240.27$
Mercury(II) chloride	Mercury(II) nitrate monohydrate	Diphenylcarbazone

Preparation of Reagent

Solution I Dissolve 100 mg 1,5-diphenylcarbazone in 100 ml ethanol (96%).

Solution II Dissolve 40 g ammonium acetate in 100 ml water and adjust to pH 3.5 with ca. 30 ml nitric acid (65%).

Spray solution Ia Dissolve 2 g mercury(II) chloride in 100 ml ethanol [1, 3, 4].

Spray solution Ib Dissolve 200 mg 1,5-diphenylcarbazone in 100 ml ethanol [1, 3, 4].

Spray solution IIa Dissolve 200 mg mercury(II) nitrate monohydrate in 100 ml nitric acid ($c = 0.1$ mol/l) [6].

Spray solution IIb Mix 4 ml solution I and 26 ml solution II [6].

Storage Solutions I and II and spray solutions Ia and Ib may be stored in the refrigerator for ca. 1 month, both spray solutions IIa and IIb may be kept for ca. 1 week.

Substances
1,5-Diphenylcarbazone
Mercury(II) nitrate monohydrate
Mercury(II) chloride
Nitric acid 0.1 mol/l Titrisol
Nitric acid (65%)
Ethanol
Ammonium acetate

Reaction

The reaction mechanism has not yet been elucidated.

Method

Variant I: The chromatogram is dried in a stream of warm air or in the drying cupboard (10 min, 120 °C), cooled to room temperature and either sprayed with spray solutions Ia and Ib one after the other (with brief drying in a stream of cold air in between) [1] or sprayed with a mixture of equal volumes of these two spray reagents until the layer begins to be transparent [2, 4].

In the case of organophosphorus insecticides [7] and usually also in the case of barbiturates and other pharmaceuticals [1, 3] white chromatogram zones are produced on a lilac background. Barbiturates sometimes appear – especially after the use of basic mobile phases – as blue-violet colored zones on a pink background [2, 5].

Variant II: The chromatogram is dried in a stream of warm air or in the drying cupboard (10 min, 120 °C), cooled to room temperature and then sprayed twice with spray solution IIa until the layer is transparent (dry in a stream of warm air after each spray step!). It is then sprayed with a small amount of spray solution IIb [6].

Barbiturates and their metabolites always appear as red-violet chromatogram zones on a white background with this variant of the reagent [6].

Note: The reaction for barbiturates according to variant I is increased in sensitivity if the chromatogram is exposed to direct sunlight or UV light after it has been sprayed; this causes the background coloration to fade almost completely and the blue zones stand out more distinctly [4].

While barbiturate metabolites are more intensely colored by the mercury(I) nitrate reagent (q.v.) the unaltered barbiturates react more sensitively to the mercury(II) diphenylcarbazone reagent; the detection limits lie between 0.05 µg (Luminal®) and 10 µg (Prominal®) per chromatogram zone [6].

The reagent can be employed on silica gel and kieselguhr layers.

Procedure Tested

Barbiturates and Metabolites [6]

Method	Ascending, one-dimensional development in a trough chamber with chamber saturation.
Layer	TLC plates Silica gel 60 F_{254} (MERCK).
Mobile phase	Chloroform – acetone (80 + 20).
Migration distance	ca. 6 cm
Running time	7 min

Detection and result: The chromatogram was dried for 5 min in a stream of hot air (120–150 °C) then treated with a stream of cold air for 5 min. It was then sprayed twice to transparency with spray solution IIa (drying in between for 3 min in a stream of hot air). It was then heated in a stream of hot air for 5 min. After cooling for 5 min in a stream of cold air it was sprayed with a small amount of spray solution IIb. Barbiturates and their metabolites appeared as violet chromatogram zones on an almost colorless background. The detection limits for barbiturates lay between 50 ng (Luminal®) and 10 µg (Prominal®, Fig. 1).

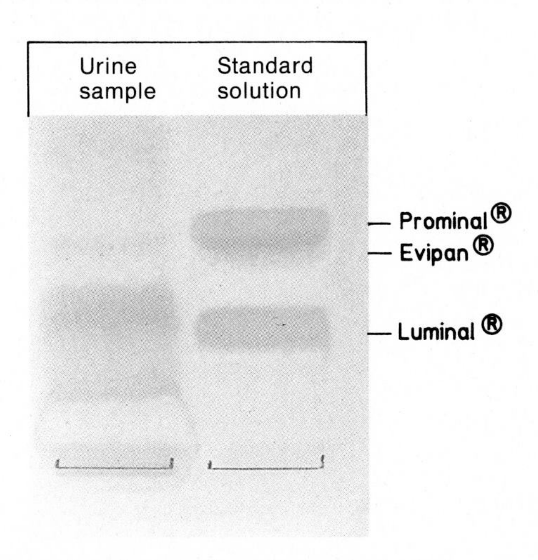

Fig. 1: Chromatograms of a urine sample and a standard solution.

References

[1] Dybowski, R., Gough, T. A.: *J. Chromatogr. Sci.* **1984,** *22*, 104–110.
[2] Owen, P., Pendlebury, A., Moffat, A. C.: *J. Chromatogr.* **1978,** *161*, 195–203.
[3] Berry, D. J., Grove, J.: *J. Chromatogr.* **1973,** *80*, 205–219.
[4] Christensen, E. K. J., Vos, T., Huizinga, T.: *Pharm. Weekbl. Ned.* **1965,** *100*, 517–531.
[5] Brunsmann, P. W. F., Paalman, A. C. A.: *Pharm. Weekbl. Ned.* **1971,** *106*, 933–937.
[6] Interschick, E., Wüst, H., Wimmer, H.: *GIT Fachz. Lab.* **1981,** *4*, 412–440.
[7] Mirashi, S. V., Kurekhar, M. P., D'Souza, F. C., Meghal, S. K.: *J. Chromatogr.* **1983,** *268*, 352–354.

2-Methoxy-2,4-diphenyl-3(2H)-furanone Reagent
(MDPF Reagent)

Reagent for:

- Amines
  e.g. colchicine [1]

$C_{17}H_{14}O_3$
$M_r = 266.30$

Preparation of Reagent

Dipping solution Dissolve 25 mg 2-methoxy-2,4-diphenyl-3(2H)-furanone in 50 ml methanol.

Storage The dipping solution may be stored for several days.

Substances 2-Methoxy-2,4-diphenyl-3(2H)-furanone
Methanol

Reaction

MDPF reacts directly with primary amines to form fluorescent products. Secondary amines yield nonfluorescent derivatives, which may be converted into fluorescent substances by a further reaction with primary amines [2].

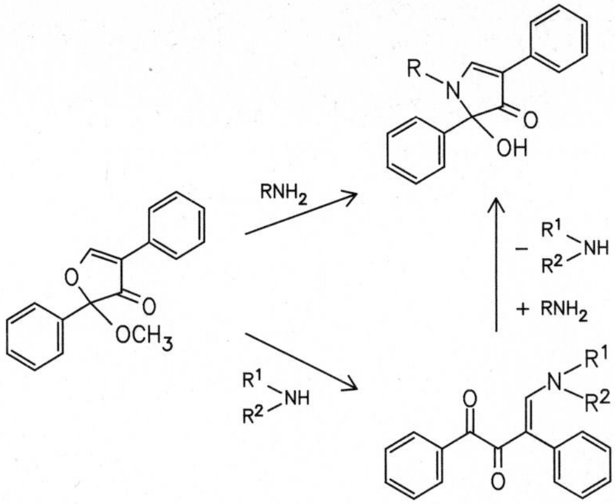

Method

The chromatograms are freed from mobile phase in a stream of warm air, immersed in the reagent solution for 4 s or sprayed evenly with it and then heated to 110 °C for 20 min.

Yellow chromatogram zones are formed on a colorless background; these exhibit yellow fluorescence in long-wavelength UV light ($\lambda = 365$ nm).

Note: The reagent can be applied to silica gel, Si 50000 and kieselguhr layers.

Procedure Tested

Colchicine [1]

Method Ascending, one-dimensional development in a trough chamber without chamber saturation.

Layer HPTLC plates Silica gel 60 (MERCK) which had been washed by developing once with chloroform – methanol (1 + 1) and then dried at 110 °C for 30 min before applying the samples.

Mobile phase Acetone – toluene – ammonia solution (25%) (40 + 15 + 5).

Migration distance 5 cm

Running time 10 min

Detection and result: The chromatogram was freed from mobile phase in a stream of warm air (45 min), immersed in the reagent solution for 4 s and then heated to 110 °C for 20 min.

Colchicine (hR_f 35 – 40) appeared as yellow fluorescent zone on a dark background in long-wavelength UV light ($\lambda = 365$ nm). The detection limit was 10 ng per chromatogram zone.

In situ quantitation: The fluorimetric analysis was carried out with excitation at $\lambda_{exc} = 313$ nm and evaluation at $\lambda_{fl} > 390$ nm (Fig. 1).

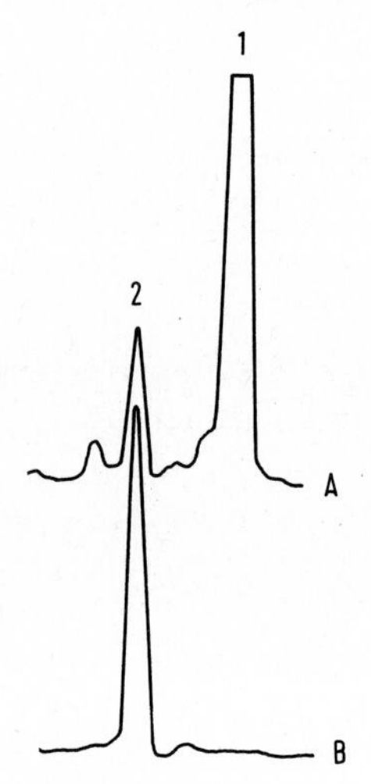

Fig. 1: Fluorescence scan of a chromatogram track with an extract of *Colchicum autumnale* (A) and of a reference track (B, 1 µg colchicine); start (1), colchicine (2).

References

[1] Müller, J.: Thesis, Fachhochschule Gießen, Fachbereich Technisches Gesundheitswesen, 1987.
[2] Nakamura, H., Tanii E., Tamura, Z.: *Anal. Chem.* **1982**, *54*, 2482–2485.

3-Methyl-2-benzothiazolinone-hydrazone Reagent (MBTH Reagent, BESTHORN's Reagent)

Reagent for:

- Carbonyl compounds [1]
- Mycotoxins
  e.g. patulin [2–10]
  moniliformine [11]
- Penicillic acid [12]

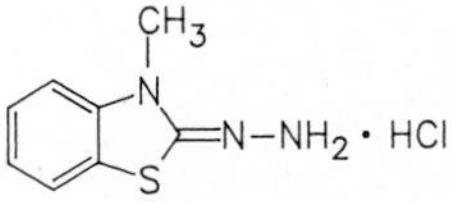

$C_8H_{10}ClN_3S$

$M_r = 215.71$

Preparation of Reagent

Dipping solution Dissolve 0.5–1 g 3-methyl-2-benzothiazolinone-hydrazone hydrochloride (BESTHORN's hydrazone) in 100 ml methanol – water (1 + 1), water [2, 4–8] or methanol [3, 11]. If precipitation occurs filter the solution before use.

Storage The dipping solution is only stable for a few days and, hence, should be freshly made up each time.

Substances 3-Methyl-2-benzothiazolinone-hydrazone hydrochloride
Methanol

Reaction (according to [13])

CH_3 N C=N−NH_2 S (1) + RC(=O)H → CH_3 N C=N−N=CH−R S (2)

(1) — Oxidation, H^+ → CH_3 N^+ C−N=NH S (3)

(2) + (3) — Oxidation → $[CH_3$ N^+ C−N=N−C(R)=N−N=C CH_3 N S S$]^+$ (4)

Method

The chromatogram is freed from mobile phase, immersed in the reagent solution for 1 s or homogeneously sprayed with it and then heated to 110 – 130 °C for up to 2 h. The chromatographic zones produced are usually blue-violet (e.g. patulin, moniliformine) on a pale background; they exhibit yellow to yellow-orange fluorescence (penicillic acid, patulin) in long-wavelength UV light ($\lambda = 365$ nm).

Note: The reagent can be employed on silica gel, kieselguhr, cellulose and polyamide layers. When left exposed to air the whole chromatogram is slowly colored blue because of the formation of the blue cation (3) (see Reaction). The detection limits for patulin, moniliformine and penicillic acid are ca. 50 ng per chromatogram zone.

Procedure Tested

Penicillic Acid [14]

Method Ascending, one-dimensional development in a trough chamber with chamber saturation.

Layer HPTLC plates Silica gel 60 F_{254} (MERCK).

Mobile phase Toluene – ethyl acetate – formic acid (60 + 30 + 10).

Migration distance 5 cm

Running time 10 min

Detection and result: The chromatogram was freed from mobile phase, immersed in reagent solution for 1 s and then heated to 130 °C for 90–120 min. Penicillic acid (hR_f 45–50) yielded yellow chromatogram zones which fluoresced yellow in long-wavelength UV light ($\lambda = 365$ nm). The detection limits were 50 to 100 ng per chromatogram zone.

In situ quantitation: Fluorimetric analysis was carried out at $\lambda_{exc} = 365$ nm and $\lambda_{fl} > 560$ nm (Fig. 1).

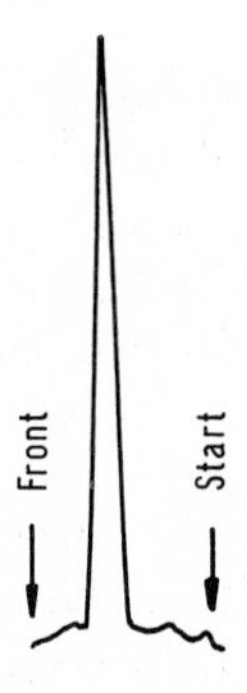

Fig. 1: Fluorescence scan of a chromatogram track of 0.5 µg penicillic acid.

References

[1] Archer, A. W.: *J. Chromatogr.* **1978**, *152*, 290–292.
[2] Reiss, J.: *J. Chromatogr.* **1973**, *86*, 190–191.
[3] Johann, H., Dose, K.: *Fresenius Z. Anal. Chem.* **1983**, *314*, 139–142.
[4] Altmayer, B., Eichhorn, K. W., Plapp, R.: *Z. Lebensm. Unters. Forsch.* **1982**, *175*, 172–174.
[5] Scott, P. M., Kennedy, B. P. C.: *J. Assoc. Off. Anal. Chem.* **1973**, *56*, 813–816.
[6] Eyrich, W.: *Chem. Mikrobiol. Technol. Lebensm.* **1975**, *4*, 17–19.
[7] Dutton, M. F., Westlake, K.: *J. Assoc. Off. Anal. Chem.* **1985**, *68*, 839–842.
[8] Gimeno, A., Martins, M. L.: *J. Assoc. Off. Anal. Chem.* **1983**, *66*, 85–91.
[9] Bergner-Lang, B., Kächele, M., Stengel, E.: *Dtsch. Lebensm.-Rundsch.* **1983**, *79*, 400–404.
[10] Leuenberger, U., Gauch, R., Baumgartner, E.: *J. Chromatogr.* **1978**, *161*, 303–309.
[11] Jansen, C., Dose, K.: *Fresenius Z. Anal. Chem.* **1984**, *319*, 60–62.
[12] Wilson, D. M., Tabor, W. H., Trucksess, M. W.: *J. Assoc. Off. Anal. Chem.* **1976**, *59*, 125–127.
[13] Sawicky, E., Hauser, T. R., Stanley, T. W.: *Anal. Chem.* **1961**, *33*, 93–96.
[14] Kany, E., Jork, H.: GDCh-training course Nr. 302 „Möglichkeiten der quantitativen Direktauswertung", Universität des Saarlandes, Saarbrücken 1987.

1,2-Naphthoquinone-4-sulfonic Acid – Perchloric Acid – Formaldehyde Reagent

Reagent for:

- Sterols
  e.g. ergosterol
  stigmasterol
  cholesterol [1]
- Alkaloids
  e.g. codeine
  morphine
  heroin
  6-monoacetyl-
  morphine [2]

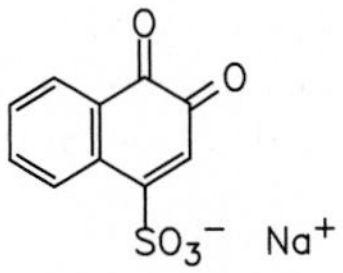

$C_{10}H_5NaO_5S$	$HClO_4$	CH_2O
$M_r = 260.20$	$M_r = 100.46$	$M_r = 30.03$
1,2-Naphthoquinone-4-sulfonic acid sodium salt	Perchloric acid	Formaldehyde

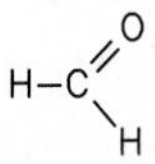

Preparation of Reagent

Dipping solution Dissolve 100 mg 1,2-naphthoquinone-4-sulfonic acid sodium salt in 40 ml ethanol and add 20 ml perchloric acid (70%), 18 ml water and 2 ml formaldehyde solution in that order.

Storage The reagent solution may be kept for ca. 4 weeks in the refrigerator.

Substances 1,2-Naphthoquinone-4-sulfonic acid sodium salt
Ethanol
Perchloric acid (70%)
Formaldehyde solution (37%)

Reaction

The reaction mechanism has not been elucidated. It is possible that formaldehyde reacts by oxidation as in the MARQUIS' reaction (see formaldehyde – sulfuric acid reagent), whereby colored salts are formed with naphthoquinone sulfonic acid.

Method

The chromatograms are freed from mobile phase (stream of warm air 5 min), immersed for 4 s in the reagent solution or sprayed homogeneously with it until they begin to be transparent and then heated to 70 °C for ca. 10 min.

Chromatogram zones are produced that are initially pink but turn blue as the heating continues; they are on a pale blue background.

Note: The color development depends on the temperature and duration of heating [1]. The detection limit for sterols and morphine alkaloids is in the lower nanogram range [1, 2].

The reagent can be employed on silica gel, kieselguhr and Si 50000 layers.

Procedure Tested

Morphine Alkaloids [2]

Method	Ascending, one-dimensional development in a trough chamber without chamber saturation.
Layer	HPTLC plates Silica gel 60 (MERCK) which had been prewashed once with chloroform – methanol (50 + 50) and then dried at 110 °C for 20 min.
Mobile phase	Methanol – chloroform – water (12 + 8 + 2).
Migration distance	6 cm
Running time	20 min

Detection and result: The chromatogram was dried in a stream of warm air for 5 min, immersed in the reagent solution for 4 s and heated to 70 °C for ca. 10 min.

Morphine (hR_f 25–30), codeine (hR_f 30–35), 6-monoacetylmorphine (hR_f 40–45) and heroin (hR_f 50–55) yielded blue chromatogram zones on a pale blue background. The detection limits were 10 to 20 ng substance per chromatogram zone.

In situ quantitation: The absorption photometric analysis was made in reflectance at $\lambda = 610$ nm (Fig. 1).

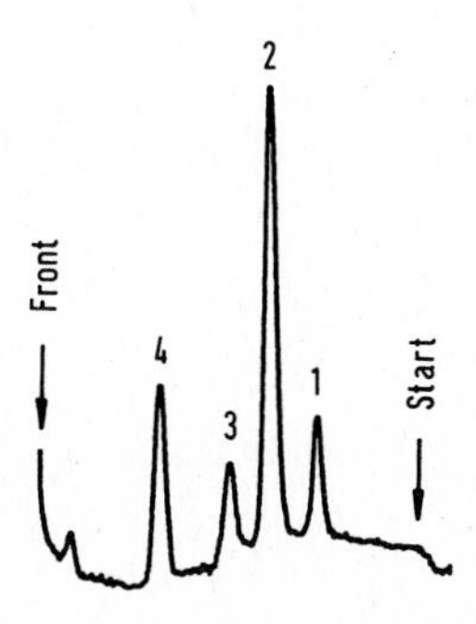

Fig. 1: Reflectance scan of the alkaloid mixture with ca. 50 ng substance per chromatogram zone. Morphine (1), codeine (2), 6-monoacetylmorphine (3), heroin (4).

References

[1] Richter, E.: *J. Chromatogr.* **1965,** *18,* 164–167.
[2] Patzsch, K., Funk, W., Schütz, H.: *GIT Fachz. Lab., Supplement 3 „Chromatographie"* **1988,** *32,* 83–91.

Ninhydrin – Collidine Reagent

Reagent for:

- Amines [1]
- Aminoglycoside antibiotics [2, 3]
- Amino acids [5–7]
- Peptides [6]

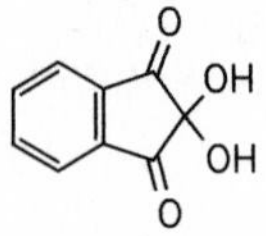

$C_9H_6O_4$ $M_r = 178.15$ Ninhydrin

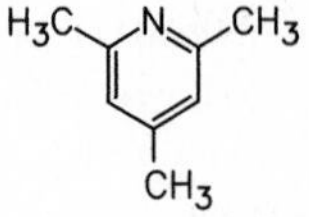

$C_8H_{11}N$ $M_r = 121.18$ Collidine

Preparation of Reagent

Dipping solution Dissolve 0.3 g ninhydrin (2,2-dihydroxy-1,3-indanedione) in 95 ml 2-propanol and add 5 ml collidine (2,4,6-trimethylpyridine) and 5 ml acetic acid (96%).

Storage If the solvents are not sufficiently pure the solution may only be kept for a short time.

Substances Ninhydrin
2,4,6-Trimethylpyridine
Acetic acid (96%)
2-Propanol

Reaction

The course of the reaction has not been fully explained; one possible route is:

Method

The chromatogram is freed from mobile phase, immersed for 1 s in the reagent solution and then heated to 95–120 °C for 5–10 min. After 15 min stabilization time mainly reddish but sometimes blue substance zones appear on a pale background.

Note: The reagent can be employed on silica gel and cellulose layers and the dipping solution can also be employed as a spray solution. The addition of collidine

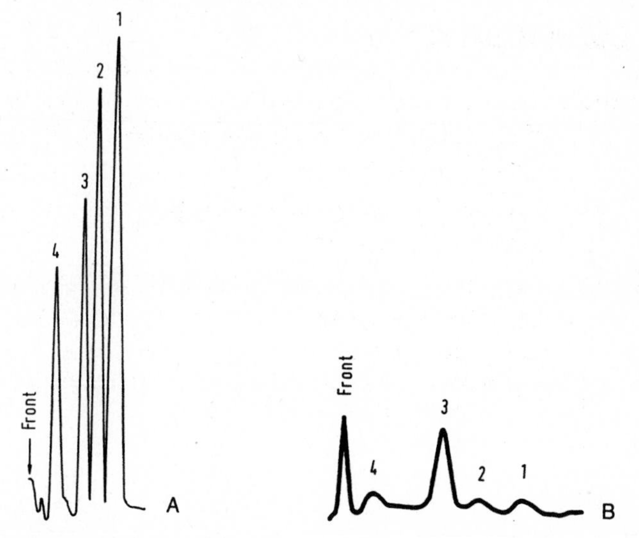

Fig. 1: Comparative recordings of the reflectance scans of a mixture of phenylethylamine (1), tyramine (2), serotonin (3) and histamine (4); A. ninhydrin reagent (q.v.), B. ninhydrin – collidine reagent.

buffers the layer and allows, for example, the differentiation of glycine and serine by color [4]. In the case of biogenic amines the reagent is much more sensitive if collidine is omitted* (Fig. 1). Ammonia vapors interfere with the reaction. If the zones fade rapidly they can be stabilized by complex formation by the addition of tin, copper [8], cobalt or cadmium salts to the reagent (cf. also copper(II) nitrate reagent for the stabilization of "ninhydrin spots").

Procedure Tested

Gentamycin Complex [2, 3, 10]

Method Ascending, one-dimensional development at 10–12 °C in a twin-trough chamber, with 25% ammonia in that part not containing mobile phase. The chamber was equilibrated for 15 min before starting development.

Layer HPTLC plates Silica gel 60 (MERCK) prewashed by developing three times with chloroform – methanol (50 + 50) with intermediate and final drying at 110 °C for 30 min.

Mobile phase Chloroform – ethanol – ammonia solution (25%) (10 + 9 + 10).
The *lower organic phase was employed.*

Migration distance ca. 5 cm

Running time ca. 20 min

Detection and result: The chromatogram was treated in a stream of cold air for 30 min in order to remove the mobile phase, it was then immersed in the reagent solution for 1 s and finally, after drying in the stream of warm air it was heated to 95 °C for 10 min. After 15 min stabilization time at room temperature blue chromatogram zones are produced on a colorless background.

Note: Under these conditions gentamycins C_2 and C_{2a} form a common zone. A separation can be achieved, for example, with methanol – 0.1 mol/l LiCl in 32% aqueous ammonia solution (5 + 25) on KC18F plates (WHATMAN) [9].

---

* Jork, H.: Private communication, Universität des Saarlandes, D-6600 Saarbrücken, 1988

In situ quantitation: The direct quantitative analysis of the blue-violet derivatives should be made in reflectance at $\lambda = 580$ nm ca. 15 min after color development. The detection limits are 50 ng substance per chromatogram zone for gentamycin C_{1a} (hR_f 35–40), $C_2 + C_{2a}$ (hR_f 40–45), and C_1 (hR_f 45–50) (Fig. 2).

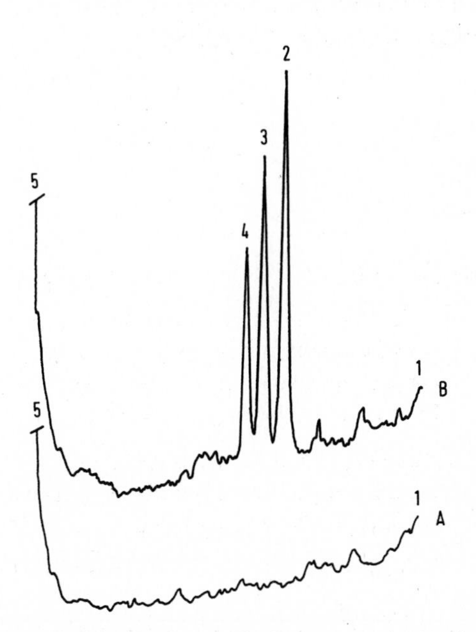

Fig. 2: Absorption scans (A) of a blank and (B) of a gentamycin standard (800 ng substance per application). Start (1), gentamycin C_{1a} (2), gentamycin $C_2 + C_{2a}$ (3), gentamycin C_1 (4), solvent front (5).

References

[1] Kirchner, J. G. in: *Thin-Layer Chromatographie.* 2nd Ed., J. Wiley & Sons, New York-Chichester-Brisbane-Toronto, 1978.

[2] Kiefer, U.: Thesis, Fachhochschule Gießen, Fachbereich Technisches Gesundheitswesen, 1984.

[3] Funk, W., Canstein, M. von, Couturier, T., Heiligenthal, M., Kiefer, U., Schlierbach, S., Sommer, D.: Proceedings of the 3rd International Symposium on Instrumental HPTLC, Würzburg. Bad Dürkheim: IfC-Verlag 1985, p. 281–311.

[4] Woidwood, A., J.: *J. Gen. Microbiol.* **1949,** *3,* 312–318.

[5] Nürnberg, E.: *Arch. Pharm.* (Weinheim) **1959,** *292/64,* 610–620.

[6] Schiltz, E., Schnackerz, K. D., Gracy, R. W.: *Anal. Biochem.* **1977,** *79,* 33–41.
[7] Karsai, T.: *Hung. Sci. Instruments* **1982,** *53,* 15–22.
[8] Datta, S., Data, S. C.: *J. Chromatogr.* **1979,** *170,* 228–232.
[9] Kunz, F. R. : Thesis, Universität des Saarlandes, Saarbrücken 1988.
[10] Ullmann, U.: *Arzneim.-Forsch.* **1971,** *21,* 263–267.

4-(4-Nitrobenzyl)pyridine Reagent (NBP Reagent)

Reagent for:

- Epoxides
  e.g. trichothecene-mycotoxins [1 – 6]
  valepotriates [7, 17]
- Olefins, acetylene derivatives [8]
- 4-Hydroxycumarin, anthraquinone [8]
- Alkylating agents [9-12]
  e.g. sulfur and nitrogen mustard derivatives,
  bis-(halogenalkyl)sulfides,
  N,N,N-(trihalogenalkyl)amines and
  N,N-bis-(halogenalkyl)alkylamines [9, 10],
  diazoalkanes and aziridines [10],
  substances with labile halogens [12]
- Pyrethroid insecticides
  e.g. S-bioallethrine [13]
- Organophosphorus insecticides [16]

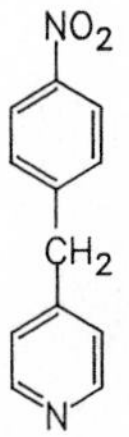

NO_2

CH_2

N

$C_{12}H_{10}N_2O_2$
$M_r = 214.23$

Preparation of Reagent

Dipping solution I Dissolve 9 g 4-(4-nitrobenzyl)pyridine (NBP) and 90 mg butylhydroxytoluene (BHT) in 90 ml chloroform.

Dipping solution II Dissolve 24 ml tetraethylene pentamine and 60 mg butylhydroxytoluene in 36 ml dichloromethane.

Spray solution I Dissolve 2 – 5 g 4-(4-nitrobenzyl)pyridine in 100 ml acetone [7 – 10].

Spray solution II Dissolve 92 mg potassium hydrogen phthalate in 100 ml water and adjust the pH to 5.0 with sodium hydroxide solution (1 mol/l). Dissolve 5 g sodium perchlorate in this solution [9].

A sodium acetate buffer solution (0.05 mol/l; pH 4.6) may be employed as an alternative [10].

Spray solution III Piperidine [10] or alternatively a mixture of triethylamine and acetone (1 + 4) or sodium hydroxide solution ($c = 10^{-5}$ mol/l; pH 8.5 – 9) [8].

Storage All solutions may be stored for several days in the refrigerator.

Substances
4-(4-Nitrobenzyl)pyridine
Tetraethylene pentamine
Chloroform
Dichloromethane
Acetone
Potassium hydrogen phthalate
Sodium perchlorate monohydrate
Sodium hydroxide solution (1 mol/l)
Piperidine
Triethylamine
Butylhydroxytoluene
(= 2,6-di-*tert*-butyl-4-methyl phenol)

Reaction

NBP reacts with epoxides according to I to yield methine dyestuffs [14] and with alkylating agents (R-X, X = e.g. halogen) according to II to yield colored pigments [15].

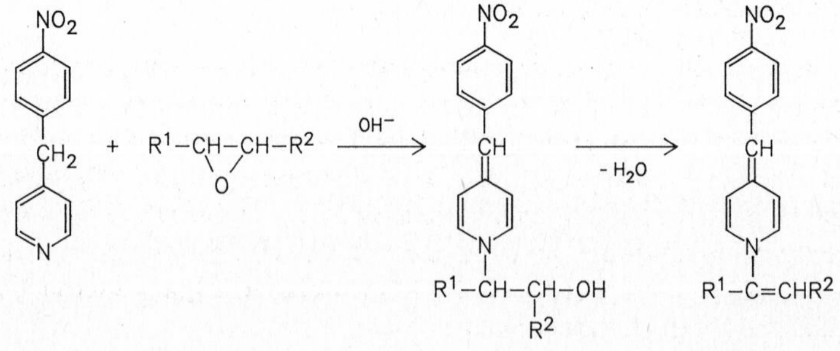

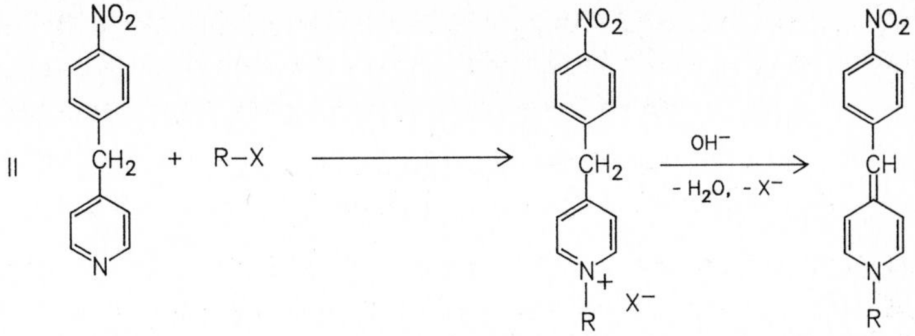

Method

1. Epoxides: The chromatograms are dried in a stream of warm air, immersed in dipping solution I for 1 – 2 s and then heated to 120 – 150 °C for 15 – 30 min. They are then cooled to room temperature and immersed for 1 – 2 s in dipping solution II [1].

2. Acetylene derivatives and substances containing labile halogen: The dried chromatograms are homogeneously sprayed with spray solution I, dried in a stream of hot air and sprayed once more with the same solution. The color can then be intensified by spraying with sodium hydroxide solution (10^{-5} mol/l) [8].

3. Alkylating agents: The chromatograms are freed from mobile phase, then sprayed homogeneously with spray solution I, dried briefly in a stream of cold air, sprayed homogeneously with spray solution II and then heated to 105 – 140 °C for 10 – 25 min. After cooling to room temperature they are then sprayed with spray solution III [9, 10].

4. Organophosphorus insecticides: The chromatograms are freed from mobile phase, immersed in dipping solution I for 10 s and exposed to a saturated acetic anhydride atmosphere for 15 s. After heating to 110 °C for 30 min the chromatogram is immersed for 10 s in dipping solution II and dried for a few minutes in a stream of cold air [16].

Trichothecenes, valepotriates, organophosphorus insecticides and alkylating agents yield blue [1, 7, 9, 16] and acetylene derivatives and substances containing labile halogens yellow to red-violet [8, 13] chromatogram zones on a colorless background. The colors produced fade after a short period [7, 10]. By subsequently dipping the dried chromatogram in liquid paraffin – n-hexane (1 + 2; containing 0.1% butylhydroxytoluene) for 2 s and drying the chromatogram once more in a stream of cold air the intensity of the colored chromatogram zones is increased and stabilized for more than 45 min.

Note: Trichothecenes and valepotriates only react when there is an epoxy group present in the molecule; their detection limits are in the range 25 – 200 ng substance per chromatogram zone [1]. The detection limits for acetylene derivatives are 100 – 800 ng substance per chromatogram zone, but not all give a positive reaction [8].

Spray solution I can also be employed as a dipping solution. Dipping solution II can be replaced by a freshly made up 10% methanolic solution of anhydrous piperazine.

The reagent can be employed on silica gel, kieselguhr, Si 50 000, cellulose, CN and RP layers. NH_2 layers are unsuitable, since no coloration is produced, as are Nano-SIL C_{18} UV_{254} plates (MACHEREY-NAGEL), since the whole plate background is colored violet.

Procedure Tested

Organophosphorus Insecticides [16]

Method Ascending, one-dimensional stepwise development in a twin trough chamber without chamber saturation.

Layer HPTLC plates Silica gel 60 (MERCK).

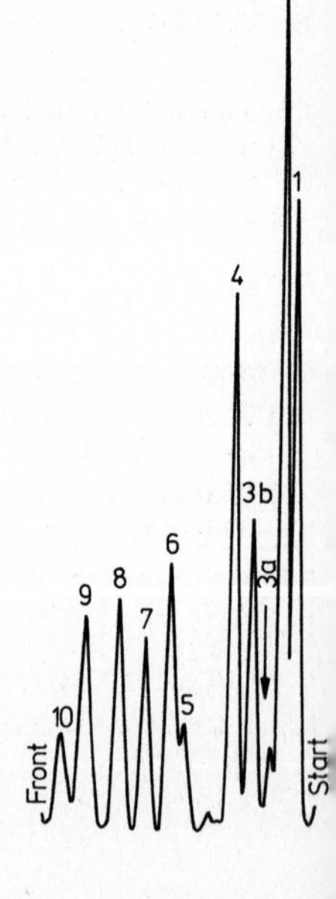

Fig. 1: Absorption scan of a chromatogram track with 150 ng of each substance per chromatogram zone: 1 = demeton-S-methylsulfone, 2 = dimethoate, 3a = transmevinphos, 3b = cis-mevinphos, 4 = demeton-S-methyl, 5 = triazophos, 6 = azinphosmethyl, 7 = azinphos-ethyl, 8 = malathion, 9 = parathion-methyl and 10 = parathionethyl.

Mobile phase Chloroform – diethylether – toluene – *n*-hexane (29.3 + 25.7 + 20 + 25).

Migration distance a. 6 cm; b. 2 cm

Running time a. 20 min; b. 5 min

Detection and result: The chromatogram was freed from mobile phase in a stream of warm air for 3 min and then immersed in dipping solution I for 10 s and exposed to a saturated acetic anhydride atmosphere for 15 s. After heating to 110 °C for 30 min the chromatogram was immersed for 10 s in dipping solution II and dried in a stream of cold air. By subsequent dipping the plate into liquid paraffin – *n*-hexane (1 + 2 v/v; containing 0.1% butylhydroxytoluene) for 2 s and drying in a stream of cold air the intensity of the blue to violet chromatogram zones is increased and stabilized for more than 45 min.

By this procedure it is possible to derivatize all of the 10 investigated organophosphorus insecticides (Fig. 1).

In situ quantitation: The direct quantitative determination of the blue-violet derivatives should be made in reflectance at $\lambda = 580$ nm.

References

[1] Takitani, S., Asabe, Y., Kato, T.: *J. Chromatogr.* **1979**, *172*, 335–342.
[2] Sakamoto, T., Swanson, S. P., Yoshizawa, T.: *J. Agric. Food Chem.* **1986**, *34*, 698–701.
[3] Schmidt, R., Ziegenhagen, E., Dose, K.: *Z. Lebensm. Unters. Forsch.* **1982**, *175*, 169–171.
[4] Corley, R. A., Swanson, S. P., Buck, W. B.: *J. Agric. Food Chem.* **1985**, *33*, 1085–1089.
[5] Bata, A., Vanyi, A., Lasztity, R.: *Acta Vet.* **1984**, *32*, 51–56.
[6] Bata, A., Teren, J., Lasztity, R.: *Acta Vet.* **1984**, *32*, 147–152.
[7] Braun, R., Dittmar, W., Machut, M., Wendland, S.: *Dtsch. Apoth. Ztg.* **1982**, *122*, 1109–1113; **1983**, *123*, 2474–2477.
[8] Schulte, K. E., Rücker, G.: *J. Chromatogr.* **1970**, *49*, 317–322.
[9] Sass, S., Stutz, M. H.: *J. Chromatogr.* **1981**, *213*, 173–176.
[10] Norpoth, K., Schriewer, H., Rauen, H. M.: *Arzneim.-Forsch.* (Drug Res.) **1971**, *21*, 1718–1721.
[11] Norpoth, K., Papatheodorou, T.: *Naturwissenschaften* **1970**, *57*, 356.
[12] Cee, A.: *J. Chromatogr.* **1978**, *150*, 290–292.
[13] Ruzo, L. O., Gaughan, L. C., Casida, J. E.: *J. Agric. Food Chem.* **1980**, *28*, 246–249.
[14] Brewer, J. H., Arnsberger, R. J.: *J. Pharm. Sci.* **1966**, *55*, 57–59.
[15] Friedman, O. M., Boger, E.: *Anal. Chem.* **1961**, *33*, 906–910.
[16] Funk, W., Cleres, L., Enders, A., Pitzer, H., Kornapp, M.: *J. Planar Chromatogr.* **1989**, *2*, 285–289.
[17] Rücker, G., Neugebauer, H., Eldin, M. S.: *Planta Med.* **1981**, *43*, 299.

Perchloric Acid Reagent

Reagent for:

- Steroids [1 – 8]
- Bile acids [9, 10]
- Polystyrenes [11]
- Antiepileptics [12 – 15]
  e.g. carbamazepine, primidone
- Psychopharmaceuticals [12, 16]
  e.g. chlorodiazepoxide, diazepam,
  nitrazepam, oxazepam
- Phenobarbital [12]
- Fatty acid esters [17]
- Nucleosides, nucleotides [18]
- Sugars [19, 20]
- Indoles
  e.g. tryptophan, tryptamine [21]

$HClO_4$
$M_r = 100.46$

Preparation of Reagent

Dipping solution Add 50 ml perchloric acid (70%) carefully to 50 ml water and after cooling dilute the mixture with 50 ml ethanol.

Spray solution I For steroids: 20% aqueous perchloric acid.

Spray solution II For bile acids: 60% aqueous perchloric acid.

Storage The reagent solutions may be stored for an extended period of time.

Substances Perchloric acid (60%)
Perchloric acid (70%)
Ethanol

Reaction

The reaction mechanism has not been elucidated.

Method

The chromatogram is freed from solvent, dipped in the reagent solution for 5–10 s and then heated to 120–150 °C for 5–10 min. (Caution: Remove perchloric acid from the back of the chromatographic plate!).

Mainly colored zones are produced on a colorless background; they are fluorescent under long-wavelength UV light ($\lambda = 365$ nm) and are suitable for quantitative analysis [9, 10].

Bile acids also yield fluorescence when an only 5% perchloric acid is employed as reagent and the chromatogram is only heated to 100 °C until coloration commences [9]. Steroids can also be detected with 2% methanolic perchloric acid [4].

Note: Heating for too long and to too high a temperature can lead to charring of the substances. For this reason heating, for example, to 80 °C for 30 min has also been recommended [11]. The dipping solution can also be employed as a spray solution.

Danger warning: Perchloric acid sprays can condense in the exhausts of fume cupboards and lead to uncontrolled explosions! Dipping is to be preferred for this reason.

The reagent can be employed on silica gel, kieselguhr and Si 50 000 layers.

Procedure Tested

Carbamazepine [13, 14]

Method Ascending, one-dimensional development in a trough chamber. After application of the samples the TLC plate was equilibrated in a chamber at 42% relative humidity for ca. 30 min and then developed immediately.

Layer	HPTLC plates Silica gel 60 (MERCK). Before application of the samples the plates were prewashed three times with chloroform – methanol (1 + 1) and dried at 110 °C for 30 min after each wash.
Mobile phase	Chloroform – acetone (16 + 3).
Migration distance	7 cm
Running time	20 min

Detection and result: The chromatogram was dried in a stream of cold air, immersed in the reagent solution for 5 s, dried in the air and then heated to 120 °C for 7 min. It could be inspected after allowing to cool for 30 min. Carbamazepine (hR_f 30–35) fluoresced blue in long-wavelength UV light ($\lambda = 365$ nm).

The fluorescence was stabilized and enhanced by a factor of 30 by dipping into a solution of liquid paraffin – chloroform – triethanolamine (10 + 60 + 10).

The detection limit for carbamazepine was 50 pg per chromatogram zone.

In situ quantitation: The direct fluorimetric analysis was carried out under long-wavelength UV light ($\lambda_{exc} = 365$ nm; $\lambda_{fl} > 430$ nm, Fig. 1).

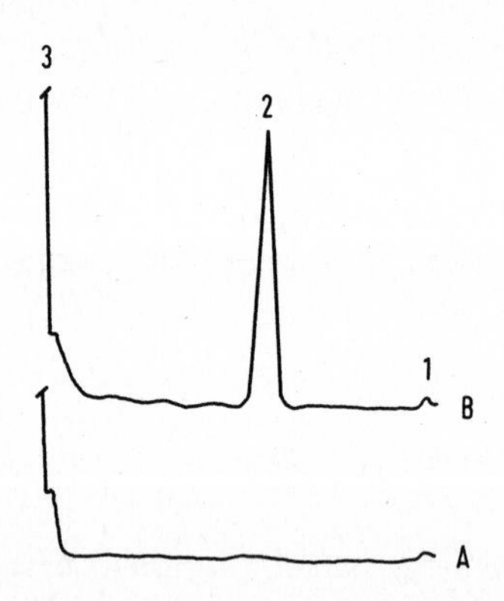

Fig. 1: Fluorescence scan of (A) a blank and (B) a carbamazepine standard with 200 ng per chromatogram zone. Start (1), carbamazepine (2), solvent front (3).

References

[1] Metz, H.: *Naturwissenschaften* **1961**, *48*, 569-570.
[2] Shafiullah, H., Khan, E. A.: *Acta Chim. Acad. Sci. Hung.* **1980**, *103*, 329-332.
[3] Shafiullah, H., Ansari, J. A.: *Acta Chim. Hung.* **1983**, *112*, 373-375.
[4] Gerner, R., Halberstadt, E.: *Z. Geburtshilfe Perinatol.* **1979**, *183*, 272-274.
[5] Ahmad, M. S., Khan, I. A.: *Acta Chimica* **1981**, *106*, 111-113.
[6] Khan, S. E. A.: *Acta Chimica* **1980**, *103*, 329-332.
[7] Ahamad, M., Khan, N., Ansari, A.: *Acta Chimica* **1984**, *115*, 193-195.
[8] Shafiullah, H., Shamsuzzaman, A.: *Acta Chimica* **1981**, *107*, 97-100.
[9] Klaus, R.: *J. Chromatogr.* **1985**, *333*, 276-287.
[10] Hara, S., Takeuchi, M.: *J. Chromatogr.* **1963**, *11*, 565-567.
[11] Otocka, E. P.: *Macromolecules* **1970**, *3*, 691-694.
[12] Faber, D. B., Man in't Veld, W. A.: *J. Chromatogr.* **1974**, *93*, 238-242.
[13] Funk, W., Canstein, M. v., Couturier, T., Heiligenthal, M., Kiefer, U., Schlierbach, S., Sommer, D. in: Proceedings of the 3rd Int. Symposium on Instrumental HPTLC Würzburg. Bad Dürkheim: IfC-Verlag 1985, p. 281-311.
[14] Canstein, M. von: Thesis, Fachhochschule Gießen, Fachbereich Technisches Gesundheitswesen, 1984.
[15] Goenechea, S., Hecke-Seibicke, E.: *Z. Klin. Chem. Klin. Biochem.* **1972**, *10*, 112-113.
[16] Besserer, K. Henzler, S., Kohler, E., Mallach, H. J.: *Arzneim. Forsch.* **1971**, *21*, 2003-2006.
[17] Asif, M., Ahmad, M. S., Mannan, A., Itoh, T., Matsumoto, T.: *J. Amer. Oil Chem. Soc.* **1983**, *60*, 581-583.
[18] Bühlmayer, P., Graf, G., Waldmeier, F., Tamm, Ch.: *Helv. Chim. Acta* **1980**, *63*, 2469-2487.
[19] Nagasawa, K., Ogamo, A., Harada, H., Kumagai, K.: *Anal. Chem.* **1970**, *42*, 1436-1438.
[20] Gangler, R. W., Gabriel, O: *J. Biol. Chem.* **1973**, *248*, 6041-6049.
[21] Nakamura, H., Pisano, J. J.: *J. Chromatogr.* **1978**, *152*, 167-174.

Peroxide Reagent (1-Naphthol – N^4-Ethyl-N^4-(2-methanesulfonamidoethyl)-2-methyl-1,4-phenylenediamine Reagent)

Reagent for:

- Peroxides [1]
  e.g. hydrogen peroxide, per acids, diacylperoxides, hydroperoxides, ketone peroxides [1]

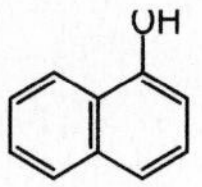

$C_{10}H_8O$
$M_r = 144.17$
1-Naphthol

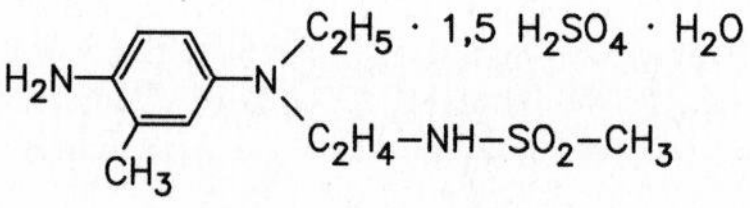

$C_{12}H_{21}N_3O_2S \cdot 1{,}5H_2SO_4 \cdot H_2O$
$M_r = 436.25$
Color developer 3

Preparation of Reagent

Dipping solution Dissolve 3 g 1-naphthol in 150 ml methanol and add 1350 ml water. Dissolve 0.5 g potassium disulfite (potassium metabisulfite) in this solution, add 20 ml glacial acetic acid and dissolve 0.5 g iron(II) sulfate · $7H_2O$ followed by 2.2 g N^4-ethyl-N^4-(2-methanesulfonamidoethyl)-2-methyl-1,4-phenylenediamine (sesquisulfate, monohydrate) (color developer 3, MERCK). Care should be taken at each step that the solution has clarified before adding further components.

Storage The dipping solution may be used for ca. 1 week. Later it slowly turns blue causing the plate background to deteriorate.

Substances 1-Naphthol
Color developer 3
Potassium disulfite
Iron(II) sulfate heptahydrate
Methanol
Acetic acid (glacial acetic acid)

Reaction

Under the influence of peroxides aromatic amines (color developer 3) react with phenols to yield quinone imines [1].

$$\text{Aromatic amine} + \text{1-Naphthol} \xrightarrow{\text{Oxidizing agent}} \text{Quinone imine dyestuff.}$$

Method

The chromatograms are freed from mobile phase, immersed in reagent solution for 1 s or sprayed evenly with it and dried in a stream of cold air. Blue chromatogram zones develop in a few minutes on a pale reddish background.

Note: The detection limits for the various peroxides are from 0.5 to 2 μg substance per chromatogram zone [1]. The reaction works best when the reagent solution is 3–5 days old; later the background absorption increases. The background coloration that is produced on drying in a stream of cold air can be largely avoided by drying the plate after dipping in the absence of oxygen, first with a moist and then with a stream of dry nitrogen [1].

The reagent can be employed on silica gel, kieselguhr, Si 50 000 and particularly sensitively [1] on cellulose layers.

Procedure Tested

Dibenzoyl Peroxide in Acne Preparations [2]

Method Ascending, one-dimensional development in a trough chamber with chamber saturation.

Layer HPTLC plates Silica gel 60 F_{254} (MERCK).

Mobile phase Toluene – dichloromethane – glacial acetic acid (50+2+1).

Migration distance 5 cm

Running time 10 min

Detection and result: The chromatogram was freed from mobile phase in a stream of cold air, immersed in the reagent solution for 1 s and then dried in a stream of cold air for 15 min. At first reddish and then after 60 min violet chromatogram zones developed on a pink background (detection limit of dibenzoyl peroxide: ca. 500 ng).

Note: The background coloration could be avoided if the dipped chromatogram was stored in a chamber first over streaming moist nitrogen gas for 15 min and

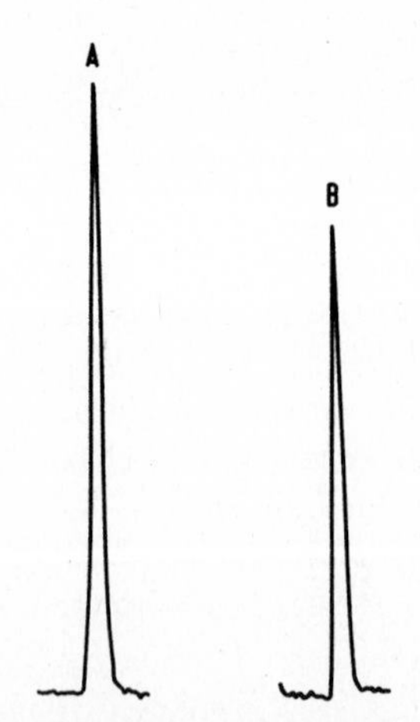

Fig. 1: Reflectance scans of 5 µg dibenzoyl peroxide ($\lambda = 620$ nm); after dipping the chromatogram was dried in (A) a stream of cold air and (B) in a stream of nitrogen.

then for a further 15 min over dry nitrogen gas. The zone intensity was then ca. 30% less than when a stream of cold air was employed for drying (Fig. 1).

In situ quantitation: The direct analysis was carried out at $\lambda = 510$ nm if the zones were red in color and at $\lambda = 620$ nm if they were violet.

References

[1] Huber, W., Fröhlke, E.: *Chromatographia* **1972**, *5*, 256–257.
[2] Jork, H., Kany, E.: GDCh-training course Nr. 300 „Einführung in die DC", Universität des Saarlandes, Saarbrücken 1987.

1,2-Phenylenediamine – Trichloroacetic Acid Reagent

Reagent for:

- α-Keto acids
  e.g. ascorbic acid,
  dehydroascorbic acid [1]
- Sugars [2]

(structure: benzene ring with two adjacent NH_2 groups)	CCl_3COOH
$C_6H_8N_2$	$C_2HCl_3O_2$
$M_r = 108.14$	$M_r = 163.39$
1,2-Phenylene-diamine	Trichloro-acetic acid

Preparation of Reagent

Dipping solution Dissolve 50 mg 1,2-phenylenediamine in 50 ml 10% ethanolic trichloroacetic acid.

Storage The reagent solution should always be made up freshly.

Substances 1,2-Phenylenediamine
Trichloroacetic acid
Ethanol

Reaction

The following reactions are to be expected with ascorbic acid and dehydroascorbic acid:

I

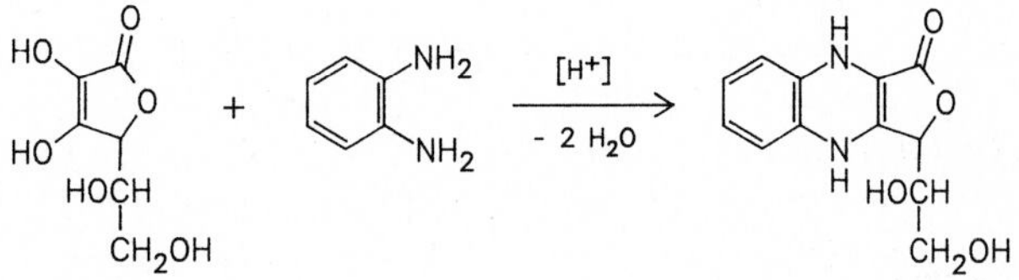

$[H^+]$, $- 2\ H_2O$

1. Ascorbic acid

II

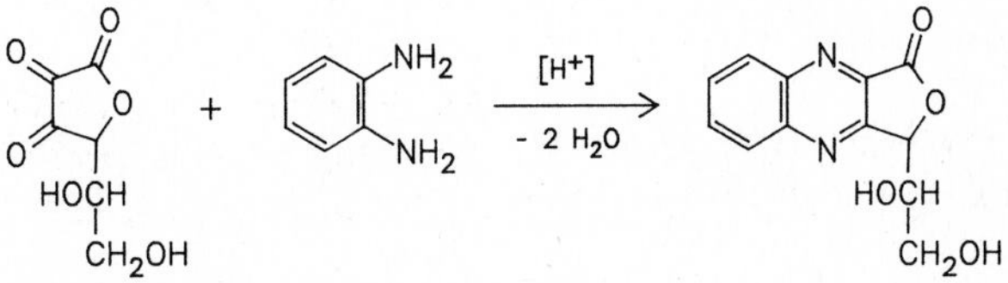

$[H^+]$, $- 2\ H_2O$

2. Dehydroascorbic acid

Method

The chromatogram is freed from mobile phase in a stream of cold air, either immersed in the reagent solution for 5 s or homogeneously sprayed with it and then heated to 110–120°C for 10–15 min [2].

Chromatogram zones are produced which exhibit a green fluorescence under long-wavelength UV light ($\lambda = 365$ nm).

Note: The chromatogram can be dipped into liquid paraffin – *n*-hexane (1 + 2) for 1 s to stabilize the fluorescence.

The reagent can be employed on silica gel, and cellulose layers.

Procedure Tested

Ascorbic Acid and Dehydroascorbic Acid [1]

Method Ascending, one-dimensional development in a twin-trough chamber with chamber saturation.

Layer HPTLC plates Silica gel 60 (MERCK). Before application of the samples the layer was prewashed once with the mobile phase and dried at 110 °C for 20 min. Before it was placed in the developing chamber the prepared HPTLC plate was preconditioned for 30 min at 0% relative humidity (over conc. sulfuric acid).

Mobile phase Acetone – toluene – formic acid (60 + 30 + 10).

Migration distance 6 cm

Running time 11 min

Detection and result: The chromatogram was dried in a stream of cold air for 2 min, immersed in the dipping solution for 5 s and then heated to 110 °C for

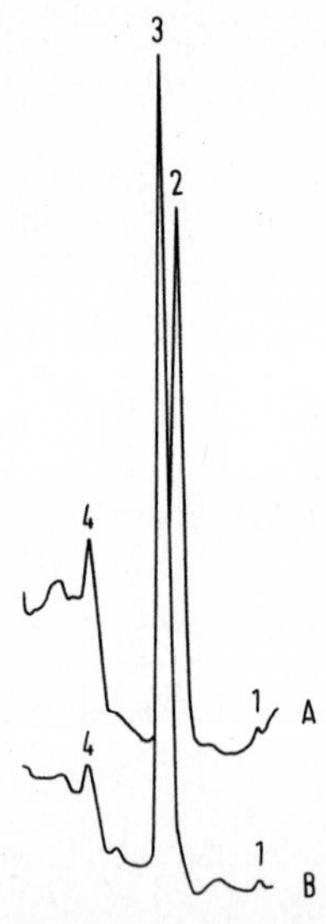

Fig. 1: Fluorescence scans of chromatogram tracks with 200 ng ascorbic acid (A) and 200 ng dehydroascorbic acid (B): Start (1), ascorbic acid (2), dehydroascorbic acid (3), β-front (4).

10 min. After cooling it was immersed for 1 s in liquid paraffin – *n*-hexane (1 + 2) to stabilize the fluorescence.

In long-wavelength UV light ($\lambda = 365$ nm) ascorbic acid (hR_f 45–50) and dehydroascorbic acid (hR_f 50–55) yielded green fluorescent zones on an orange-colored background. The detection limits were less than 5 ng substance per chromatogram zone.

In situ quantitation: The fluorimetric determination was carried out under long-wavelength UV light ($\lambda_{exc} = 365$ nm, $\lambda_{fl} = 546$ nm, monochromatic filter; Fig. 1).

References

[1] Schnekenburger, G.: Thesis, Fachhochschule Gießen, Fachbereich Technisches Gesundheitswesen, Gießen 1987.

[2] Fengel, D., Przyklenk, M.: *Sven. Papperstidn.* **1975**, *78*, 17–21.

Phosphomolybdic Acid Reagent

Reagent for:

- Reducing substances
  e.g. antioxidants [1–3], ascorbic acid [4],
  isoascorbic acid [4, 5], vitamin E [1]
- Steroids [6–9]
- Bile acids, bile acid conjugates [9, 10]
- Lipids [11–14], phospholipids [11, 12]
- Fatty acids [15] or their methyl esters [16]
- Triglycerides [14, 15]
- Subst. phenols [5]
- Indole derivatives [5]
- Prostaglandins [17]
- Components of essential oils
  e.g. carvone, agarofuran [18]
- Morphine [5]

$H_3(P(Mo_3O_{10})_4)$

$M_r = 1825.28$

Preparation of Reagent

Dipping solution Dissolve 250 mg phosphomolybdic acid (molybdatophosphoric acid) in 50 ml ethanol.

Spray solution Dissolve 2–20 g phosphomolybdic acid in 100 ml ethanol [1–3, 6–11, 13, 16], in 2-propanol – methanol (70 + 30) [14] or in water [5, 12].

Storage The solutions may be stored in the dark for ca. 10 days [14].

Substances Molybdatophosphoric acid hydrate
Ethanol

Reaction

A large number of organic substances can be oxidized with phosphomolybdic acid, whereby a portion of the Mo(VI) is reduced to Mo(IV), which forms blue-grey mixed oxides with the remaining Mo(VI).

$$2MoO_3 \xrightarrow[\text{Reduction}]{} MoO_2 \cdot MoO_3$$

Method

Dry the chromatogram in a stream of warm air and immerse for 2–3 s in the reagent solution or spray the layer with it until this acquires an even yellow coloration and dry in a stream of warm air (ca. 2 min).

Blue zones appear immediately or after a few minutes on a yellow background. The background can then be lightened [7] by placing the chromatogram in a twin-trough chamber whose second trough contains 25% ammonia solution.

Note: Occasionally in order to achieve optimal color development a brief heating is to be recommended after dipping or spraying (normally at 105–120 °C for 10 min, or in the case of saturated lipids at 150–180 °C). This also causes sugars to yield blue derivatives. Heating for too long can result in the background being darkened [11]. The detection sensitivity can be improved by adding 4 ml conc. hydrochloric acid to every 100 ml reagent solution.

The detection limit per chromatogram zone is 50–200 ng for lipids [11], 200–400 ng for antioxidants [3] and several ng for ascorbic acid.

The phosphomolybdic acid reagent can be employed on silica gel, aluminium oxide, polyamide, RP-2, RP-18 and cellulose phases and also on silver nitrate-impregnated silica gel [13].

Procedure Tested

Lecithin, Sphingomyelin [19]

Method Ascending, one-dimensional development in a linear chamber (CAMAG).

Layer HPTLC plates Silica gel 60 (MERCK).
Before applying the sample it was best to prewash the layer with mobile phase (Fig. 1) and then activate at 120 °C for 15 min.

Mobile phase Chloroform – methanol – water (30 + 10 + 1.4).

Migration distance 5 cm

Running time 10 min

Detection and result: The chromatogram was dried in a stream of cold air for 2 min and sprayed three times with the spray reagent until it began to appear transparent. The plate was dried in cold air after each spray step and finally heated to 120 °C for 15 min. Lecithin (hR_f 15) and sphingomyelin (hR_f 5 – 10) appeared as dark blue zones on a yellow background (Fig. 1).

In situ quantitation: The photometric analysis was performed in reflectance at $\lambda = 650$ nm.

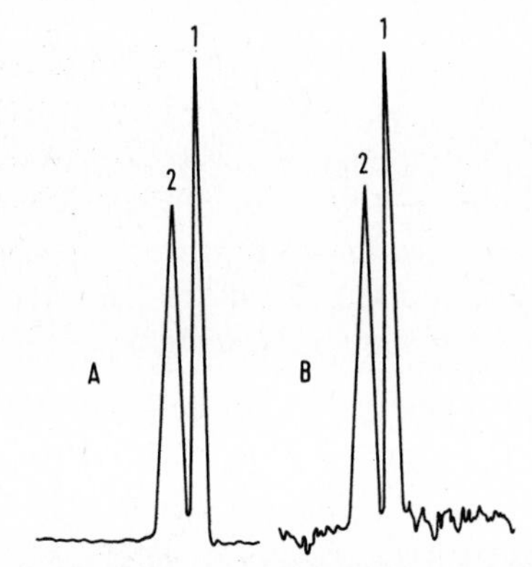

Fig. 1: Reflectance scan of chromatogram tracks (A: layer prewashed with mobile phase, B: with methanol) with 750 ng each substance per chromatogram zone. Sphingomyelin (1), lecithin (2).

References

[1] Seher, A.: *Fette, Seifen, Anstrichm.* **1959,** *61,* 345-351.
[2] Heide, R. F. van der: *J. Chromatogr.* **1966,** *24,* 239-243.
[3] Laub, E., Woller, R.: *Dtsch. Lebensm. Rundsch.* **1976,** *72,* 276–279.
[4] Kovatscheva, E., Popova, J., Kratschanova, M., Ivanova, V.: *Nahrung* **1983,** *27,* 9-13.

[5] Reio, L.: *J. Chromatogr.* **1958,** *1,* 338–373: **1960,** *4,* 458–476.
[6] Jarc, H., Ruttner, O., Krocza, W.: *Fleischwirtschaft* **1976,** *9,* 1326–1328.
[7] Neidlein, R., Koch, E.: *Arch. Pharm.* (Weinheim) **1980,** *313,* 498–508.
[8] Domnas, A. J., Warner, S. A., Johnson, S. L.: *Lipids* **1983,** *18,* 87–89.
[9] Raedsch, R., Hofmann A. F., Tserng, K. Y.: *J. Lipid Res.* **1979,** *20,* 796–800.
[10] Cass, O. W., Cowen, A. E., Hofmann A. F., Coffin, S. B.: *J. Lipid Res.* **1975,** *16,* 159–160.
[11] Sherma, J., Bennett, S.: *J. Liq. Chromatogr.* **1983,** *6,* 1193–1211.
[12] Vinson, J. A., Hooyman, J. E.: *J. Chromatogr.* **1977,** *135,* 226–228.
[13] Schieberle, P., Grosch, W.: *Z. Lebensm. Unters. Forsch.* **1981,** *173,* 192–198: 199–203.
[14] Studer, A., Traitler, H.: *J. High Resolut. Chromatogr. Chromatogr. Commun.* **1985,** *8,* 19–22.
[15] Bringi, N. V., Padley, F. B., Timms, R. E.: *Chem. Ind.* (London) **1972,** *20,* 805–806.
[16] Garssen, G. J., Vliegenthart, J. F. G., Boldingh, J.: *Biochem. J.* **1972,** *130,* 435–442.
[17] Goswami, S. K., Kinsella, J. E.: *J. Chromatogr.* **1981,** *209,* 334–336.
[18] Jork, H.: *Planta Med.* **1979,** *37,* 137–142.
[19] Miller, H., Wimmer, H. in: Jork, H., Wimmer, H.: *Quantitative Auswertung von Dünnschicht-Chromatogrammen.* Darmstadt: GIT, 1982. III/3–82.

o-Phthalaldehyde Reagent (OPT, OPA)

Reagent for:

- Primary amines [1–6]
  e.g. aminoglycoside antibiotics
- Amino acids [1, 2, 7–9]
- Peptides [4, 8, 9]
- Imidazole derivatives [1]
- Indole derivatives
  e.g. pindolol [10]
  hydroxyindolylacetic acid [11–14]
  bufotenine [14, 15]
  serotonin, 5-hydroxytryptamine [11, 14–17]
- Ergot alkaloids [18, 19]

$HS-CH_2-CH_2-OH$

$C_8H_6O_2$
$M_r = 134.14$
o-Phthalaldehyde

C_2H_6OS
$M_r = 78,13$
2-Mercaptoethanol

Preparation of Reagent

Dipping solution Make 0.1 g *o*-phthalaldehyde (phthaldialdehyde, OPA) and 0.1 ml 2-mercaptoethanol (2-hydroxy-1-ethanethiol) up to 100 ml with acetone.

Storage The reagent solution is stable for several days when stored in the dark at room temperature.

Substances Phthaldialdehyde
2-Mercaptoethanol
Acetone

Reaction

In the presence of 2-mercaptoethanol *o*-phthalaldehyde reacts with primary amines to yield fluorescent isoindole derivatives [20]:

$$C_6H_4(CHO)_2 + H_2N{-}R + HS{-}CH_2{-}CH_2{-}OH \longrightarrow \text{isoindole}(N{-}R)(S{-}CH_2{-}CH_2{-}OH)$$

o-Phthalaldehyde Amine 2-Mercaptoethanol Fluorescing derivative

Method

Immerse the dried chromatogram for 1 s in the reagent solution and then heat to 40 – 50 °C in the drying cupboard for 10 min.

Substance zones are produced that mainly yield blue fluorescence under long-wavelength UV light ($\lambda = 365$ nm) (indoles occasionally fluoresce yellow [15]), colored zones are also produced occasionally. The fluorescence is stabilized by immersing in 20% methanolic polyethylene glycol solution [5].

For the detection of the ergot alkaloids 0.2 g *o*-phthalaldehyde in 100 ml conc. sulfuric acid (!) [15] or buffer solution is employed as spray solution. Cysteine [11, 12, 15, 19] or sulfurous acid [17] is occasionally substituted for mercapto-ethanol.

Note: The reagent can be employed on silica gel, alumina and silica gel plates but not on amino or polyamide layers. The dipping solution can also be employed as a spray solution.

If the reaction proves difficult the TLC plate should first be dipped in 1% solution of triethylamine in acetone or in a solution of 1 to 2 drops sodium hydroxid solution ($c = 10$ mol/l) in methanol to optimize the pH for the reaction. This effect can also be achieved by employing borate buffer, pH = 11, instead of acetone in the spray reagent [10, 11].

Since the fluorescence intensity of the zones on silica gel layers is reduced after a few minutes the determination of aromatic amino acids is usually performed on cellulose layers, where the fluorescence remains stable for a longer period of time [8]. Under these conditions proline can still be detected after heating the plate to 110 °C for 1 h, while the other zones fade.

The detection limits for amino acids and peptides are between 50 and 200 pmol per chromatogram zone [9], 400 pg for 5-hydroxyindolylacetic acid [11] and 300 pg for dihydroxyergotoxin [19].

Procedure Tested

Gentamycin C Complex [21]

Method Ascending, one-dimensional development in a twin-trough chamber (CAMAG) with 5 ml ammonia solution (25%) in the trough free from mobile phase. Chamber saturation: ca. 15 min; development at 10 – 12 °C.

Layer HPTLC plates Silica gel 60 (MERCK), prewashed by developing three times with chloroform – methanol (1 + 1) and drying at 110 °C for 30 min after each development.

Mobile phase Chloroform – ethanol – ammonia solution (25%) (10 + 9 + 10); the *lower organic phase* was used for chromatography.

Migration distance ca. 5 cm

Running time ca. 20 min

Detection and result: The chromatogram was freed from mobile phase for ca. 45 min in a current of cold air, immersed for 1 s in the reagent solution and dried in the air.

Alongside gentamycin C_1 (hR_f 45 – 50) and gentamycin C_{1a} (hR_f 35 – 40), gentamycins C_2 and C_{2a} formed a common zone (hR_f 40 – 45; Fig. 1).

Quantitation could be performed fluorimetrically: $\lambda_{exc} = 313$ nm, $\lambda_{fl} > 390$ nm. The chromatogram was first immersed for 1 s in a solution of liquid paraffin – *n*-hexane (1 + 2) to stabilize the fluorescence. The detection limit was ca. 30 ng per chromatogram zone for each substance.

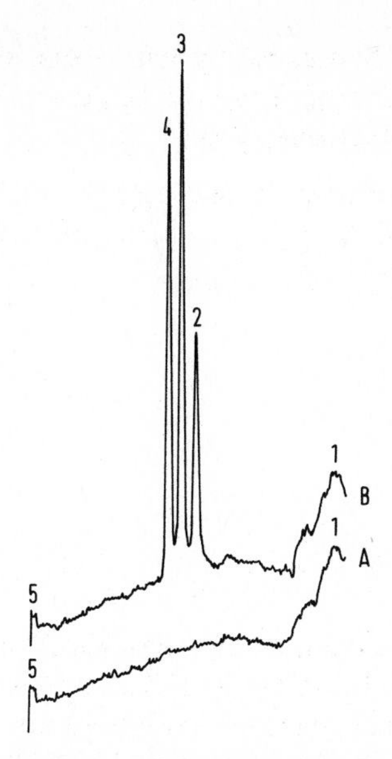

Fig. 1: Fluorescence scan of (A) a blank track and (B) a standard gentamycin mixture (800 ng C complex per application). Start (1), gentamycin C_{1a} (2), C_2/C_{2a} (3), C_1 (4), solvent front (5).

References

[1] Aures, D., Fleming, R., Hakanson, R.,: *J. Chromatogr.* **1968,** *33,* 480–493.
[2] Turner, D., Wightman, S. L.: *J. Chromatogr.* **1968,** *32,* 315–322.
[3] Shelley, W. B., Juhlin, L.: *J. Chromatogr.* **1966,** *22,* 130–138.
[4] Edvinsson, L., Håkanson, R., Rönnberg, A. L., Sundler, F.: *J. Chromatogr.* **1972,** *67,* 81–85.
[5] Gübitz, G.: *Chromatographia,* **1979,** *12,* 779–781.
[6] Stahl, E., Zimmer, C., Juell, S.: *Z. Lebensm. Unters. Forsch.* **1982,** *175,* 88–92.
[7] Davies, H. M., Miflin, B. J.: *J. Chromatogr.* **1978,** *153,* 284–286.
[8] Lindeberg, E. G. G.: *J. Chromatogr.* **1976,** *117,* 439–441.
[9] Schiltz, E., Schnackerz, K. D., Gracy, R. W.: *Anal. Biochem.* **1977,** *79,* 33–41.
[10] Spahn, H., Prinoth, M., Mutschler, E.: *J. Chromatogr.* **1985,** *342,* 458–464.
[11] Zahn, H.: *Ärztl. Lab.* **1984,** *30,* 279–283.
[12] Niederwieser, A., Giliberti, P.: *J. Chromatogr.* **1971,** *61,* 65–99.
[13] Korf, J., Valkenburgh-Sikkema, T.: *Clin. Chim. Acta* (Amsterdam) **1969,** *26,* 301–306.
[14] Narasimhachari, N., Plaut, J.: *J. Chromatogr.* **1971,** *57,* 433–437.
[15] Narasimhachari, N., Lin, R.-L., Plaut, J., Leiner, K.: *J. Chromatogr.* **1973,** *86,* 123–131.

[16] Garcia-Moreno, C., Rivas-Gonzalo, J., Pena-Egido, M., Marine-Font, A: *J. Assoc. Off. Anal. Chem.* **1983**, *66*, 115-117.
[17] Geissler, H. E., Mutschler, E.: *Arzneim. Forsch.* **1976**, *26*, 75-78.
[18] Szabo, A., Karacsony, E. M.: *J. Chromatogr.* **1980**, *193*, 500-503.
[19] Prosek, M., Katic, M., Korsic, J., in: *Chromatography in Biochemistry, Medicine and Environmental Research.* Amsterdam: Elsevier, 1983, Vol. 1, p. 27-36.
[20] Skaaden, T., Greibrokk, T.: *J. Chromatogr.* **1982**, *247*, 111-122.
[21] Kiefer, U.: Thesis, Fachhochschule Gießen, Fachbereich Technisches Gesundheitswesen, 1984.

Picric Acid – Perchloric Acid Reagent

Reagent for:

- Δ^5-3β-Hydroxysteroids [1]
- Sterols
  e.g. cholesterol [1, 2]
  coprostanol [2]

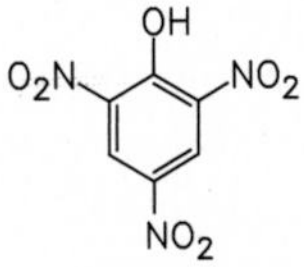

OH
O_2N NO_2
NO_2

$C_6H_3N_3O_7$
$M_r = 229.11$
Picric acid

$HClO_4$
$M_r = 100.47$
Perchloric acid

Preparation of Reagent

Dipping solution Dissolve 100 mg picric acid (2,4,6-trinitrophenol) in 36 ml acetic acid (96%) and carefully add 6 ml perchloric acid (70%).

Storage The dipping solution should be freshly made up before use.

Substances Picric acid
Acetic acid (96%)
Perchloric acid (70%)

Reaction

Presumably perchloric acid oxidizes the steroids at ring A and these then form charge transfer complexes with picric acid.

Method

After drying in a stream of cold air the chromatogram is immersed in the reagent solution for 1 s and heated to 70–75 °C for 3–5 min (until the color develops optimal intensity). Red-colored zones are usually formed on a white background; occasionally, however, yellow to brown chromatogram zones, which gradually fade, are formed on a white to pale yellow background [1].

Note: The reagent can be employed on silica gel and cellulose layers. The coloration of the stained chromatogram zones is dependent on the temperature and duration of heating. For instance, cholesterol appears bluish-pink after heating to 75–80 °C for 3–5 min [1], but yellow to brown-colored after heating for 20–30 min (cf. "Procedure Tested").

Warning of danger: The perchloric acid-containing reagent should not be employed as a spray solution for reasons of safety.

Procedure Tested

Cholesterol, Coprostanol [2]

Method	Ascending, one-dimensional development in a trough chamber.
Layer	HPTLC plates Silica gel 60 F_{254} (MERCK). The plates were prewashed by developing once to the upper edge with chloroform – methanol (1 + 1) and then activated at 110 °C for 30 min.
Mobile phase	Cyclohexane – diethyl ether (10 + 10).
Migration distance	6 cm
Running time	ca. 15 min

Detection and result: The developed chromatogram was dried in a stream of cold air, immersed in the reagent solution for 1 s and heated to 80 °C for 20–30 min (until optimal color development occurred). Yellow to brown-colored zones were produced on a pale yellow-colored background; these were suitable for quantitative analysis. The detection limits for cholesterol (hR_f 20–25) and coprostanol 25–30) were a few nanograms per chromatogram zone.

In situ quantitation: The absorption-photometric analysis was carried out at $\lambda = 378$ nm.

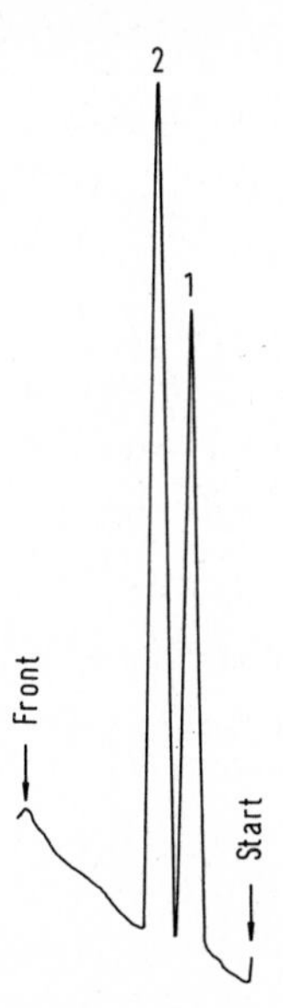

Fig. 1: Absorption scan of a chromatogram track with 250 ng cholesterol and 500 ng coprostanol per chromatogram zone. Cholesterol (1), coprostanol (2).

References

[1] Eberlein, W. R.: *J. Clin. Endocrinol* **1965,** *25,* 288–289.
[2] Schade, M.: Thesis, Fachhochschule Gießen, Fachbereich Technisches Gesundheitswesen, 1986.

Pinacryptol Yellow Reagent

Reagent for:

- Anionic active and nonionogenic detergents [1 – 7]
- Organic anions of aliphatic phosphates, phosphonates, sulfates, sulfonates and sulfamates [8, 9]
- Sweeteners e.g. cyclamate, saccharin, dulcin [10]

$C_{21}H_{22}N_2O_7S$
$M_r = 446.48$

Preparation of Reagent

Dipping solution Dissolve 100 mg pinacryptol yellow in 100 ml ethanol (95%) with gentle heating.

Spray solution Dissolve 100 mg pinacryptol yellow in 100 ml water with boiling [2, 3] or in ethanol (95%) with heating [8, 10].

Storage The reagent solution may be stored for several months in the dark.

Substances Pinacryptol yellow
Ethanol

Reaction

The reaction has not been elucidated. An exchange of counterions is probable in the case of anionic compounds [8].

Method

The chromatogram is freed from mobile phase in a stream of warm air and either immersed for 2 s in the dipping solution or homogeneously sprayed with it until the layer begins to be transparent. In the case of detergents the chromatograms are evaluated while still moist [3], in the case of sweeteners after drying for 10 min in the dark [10].

Fluorescent chromatogram zones are produced on a dark or fluorescent background under long-wavelength ($\lambda = 365$ nm) and occasionally short-wavelength UV light ($\lambda = 254$ nm).

Note: Detergents fluoresce blue, yellow or orange [1, 3, 5, 6]. On cellulose layers aliphatic organic anions with chain lengths of at least 3 C atoms yield yellow to orange fluorescence on a pale green background; increasing chainlength of the organic part of the molecule has a positive effect and hydroxyl groups a negative effect on the fluorescence intensity [8]. Inorganic anions, carboxylic acids and amino acids do not react [8]. Cyclamate fluoresces orange and saccharin and dulcin can be recognized as orange and dark violet zones respectively [10].

The reported detection limits per chromatogram zone are 5 – 50 µg for detergents [3, 5], 0.025 – 5 µg for aliphatic phosphate, phosphonate, sulfate, sulfonate and sulfamate ions [8] and 0.2 – 1 µg for sweeteners [10].

The reagent can be employed on silica gel, cellulose, polyamide and alumina layers.

Procedure Tested

Sweeteners [10, 11]

Method	Ascending, one-dimensional double development in a trough chamber with chamber saturation.
Layer	HPTLC plates Cellulose (MERCK).
Mobile phase	Ethyl acetate – acetone – ammonia (25%) (6 + 6 + 1).
Migration distance	2×8 cm
Running time	2×10 min with 5 min intermediate drying in a stream of warm air.

Detection and result: The HPTLC plate was freed from mobile phase, immersed in the reagent solution for 2 s, dried for a few seconds in a stream of cold air and then immediately stored in the dark for 10 min.

In long-wavelength UV light ($\lambda = 365$ nm) cyclamate (hR_f 5–10) appeared as a light orange fluorescent zone while saccharin (hR_f 15–20) and acesulfame (hR_f 25–30) yielded violet and brownish zones respectively on a pale blue fluorescent background. Under the chromatographic conditions chosen dulcin lay in the region of the front. The fluoresence colors are concentration-dependent. At $\lambda =$ 254 nm the sweeteners were detectable as weak light fluorescent zones.

In situ quantitation: The fluorimetric determination was carried out at λ_{exc} = 313 nm and λ_{fl} = 365 nm (monochromatic filter M 365; Fig. 1).

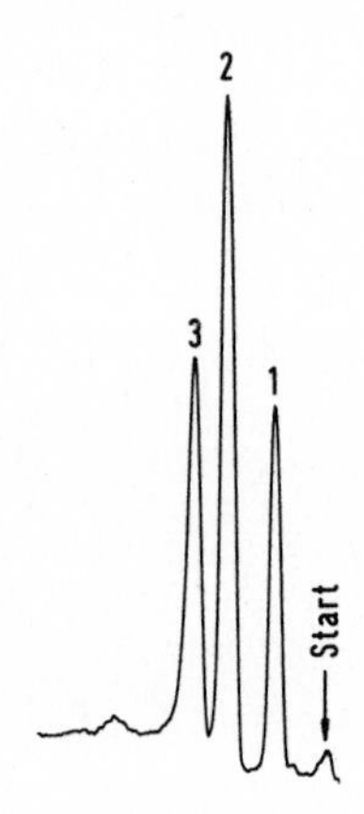

Fig. 1: Fluorescence scan of a chromatogram track with 3 µg each of acesulfame and saccharin and 2.25 µg of cyclamate per chromatogram zone. Cyclamate (1), saccharin (2), acesulfame (3).

References

[1] König, H.: *Fresenius Z. Anal. Chem.* **1971**, *254*, 337–345.
[2] Köhler, M., Chalupka, B.: *Fette, Seifen, Anstrichm.* **1982**, *84*, 208–211.
[3] Matissek, R., Hieke, E., Baltes, W.: *Fresenius Z. Anal. Chem.* **1980**, *300*, 403–406.
[4] Bey, K.: *Fette, Seifen, Anstrichm.* **1965**, *67*, 217–221.
[5] Matissek, R.: *Tenside Detergents* **1982**, *19*, 57–66.
[6] Read, H.: Proc. Int. Symp. Instr. High Perform. Thin-Layer Chromatogr., 3rd, Würzburg 1985, p. 157–171.

[7] Longman, G. F.: *The Analysis of Detergents and Detergent Products.* J. Wiley & Sons, Ltd., Chichester 1975.
[8] Nagasawa, K., Ogamo, A., Anryu, K.: *J. Chromatogr.* **1972,** *67*, 113–119.
[9] Takeshita, R., Jinnai, N., Yoshida, H.: *J. Chromatogr.* **1976,** *123*, 301–307.
[10] Nagasawa, K., Yoshidome, H., Anryu, K.: *J. Chromatogr.* **1970,** *52*, 173–176.
[11] Klein, I., Müller, E., Jork, H.: GDCh-training course Nr. 301 „Dünnschicht-Chromatographie für Fortgeschrittene", Universität des Saarlandes, Saarbrücken 1988.
[12] Borecky, J.: *J. Chromatogr.* **1959,** *2*, 612–614.

Potassium Hexacyanoferrate(III) – Ethylenediamine Reagent

Reagent for:

- Catecholamines [1–4]
- 3,4-Dihydroxyphenyl-acetic acid [5]
- 3-Hydroxytyramine [5]

$K_3(Fe(CN)_6)$
$M_r = 329.26$
Potassium hexacyano-ferrate(III)

$H_2N\text{-}CH_2CH_2\text{-}NH_2$
$C_2H_8N_2$
$M_r = 60.10$
Ethylenediamine

Preparation of Reagent

Dipping solution Mix 20 ml methanol and 5 ml ethylenediamine and then add 5 ml of a solution of 0.5 g potassium hexacyanoferrate(III) in 100 ml water.

Storage The solution may be kept in the refrigerator for 1 week; however, it should always be freshly prepared for in-situ quantitation [3].

Substances Methanol
Ethylenediamine
Potassium hexacyanoferrate(III)

Reaction

On oxidation by potassium hexacyanoferrate(III) adrenaline is converted into adrenochrome which then condenses with ethylenediamine:

Adrenaline + Ethylene–diamine $\xrightarrow[-2\,H_2O]{Fe^{3+}\ Fe^{2+}}$ Condensation product

Method

The chromatograms are dried thoroughly in a stream of warm air, immersed in the reagent solution for 1 s, dried in the air and heated in the drying cupboard at 65 °C for 30 min.

Fluorescent zones, which are suitable for quantitation [2], are visible in long-wavelength UV light ($\lambda \geq 400$ nm).

Note: The dipping solution may also be used as a spray solution [2]. Catecholamines are only separable on silica gel layers as their triacetyl derivatives but they can be separated underivatized on cellulose layers [4].

Procedure Tested

Adrenaline, Noradrenaline, Dopamine, Dopa as Triacetyl Derivatives [6]

Method	Ascending, one-dimensional development in a trough chamber. After application of the sample the layer was equilibrated for 30 min in a conditioning chamber at 18% relative humidity and then developed immediately.
Layer	HPTLC plates Silica gel 60 (MERCK).
Mobile phase	Acetone – dichloromethane – formic acid (50 + 50 + 1).
Migration distance	8 cm
Running time	20 min

Detection and result: The chromatogram was dried in a stream of warm air, immersed in the freshly prepared reagent solution for 1 s and then heated to 80 °C for ca. 15 min. Blue-yellow fluorescent zones were visible under long-wavelength UV light ($\lambda = 365$ nm).

In situ quantitation: The fluorimetric analysis was performed at $\lambda_{exc} = 405$ nm and the fluorescent light was measured at $\lambda_{fl} > 460$ nm. The detection limits were 200 pg dopamine (hR_f 65–70), 400 pg adrenaline (hR_f 55–60) or noradrenaline (hR_f 45–50) and 1–2 ng dopa (hR_f 80–85) per chromatogram zone (Fig. 1).

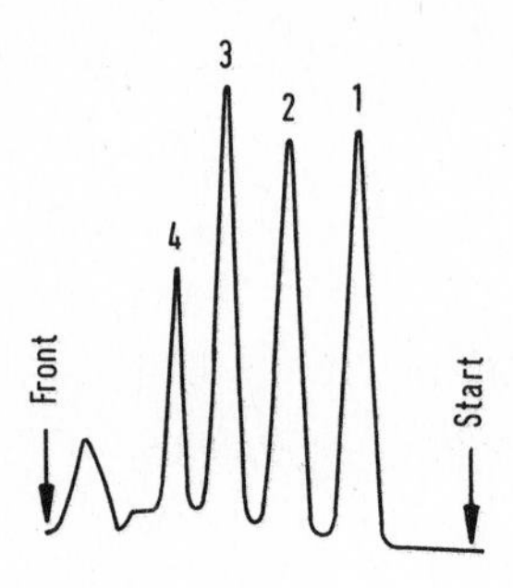

Fig. 1: Fluorescence scan of the catecholamine derivatives (each ca. 10 ng) of noradrenaline (1), adrenaline (2), dopamine (3), dopa (4).

References

[1] Ellman, G. L.: *Nature* (London) **1958,** *181,* 768-769
[2] Geissler, H. E., Mutschler, E.: *J. Chromatogr.* **1971,** *56,* 271-279.
[3] Gerardy, J., Quinaux, N., Dresse, A.: *Experientia* **1971,** *27,* 112-113.
[4] Takahashi, S., Gjessing, L. R.: *Clin. Chim. Acta* **1972,** *36,* 369-378.
[5] Clotten, R., Clotten, A.: *Hochspannungs-Elektrophorese.* Stuttgart: Thieme, 1962.
[6] Wallenstein, B.: Thesis, Universität Gießen, Institut für Pflanzenbau und Pflanzenzüchtung I, 1985.

Potassium Hexacyanoferrate(III) – Sodium Hydroxide Reagent

Reagent for:

- Vitamin B_1 and its phosphates [1 – 2]
- Adrenaline, noradrenaline, isoprenaline [3]

$K_3(Fe(CN)_6)$

$M_r = 329.25$

Preparation of Reagent

Dipping solution First dissolve 10 mg potassium hexacyanoferrate(III) and then 1 g sodium hydroxide pellets in 7 ml water and then dilute the solution with 20 ml ethanol.

Storage The dipping solution should be prepared fresh before use, since it is unstable.

Substances Potassium hexacyanoferrate(III)
Sodium hydroxide pellets
Ethanol

Reaction

Potassium hexacyanoferrate(III) forms, for example, fluorescent thiochrome with vitamin B_1:

$$[\ldots]^{2+}\ 2\ Cl^- \xrightarrow[K_3[Fe(CN)_6]]{OH^-} \ldots$$

Method

The developed chromatograms are dried in a stream of warm air for 1 min, then after cooling they are immersed in the reagent solution for 1 s and dried again in a stream of warm air for 30 s.

The dipping solution can also be used as a spray solution; other concentrations are also reported in the literature [1–3].

Vitamin B_1 and its phosphates yield bluish to bright blue fluorescent zones in long-wavelength UV light ($\lambda = 365$ nm). But adrenaline, noradrenaline, dopa and isoprenaline yield red colors and dopamine blue. Noradrenaline yields an orange-colored fluorescence in long-wavelength UV light ($\lambda = 365$ nm).

Note: The reagent may be applied to silica gel, kieselguhr, Si 50000 and cellulose layers.

Procedure Tested

Vitamin B_1 [2]

Method	Ascending, one-dimensional development in a twin-trough chamber with chamber saturation. After application of the samples the layer was equilibrated for 15 min in the mobile phase-free part of the twin-trough chamber.
Layer	HPTLC plates Silica gel 60 (MERCK) which were prewashed once to the upper edge of the plate with chloroform – methanol (1 + 1) and then dried at 110°C for 30 min.
Mobile phase	Methanol – ammonia solution (25%) – glacial acetic acid (8 + 1 + 1).
Migration distance	6 cm
Running time	ca. 20 min

Detection and result: The chromatogram was dried in a stream of warm air for 1 min, after cooling it was immersed for 1 s in the reagent solution. After redrying in a stream of warm air it was dipped into a mixture of chloroform – liquid paraffin – triethanolamine (6 + 1 + 1) for 1 s to enhance (by a factor of 2) and stabilize the fluorescence and dried again in a stream of warm air for 30 s.

Under long-wavelength UV light ($\lambda = 365$ nm) thiamine (hR_f 40 – 45) appeared as a bluish fluorescent zone which could be employed for quantitative analysis (Fig. 1). The detection limit was 500 pg vitamin B_1 per chromatogram zone.

In situ quantitation: The fluorimetric quantitation took place under long-wavelength UV light ($\lambda_{exc} = 365$ nm, $\lambda_{fl} > 430$ nm).

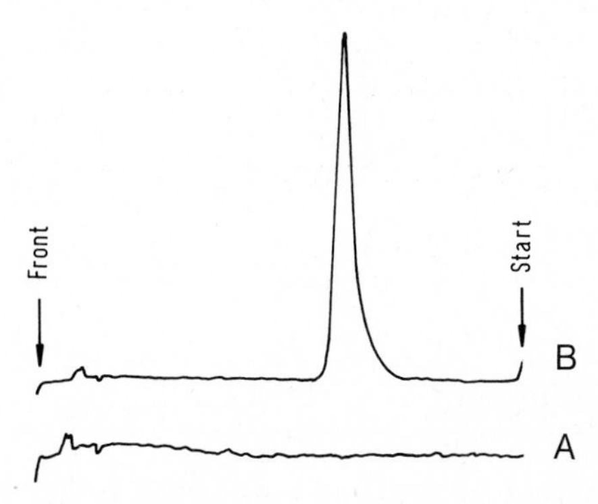

Fig. 1 Fluorescence scan of a blank track (A) and of a chromatogram track with 5 ng thiamine per chromatogram zone (B).

References

[1] Strohecker, R., Henning, H. M.: *Vitamin-Bestimmungen.* Weinheim: Verlag Chemie, 1963, p. 68 ff.
[2] Derr, P.: Thesis, Fachhochschule Gießen, Fachbereich Technisches Gesundheitswesen, 1985.
[3] Beckett, A. H., Beaven, M. A., Robinson, A. E.: *J. Pharm. Pharmacol.* **1960,** *12,* 203T–216T.

Pyrocatechol Violet Reagent

Reagent for:

- Metal ions (cations) [1 – 4]
  e.g. tin in organometallic compounds (stabilizers) [1 – 3]
  molybdenum, tungsten [4]

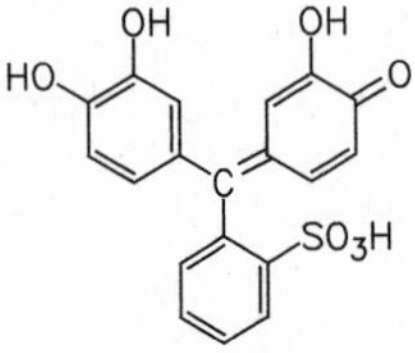

$C_{19}H_{14}O_7S$
$M_r = 386.38$

Preparation of Reagent

Dipping solution Dissolve 100 mg pyrocatechol violet (pyrocatecholsulfophthalein) in 100 ml ethanol.

Storage The reagent should always be freshly made up.

Substances Pyrocatechol violet
Ethanol (96%)

Reaction

Pyrocatechol violet forms colored complexes with a variety of metal ions, the complexes are stable in differing pH ranges.

Method

The chromatogram is freed from mobile phase, dipped for 1 s in the reagent solution or sprayed evenly with it until the plate begins to be transparent and dried in a stream of cold air.

Colored zones are formed (tin: violet-red to blue) on a yellow ochre background [3].

Note: The dipping solution can also be employed as a spray reagent.

Organotin compounds must first be decomposed before using the reagent [1–3]. This can be done by first heating the chromatogram to 110 °C for 30 min and then exposing it to saturated bromine vapors for 60 min (7–9 drops bromine in a chamber), the excess bromine is then allowed to evaporate in the air (10 min) and the plate is treated with the dyeing reagent [1, 3]. Quantitative determinations should be made at $\lambda = 580$ nm, the detection limits for organotin compounds are in the 10–20 ng per chromatogram zone range [3].

The reagent can be employed on silica gel, kieselguhr, Si 50 000 and cellulose layers.

Procedure Tested

Organotin Compounds [3, 5]

Method	Ascending, one-dimensional development in a trough chamber with chamber saturation.
Layer	HPTLC plates Silica gel 60 (MERCK).
Mobile phase	Methyl isobutyl ketone – pyridine – glacial acetic acid (97.5 + 1.5 + 1).
Migration distance	5 cm
Running time	10 min

Detection and result: The chromatogram was freed from mobile phase (heated to 110 °C for 30 min) and then exposed to bromine vapor for 1 h in a chamber, after blowing off excess bromine from the layer it was immersed for 1 s in the reagent solution. On drying in air dibutyltin dilaurate (hR_f 25–30), dibutyltin dichloride (hR_f 25–30), dioctyltin oxide (hR_f 40), tributyltin oxide (hR_f 80), tributyltin chloride (hR_f 80) and tetrabutyltin (hR_f 85–90) produced persistent blue zones on a yellow ochre background (Fig. 1).

In situ quantitation: The analysis is performed at $\lambda = 580$ nm. The detection limits for the di- and trialkyltin compounds are 10–20 ng and for tetrabutyltin 500 ng per chromatogram zone (Fig. 2).

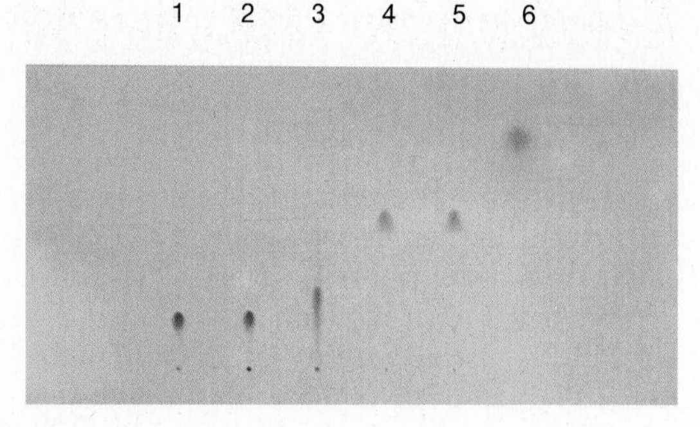

Fig. 1: Chromatogram of organotin compounds. Dibutyltin dilaurate (1), dibutyltin dichloride (2), dioctyltin oxide (3), tributyltin oxide (4), tributyltin chloride (5), tetrabutyltin (6).

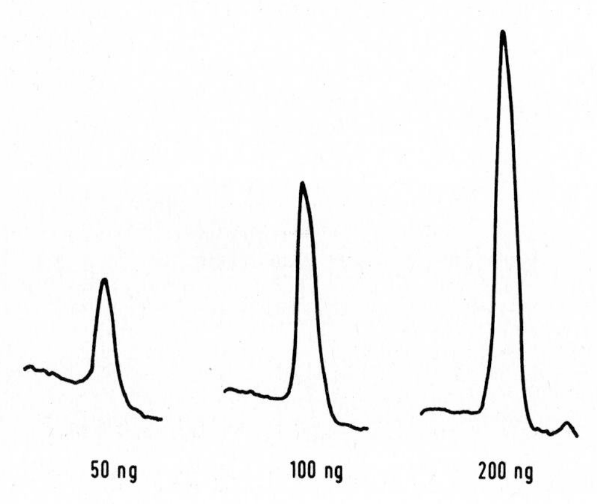

Fig. 2: Absorption scans of chromatograms with 50, 100 and 200 ng dibutyltin dichloride per chromatogram zone.

References

[1] Herold, B., Droege, K. H.: *Fresenius Z. Anal. Chem.* **1978,** *289,* 285–286.

[2] Ohlsson, S. V., Hintze, W. W.: *J. High Resolut. Chromatogr. Chromatogr. Commun.* **1983,** *6,* 89–94.

[3] Woidich, H., Pfannhauser, W., Blaicher, G.: *Dtsch. Lebensm. Rundsch.* **1976,** *72,* 421–422.

[4] Kitaeva, L. P., Volynets, M. P., Suvorova, S. N.: *J. Anal. Chem.* (USSR) **1979,** *34,* 712–717: **1980,** *35,* 199–209.

[5] Kany, E., Jork, H.: GDCh-training course Nr. 300 „Einführung in die Dünnschicht-Chromatographie", Universität des Saarlandes, Saarbrücken 1987.

Rhodamine B Reagent

Reagent for:

- Lipophilic substances, lipids [1 – 3]
  e.g. triglycerides [3], gangliosides [2]
  fatty acids and fatty acid methyl esters [3, 4]
  fatty alcohol dinitrobenzoates [5]
- Prenols, prenylquinones, prenyl vitamins [6]
- Polyphenols [7]
- Flavonols [8]
- Detergents [9]

$(C_2H_5)_2N$ ... $N(C_2H_5)_2$ Cl^- ... COOH

$C_{28}H_{31}ClN_2O_3$
$M_r = 479.02$

Preparation of Reagent

Dipping solution Dissolve 50 mg rhodamine B in 200 ml water.

Spray solution Dissolve 10 to 500 mg rhodamine B in 100 ml ethanol [3, 7], methanol [5] or water [2].

Storage Both reagent solutions may be stored over a long period.

Substances Rhodamine B (C.I. 45170)
Ethanol

Reaction

Because it contains amino and carboxylic groups rhodamine B tends to form zwitter ions which easily associate and can accumulate in lipophilic chromatogram zones.

Method

The chromatograms are freed from mobile phase and immersed in the reagent solution for 1 s or evenly sprayed with it. Red-violet zones are usually formed on a pink background; they fluoresce a deeper red than their surroundings in long-wavelength UV light ($\lambda = 365$ nm).

Note: Rhodamine B is a universal reagent that can be used on silica gel, talc, starch [5] and cellulose layers, just as on urea [1] or silver nitrate-impregnated [7] phases. Liquid paraffin-impregnated silica gel and RP layers are less suitable, since the background to the chromatographic zones is also intensely colored. It is often possible to increase the detection sensitivity by placing the plate in an atmosphere of ammonia after it has been sprayed or dipped, alternatively it can be oversprayed with sodium or potassium hydroxide solution.

Procedure Tested

Fatty Acids [10]

Method Ascending, one-dimensional development in a trough chamber with chamber saturation.

Layer TLC plates Kieselguhr (MERCK), which were impregnated before application of the samples by a single development in a 1% solution of viscous paraffin oil in petroleum ether.

Mobile phase Acetic acid – acetone – water (40 + 20 + 4).

Migration distance 17 cm

Running time 60 min

Detection and result: The chromatogram was freed from mobile phase and immersed in the reagent solution for 1 s. Arachidic acid (hR_f 15–20), stearic acid (hR_f 30–35), palmitic acid (hR_f 50–55), myristic acid (hR_f 60–65) and lauric acid (hR_f 70–75) appeared as pink zones on a reddish background.

The reaction was not particularly sensitive on paraffin-impregnated kieselguhr layers because of background coloration. For quantitation it was better to use the five-fold more sensitive rhodamine 6G reagent (q.v.).

References

[1] Rincker, R., Sucker, H.: *Fette Seifen Anstrichm.* **1972,** *74,* 21-24.
[2] Merat, A., Dickerson, J. W.: *J. Neurochem.* **1973,** *20,* 873-880.
[3] Vinson, J. A., Hooyman, J. E.: *J. Chromatogr.* **1977,** *135,* 226-228.
[4] Haeffner, E. W.: *Lipids* **1970,** *5,* 430-433.
[5] Perisic-Janjic, N., Canic, V., Lomic, S., Baykin, D.: *Fresenius Z. Anal. Chem.* **1979,** *295,* 263-265.
[6] Lichtenthaler, H., Boerner, K.: *J. Chromatogr.* **1982,** *242,* 196-201.
[7] Thielemann, H.: *Mikrochim. Acta* (Vienna) **1972,** 672-673; und: *Z. Chem.* **1972,** *12,* 223.
[8] Halbach, G., Görler, K.: *Planta Med.* **1971,** *19,* 293-298.
[9] Read, H. in: Proceedings of the 3rd International Symposium on Instrumental HPTLC, Würzburg. Bad Dürkheim: IfC-Verlag, 1985, p. 157-171.
[10] Jork, H., Kany, E.: GDCh-training course Nr. 301 „Dünnschicht-Chromatographie für Fortgeschrittene“, Universität des Saarlandes, Saarbrücken 1986.

Rhodamine 6G Reagent

Reagent for:

- Lipophilic substances, lipids [1 – 11]
  e.g. hydrocarbons [1]
  fatty acids and fatty acid esters [2, 3]
  ubiquinones [4]
  gangliosides [5]
  steroids, sterols [3, 6]
  triterpene alcohols [7]
  diglycerides, triglycerides [3, 8]
  phospholipids [3]

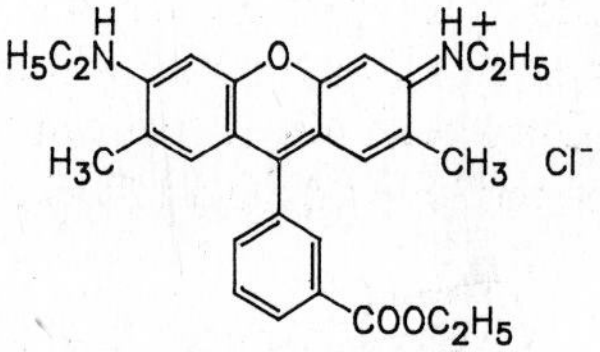

$C_{28}H_{31}ClN_2O_3$
$M_r = 479.02$

Preparation of Reagent

Dipping solution Dissolve 50 mg rhodamine 6G in 100 ml ethanol (96%).

Spray solution Dissolve 100 mg rhodamine 6G in 100 ml ethanol (96%).

Storage Both reagent solutions may be stored for an extended period.

Substances Rhodamine 6G (C.I. 45160)
Ethanol

Reaction

Rhodamine 6G accumulates in lipophilic chromatogram zones giving rise to a stronger fluorescence than in their surroundings.

Method

The chromatograms are freed from mobile phase and immersed in the reagent solution for 1 s or evenly sprayed with it. Pink-colored chromatogram zones are usually formed on a red-violet background, these fluoresce deeper yellow-orange than their environment in long-wavelength UV light ($\lambda = 365$ nm).

Note: Rhodamine 6G is a universal reagent which can also be incorporated in the TLC layers [4, 9] or added to the mobile phase [4]. The spray reagent can also be made up in water [8], acetone [4, 6] or ammonia solution (c = 2.5 mol/l) [5]. The visual detection limit is most favorable when the water from the mobile phase or the detection reagent has not completely evaporated from the layer. This can be recognized by the fact that the background fluorescence has not turned from red to pink [4].

It is often possible to increase the detection sensitivity in visible light by exposing the dipped or sprayed chromatogram to ammonia vapors; it can also be sprayed with caustic soda or potash solution. When this is done the fluorescence intensity is reduced on silica gel layers and increased on RP ones.

The reagent can be employed on silica gel, kieselguhr, cellulose and Florisil layers; these can also be impregnated, if desired, with silver nitrate.

Procedure Tested

Fatty Acids [10]

Method Ascending, one-dimensional development in a trough chamber.

Layer HPTLC plates RP-18 without fluorescence indicator (MERCK).

Mobile phase Acetone – acetonitril – 0.1 mol/l aqueous lithium chloride solution (10 + 10 + 1).

Migration distance 7 cm

Running time 8 min

Detection and result: The chromatogram was freed from mobile phase and immersed for 1 s in the reagent solution. Arachidic acid (hR_f 35–40), stearic acid

(hR_f 40–45), palmitic acid (hR_f 45–50), myristic acid (hR_f 55–60) and lauric acid (hR_f 60–65) appeared as pink-colored zones on a red-violet background; these fluoresced a deeper yellow-orange than their environment in long-wavelength UV light ($\lambda = 365$ nm) (Fig. 1).

In situ quantitation: The fluorimetric analysis was carried out in long-wavelength UV light ($\lambda_{exc} = 365$ nm; $\lambda_{fl} > 560$ nm). The detection limit for fatty acids was ca. 100 ng per chromatogram zone.

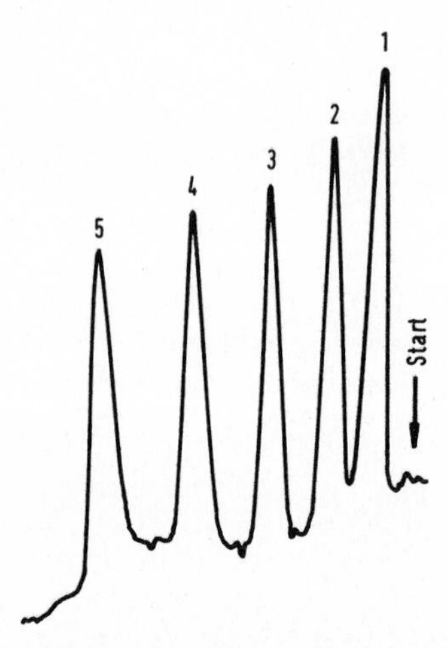

Fig. 1: Fluorescence scan of a fatty acid mixture with 500 ng substance per chromatogram zone. Arachidic acid (1), stearic acid (2), palmitic acid (3), myristic acid (4), lauric acid (5).

References

[1] Nagy, S. Nordby, H. E.: *Lipids* **1972,** *7,* 722 - 727.
[2] Ellington, J. J., Schlotzhauer, P. F., Schepartz, A. I.: *J. Amer. Oil Chem. Soc.* **1978,** *55,* 572–573.
[3] Nicolosi, R. J., Smith, S. C., Santerre, R. F.: *J. Chromatogr.* **1971,** *60,* 111–117.
[4] Rokos, J. A. S.: *J. Chromatogr.* **1972,** *74,* 357–358.
[5] Traylor, T. D., Hogan, E. L.: *J. Neurochem.* **1980,** *34,* 126–131.
[6] Garg, V. K., Nes, W. R.: *Phytochemistry* **1984,** *23,* 2925–2929.
[7] Boskou, D., Katsikas, H.: *Acta Aliment.* **1979,** *8,* 317–320.
[8] Devor, K. A., Mudd, J. B.: *J. Lipid Res.* **1971,** *12,* 396–402.
[9] Whistance, G. R., Dillon, J. F., Threlfall, D. R.: *Biochem. J.* **1969,** *111,* 461–472.

[10] Jork, H., Kany, E.: GDCh-training course 300 „Einführung in die Dünnschicht-Chromatographie" Universität des Saarlandes, Saarbrücken 1986.
[11] Abdelkader, A. B., Cherif, A., Demandre, C., Mazliak, P.: *Eur. J. Biochem.* **1973**, *32,* 155-165.

Silver Nitrate – Sodium Hydroxide Reagent

Reagent for:

- Carbohydrates
  e.g. mono- and oligosaccharides [1 – 8]
  sugar alcohols [4, 9]

$AgNO_3$	$NaOH$
$M_r = 169.87$	$M_r = 40.0$
Silver nitrate	Sodium hydroxide

Preparation of Reagent

Dipping solution I Make 1 ml of a saturated aqueous silver nitrate solution up to 200 ml with acetone. Redissolve the resulting precipitate by adding 5 ml water and shaking.

Dipping solution II Dissolve 2 g sodium hydroxide pellets in 2 ml water with heating and make up to 100 ml with methanol.

Storage Dipping solution I should be prepared freshly each day, dipping solution II may be stored for several days.

Substances Silver nitrate
Sodium hydroxide pellets
Acetone
Methanol

Reaction

The ionic silver in the reagent is reduced to metallic silver by reducing carbohydrates.

Method

The chromatograms are freed from mobile phase in a stream of warm air, immersed in solution I for 1 s or sprayed evenly with it, then dried in a stream of cold air, immersed in solution II for 1 s or sprayed with it and finally heated to 100 °C for 1 – 2 min.

Occasionally even without warming brownish-black chromatogram zones are produced on a pale brown background.

Note: The background can be decolorized by spraying afterwards with 5% aqueous ammonia solution and/or 5 – 10% sodium thiosulfate in 50% aqueous ethanol [2, 3]. The sodium hydroxide may be replaced by potassium hydroxide in dipping solution II [4].

The reagent may be employed on silica gel, cellulose and cellulose acetate layers.

Procedure Tested

Sugar Alcohols [9]

Method	Ascending, one-dimensional development in a trough chamber with chamber saturation (15 min).
Layer	HPTLC plates Si 50 000 (MERCK). Before application of the sample the plates were prewashed once by developing in chloroform – methanol (50 + 50) and then dried at 110 °C for 30 min.
Mobile phase	1-Propanol – water (18 + 2).
Migration distance	7 cm
Running time	1 h

Detection and result: The chromatogram was dried in a stream of cold air, immersed in dipping solution I for 1 s, then dried in a stream of cold air, immersed in dipping solution II for 1 s and finally heated to 100 °C for 1 – 2 min.

The sugar alcohols sorbitol (hR_f 15 – 20), mannitol (hR_f 20 – 25) and xylitol (hR_f 30 – 35) appeared as light to dark brown chromatogram zones on a beige-colored background. The detection limits were 1 ng substance per chromatogram zone.

In situ quantitation: The photometric determination of absorption in reflectance was carried out at a mean wavelength of $\lambda = 530$ nm (Fig. 1).

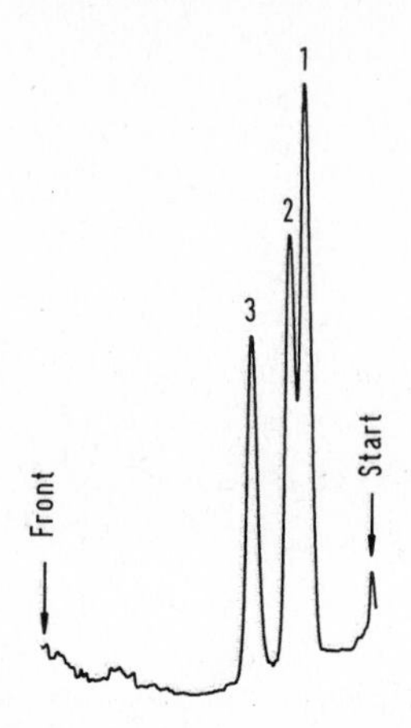

Fig. 1: Reflectance scan of a chromatogram track with 50 ng substance per chromatogram zone: sorbitol (1), mannitol (2), xylitol (3).

References

[1] Cerny, I., Trnka, T., Cerny, M.: *Collect. Czech. Chem. Commun.* **1983**, *48*, 2386–2394.
[2] Miyamoto, I., Watanabe, H., Nagase, S.: *Anal. Biochem.* **1981**, *110*, 39–42.
[3] Spitschan, R.: *J. Chromatogr.* **1971**, *61*, 169–172.
[4] Samuels, S., Fisher, C.: *J. Chromatogr.* **1972**, *71*, 291–296 and 297–306.
[5] Lee, E. Y. C., Carter, J. H.: *Arch. Biochem. Biophys.* **1973**, *154*, 636–641.
[6] Shiomi, N.: *Phytochemistry* **1981**, *20*, 2581–2583.
[7] Kanfer, J. N.: *Lipids* **1972**, *7*, 653–655.
[8] Bach, G., Berman, E. R.: *Biochim. Biophys. Acta* **1971**, *252*, 461–471.
[9] Netz, S., Funk, W.: *J. Planar Chromatogr.* **1989**, in press.

Sulfuric Acid Reagent

Reagent for:

- Steroids [1 – 3]
  e.g. estrogens, androgens, anabolics [1 – 4]
  bile acids [2, 5, 6]
  cholesterol, cholesteryl esters [7]
- Steroid conjugates [6, 8, 9]
- Sapogenins, saponins [2, 10]
- Spironolactone, canrenone [11]
- Cardiac glycosides [2]
- Alkaloids [2]
- Gibberellins [12]
- Prostaglandins [13]
- Lipids
  e.g. fatty acids [13], ceramides [14]
  phospholipids [15]
- Phenothiazines [16]
- Triphenodioxazines [17]
- Mycotoxins
  e.g. aflatoxins, trichothecenes [12 – 20]
- Antibiotics [21, 22]
- Vitamin A acid [23]

H_2SO_4
$M_r = 98.08$

Preparation of Reagent

Dipping solution Add 10 ml sulfuric acid (95–97%) cautiously to 85 ml water while cooling with ice and add 5 ml methanol after mixing thoroughly.

Spray solution Add 5–10 ml sulfuric acid cautiously under cooling to 85 to 95 ml acetic anhydride [1], ethanol [1], butanol [5] or methanol [11].

Storage The dipping solution may be kept at 4 °C for an extended period.

Substances Sulfuric acid (95–97%)
Acetic anhydride
Methanol
Ethanol absolute

Reaction

The reaction mechanism has not yet been elucidated.

Method

The chromatogram is dried in a stream of warm air for 10 min, immersed in the dipping solution for 1–2 s or evenly sprayed with the spray solution, dried in a stream of warm air and then heated to 95–140 °C for 1–20 min.

Under long-wavelength UV-light ($\lambda = 365$ nm) characteristic substance-specific yellow, green, red or blue fluorescent chromatogram zones usually appear which are often recognizable in visible light [7] – sometimes even before heating [2] – as colored zones on a colorless background and which are suitable for fluorimetric analysis [1].

Note: Sulfuric acid is a universal reagent, with which almost all classes of substance can be detected by charring at elevated temperatures (150–180 °C). The production of colored or fluorescent chromatogram zones at lower temperatures (< 120 °C) and their intensities are very dependent on the duration of heating, the temperature [2] and on the solvent employed [1]. For instance, the detection of prostaglandins is most sensitive after heating to 80 °C for only 3–5 min [13].

The following detection limits (substance per chromatogram zone) have been reported: steroids (1–10 ng) [1, 2], steroid conjugates (< 50 ng) [8], prostaglandins (< 1 ng) [13], phenothiazines (2 µg) [16].

The reagent can be used on silica gel, kieselguhr, Si 50000 and RP layers.

Procedure Tested

Cis-/trans-Diethylstilbestrol, Ethinylestradiol [24]

Method Ascending, one-dimensional development in a twin-trough chamber with the spare trough containing 2 ml (25%) ammonia solution. Before development the HPTLC plate was preconditioned with eluent vapors for ca. 60 min.

Layer HPTLC plates Silica gel 60 (MERCK); prewashed by triple development with chloroform – methanol (50 + 50) and then heated to 110 °C for 30 min.

Mobile phase Chloroform – methanol (19 + 1).

Migration distance 5 cm

Running time ca. 10 min

Detection and result: The chromatogram was freed from mobile phase and ammonia vapors and immersed twice for 10 s in reagent solution, with intermediate drying in a stream of cold air, and then heated to 95 °C for 12 min. The steroid derivatives were then visible under long-wavelength UV light ($\lambda = 365$ nm) as light blue fluorescent zones on a dark background: cis-diethylstilbestrol (hR_f 15–20), trans-diethylstilbestrol (hR_f 40–45) and ethinylestradiol (hR_f 50–55).

In situ quantitation: The chromatogram was immersed twice for ca. 1 s with brief intermediate drying in a mixture of chloroform – liquid paraffin and triethanolamine (60 + 10 + 10) to stabilize the fluorescence (for ca. 24 h) and increase its intensity (by a factor of ca. 3). The analysis was made in UV light

(λ_{exc} = 313 nm, λ_{fl} > 390 nm); the detection limit was 500 pg per chromatogram zone (Fig. 1).

Note: The result of the analysis is extremely dependent on the acid concentration employed, the immersion time and the subsequent temperature and duration of heating.

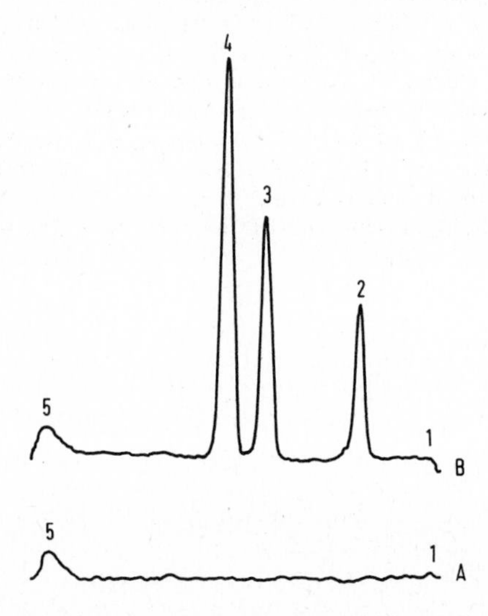

Fig. 1: Fluorescence scan of a blank (A) and of a diethylstilbestrol-ethinylestradiol mixture (B), each with 10 ng substance per chromatogram zone. Start (1), cis-diethylstilbestrol (2), trans-diethylstilbestrol (3) ethonsylestradiol (4), front (5).

References

[1] Verbeke, R.: *J. Chromatogr.* **1979,** *177,* 69–84.
[2] Heftman, E., Ko, S.-T., Bennett, R. D.: *J. Chromatogr.* **1966,** *21,* 490–494.
[3] Renwick, A. G. C., Pound, S. M., O'Shannessy, D. J.: *J. Chromatogr.* **1983,** *256,* 375–377.
[4] Smets, F., Vandewalle, M.: *Z. Lebensm. Unters. Forsch.* **1982,** *175,* 29–30.
[5] Anthony, W. L., Beher, W. T.: *J. Chromatogr.* **1964,** *13,* 567–570.
[6] Touchstone, J. C., Levitt, R., E., Soloway, R. O., Levin, S. S.: *J. Chromatogr.* **1979,** *178,* 566–570.
[7] Jatzkewitz, H., Mehl, E.: *Hoppe-Seyler's Z. Physiol. Chem.* **1960,** *320,* 251–257.
[8] Watkins, T. R., Smith, A., Touchstone, J. C.: *J. Chromatogr.* **1986,** *374,* 221–222.
[9] Batta, A. K., Shefer, S., Salen, G.: *J. Lipid Res.* **1981,** *22,* 712–714,

[10] Nakayama, K., Fujino, H., Kasai, R., Tanaka, O., Zhou, J.: *Chem. Pharm Bull.* **1986,** *34,* 2209-2213.
[11] Van der Merwe, P. J., Müller, D. G., Clark, E. C.: *J. Chromatogr.* **1979,** *171,* 519-521.
[12] Jones, D. F., McMillan, J., Radley, M.: *Phytochemistry,* **1964,** *2,* 307-314.
[13] Chiarugi, V., Ruggiero, M., Ricoveri, W.: *J. Chromatogr.* **1983,** *280,* 400-403.
[14] Do, U. H., Pei, P. T.: *Lipids* **1981,** *16,* 855-862.
[15] Sherma, J., Bennett, S.: *J. Liq. Chromatogr.* **1983,** *6,* 1193-1211.
[16] Steinbrecher, K.: *J. Chromatogr.* **1983,** *260,* 463-470.
[17] Ojha, K. G., Jain, S. K., Gupta, R. R.: *Chromatographia* **1979,** *12,* 306-307.
[18] Schmidt, R., Ziegenhagen, E., Dose, K.: *Z. Lebensm. Unters. Forsch.* **1982,** *175,* 169-171.
[19] Serralheiro, M. L., Quinta, M. L.: *J. Assoc. Off. Anal. Chem.* **1985,** *68,* 952-954.
[20] Gimeno, A.: *J. Assoc. Off. Anal. Chem.* **1979,** *62,* 579-585.
[21] Joshi, Y. C., Shukla, S. K., Joshi, B. C.: *Pharmazie,* **1979,** *34,* 580-582.
[22] Menyhart, M., Szaricskai, F., Bognar, R.: *Acta Chim. Hung.* **1983,** *113,* 459-467.
[23] De Paolis, A. M.: *J. Chromatogr.* **1983,** *258,* 314-319.
[24] Sommer, D.: Thesis, Fachhochschule Gießen, Fachbereich Technisches Gesundheitswesen, 1984.

Tetracyanoethylene Reagent (TCNE Reagent)

Reagent for:

- Aromatic hydrocarbons and heterocyclics [1, 2]
- Aromatic amines and phenols [2, 3]
- Indole derivatives [3, 4, 6]
- Carbazoles [3]
- Phenothiazines [5]

$(NC)_2C{=}C(CN)_2$

C_6N_4

$M_r = 128.09$

Preparation of Reagent

Dipping solution Dissolve 0.5 g tetracyanoethylene (TCNE) in 100 ml dichloromethane.

Spray solution Dissolve 0.5 – 1 g TCNE in ethyl acetate [3], dichloromethane [4] or acetonitrile [5].

Storage The reagent solutions may be kept for several days.

Substances Tetracyanoethylene
Dichloromethane

Reaction

Tetracyanoethylene yields a colored π-complex with aromatic compounds; in the case of aromatic amines, phenols and indoles these then react to yield the corresponding tricyanovinyl derivatives [3, 4].

Indole TCNE Tricyanovinyl derivative

π-Complex

- HCN

Method

The chromatograms are freed from mobile phase in a stream of warm air, immersed for 1 s in the dipping solution or sprayed evenly with the spray solution and dried for a few minutes in a stream of cold air [4, 6] or at 80 °C.

Variously colored chromatogram zones are formed on a pale yellow background.

Note: For some of the substances the intensities of coloration are only stable for ca. 2 h; in the case of phenols the coloration intensifies during this time [2]. The detection limits for indole derivatives lie in the lower nanogram range.

The reagent can be employed on silica gel, kieselguhr, Si 50000 and cellulose layers.

Procedure Tested

Indole and Indole Derivatives [6]

Method Ascending, one-dimensional development in a trough chamber with chamber saturation.

Layer	HPTLC plates Silica gel 60 (MERCK).
Mobile phase	Toluene – ethyl acetate (50 + 15).
Migration distance	7.5 cm
Running time	17 min

Detection and result: The chromatogram was freed from mobile phase in a stream of warm air, immersed in reagent solution for 1 s and dried in a stream of cold air.

5-Nitroindole (hR_f 40–45), 5-chloroindole (hR_f 60–65) and indole (hR_f 70–75) yielded orange-yellow chromatogram zones on a pale yellow background. If the chromatogram was exposed to ammonia vapor for 15 s the color was intensified. The detection limits were 10 ng substance per chromatogram zone.

In situ quantitation: The photometric scanning in reflectance was carried out at a mean wavelength of λ = 460 nm ($\lambda_{max\,(5\text{-nitroindole})}$ = 450 nm, $\lambda_{max\,(5\text{-chloroindole})}$ = 460 nm, $\lambda_{max\,(indole)}$ = 480 nm; Fig. 1).

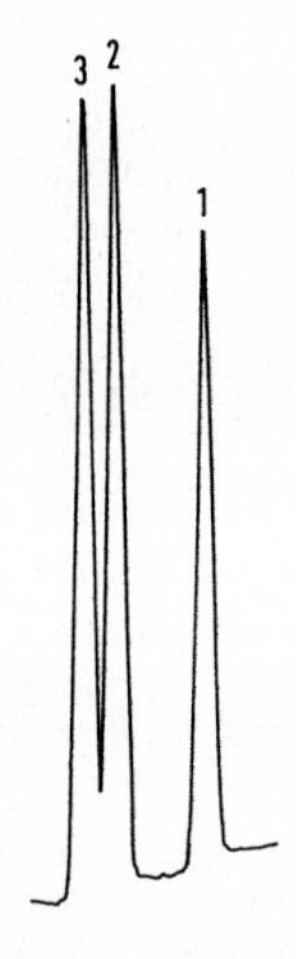

Fig. 1: Reflectance scan of a chromatogram track with 80 ng substance per chromatogram zone: 5-nitroindole (1), 5-chloroindole (2) and indole (3).

References

[1] Kucharczyk, N., Fohl, J., Vymetal, J.: *J. Chromatogr.* **1963**, *11*, 55-61.
[2] Janak, J.: *J. Chromatogr.* **1964**, *15*, 15-28.
[3] Macke, G. F.: *J. Chromatogr.* **1968**, *36*, 537-539.
[4] Heacock, R. A., Forrest, J. E., Hutzinger, O.: *J. Chromatogr.* **1972**, *72*, 343-350.
[5] Forrest, J. E., Heacock, R. A.: *J. Chromatogr.* **1973**, *75*, 156-160.
[6] Heiligenthal, M., Funk, W.: Private communication, Fachhochschule Gießen, Fachbereich Technisches Gesundheitswesen, Gießen 1988; *J. Planar Chromatogr.* **1989**, in press.

Trichloroacetic Acid Reagent

Reagent for:

- Steroids [1]
- Alkaloids from
  e.g. veratrum [1],
  colchicum [2]
- Digitalis glycosides [3, 4]
- Vitamin D_3 [5]
- Benzodiazepin-2-one derivatives [6]

CCl_3-COOH
$C_2HCl_3O_2$
$M_r = 163.39$

Preparation of Reagent

Dipping solution Dissolve 5 g trichloroacetic acid in 50 ml chloroform [2] or ethanol.

Storage The dipping solution may be kept for at least 1 week at room temperature.

Substances Trichloroacetic acid
Chloroform

Reaction

The reaction mechanism has not yet been elucidated.

Method

After drying in a stream of cold air the chromatograms are immersed for 1 s in the reagent solution or sprayed with it and then heated at 120°C for 10 min.

Mainly light blue fluorescent zones appear under long-wavelength UV light ($\lambda = 365$ nm).

Note: A few drops of 30% hydrogen peroxide solution are added to this reagent when detecting digitalis glycosides [4, 7]. Digitalis glycosides of the A series fluoresce yellow-brown, those of the B series brilliant blue and those of the C series pale blue [7]. The fluorescence can be stabilized and intensified by dipping the plate into a solution of liquid paraffin – *n*-hexane (1 + 2).

The reagent can be employed on silica gel, kieselguhr, Si 50000 and cellulose layers.

Procedure Tested

Steroids [1]

Method	Ascending, one-dimensional development in a trough chamber. The layer was conditioned for 30 min at 0% relative humidity after the samples had been applied.
Layer	HPTLC plates Silica gel 60 F_{254} (MERCK), which had been prewashed by developing once to the upper edge of the plate with chloroform – methanol (50 + 50) and then activated at 110°C for 30 min.
Mobile phase	Cyclohexane – diethyl ether (50 + 50).
Migration distance	6 cm
Running time	ca. 15 min

Detection and result: The chromatogram was dried in a stream of cold air and then intensively irradiated with UV light ($\lambda = 365$ nm) for 2 min and then immersed in the reagent solution for 1 s. It was finally heated to 120°C for 10 min and after cooling dipped into liquid paraffin – *n*-hexane (1 + 2) to intensify and stabilize the fluorescence. Light blue fluorescent zones were produced under long-wavelength UV light ($\lambda = 365$ nm) by cholesterol (hR_f 20–25), coprostanol (hR_f 25–30), 4-cholesten-3-one (hR_f 40–45), 5α-cholestan-3-one (hR_f: 60) and coprostanone (hR_f 70).

In situ quantitation: The fluorimetric analysis was carried out under long-wavelength UV light ($\lambda_{exc} = 365$ nm, $\lambda_{fl} > 430$ nm; Fig. 1). The detection limit was several nanograms substance per chromatogram zone.

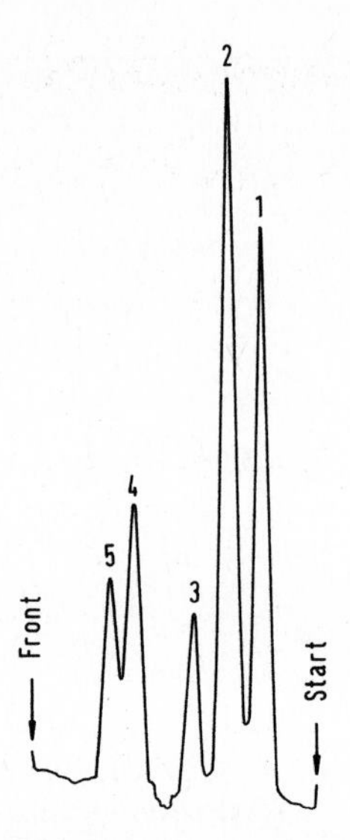

Fig. 1: Fluorescence scan of a chromatogram track with 250 ng of each substance per chromatogram zone (exception coprostanol with 500 ng). Cholesterol (1), coprostanol (2), 4-cholesten-3-one (3), 5α-cholestan-3-one (4), and coprostanone (5).

References

[1] Schade, M.: Thesis, Fachhochschule Gießen, Fachbereich Technisches Gesundheitswesen, 1986.

[2] Müller, J.: Thesis, Fachhochschule Gießen, Fachbereich Technisches Gesundheitswesen, 1987.

[3] Horvath, P., Szepesi, G., Hoznek, M., Vegh, Z., Mincsovics, E.: Proceedings of the 2nd International Symposium on Instrumental HPTLC, Interlaken. Bad Dürkheim: IfC-Verlag, 1982, p. 295–304.

[4] Balbaa, S. I., Hilal, S. H., Haggag, M. Y.: *Planta Med.* **1970,** *18,* 254–259.

[5] Zeller, M.: Thesis, Fachhochschule Gießen, Fachbereich Technisches Gesundheitswesen, 1986.

[6] Steidinger, J., Schmid, E.: *Arzneim. Forsch.* **1970,** *20,* 1232–1235.

[7] Aldrich, B. J., Frith, M. L., Wright, S. E.: *J. Pharm. Pharmacol.* **1956,** *8,* 1042–1049.

Trinitrobenzenesulfonic Acid Reagent (TNBS Reagent)

Reagent for:

- Amino acids [1]
- Aminoglycoside antibiotics [2, 3]
  e.g. neomycin, gentamycin

$C_6H_3N_3O_9S$
$M_r = 293.2$

Preparation of Reagent

Dipping solution Dissolve 100 mg 2,4,6-trinitrobenzenesulfonic acid in a mixture of 20 ml acetone, 20 ml ethanol and 10 ml water.

Storage The reagent should always be freshly made up.

Substances 2,4,6-Trinitrobenzenesulfonic acid
Acetone
Ethanol

Reaction

On heating primary amines form colored MEISSENHEIMER complexes with trinitrobenzenesulfonic acid.

$$R\text{-}NH_2 + HO_3S\text{-}C_6H_2(NO_2)_3 \xrightarrow{H_2O,\ \Delta T} \text{Meissenheimer complex } (R\text{-}HN,\ HO_3S,\ (NO_2)_3)$$

Method

The chromatograms are freed from mobile phase, immersed in the reagent solution for 1 s or sprayed evenly with it, dried in a stream of warm air and heated to 100 °C for 5 min.

Deep yellow-colored zones are formed on a pale yellow background; these remain visible for ca. 2 days.

Note: Ammonia interferes with the reaction and must be removed from the layer completely before application of the reagent.
The reagent can be employed on silica gel and kieselguhr layers. NH_2 layers are not suitable.

Procedure Tested

Gentamycin C Complex [2]

Method Ascending, one-dimensional development at 10 – 12 °C in a twin-trough chamber with 5 ml ammonia solution (25%) in the second trough; chamber saturation for 15 min.

Layer HPTLC plates Silica gel 60 (MERCK) which had been prewashed by developing three times with chloroform – methanol (1 + 1) and then dried at 110 °C for 30 min.

Mobile phase Chloroform – ammonia solution (25%) – ethanol (10 + 10 + 9); the *lower organic phase* was employed.

Migration distance ca. 5 cm

Running time ca. 20 min

Detection and result: The HPTLC plates were freed from mobile phase (ca. 30 min in a stream of cold air, the ammonia must be removed as completely as possible), immersed in the reagent for 1 s, dried in a stream of warm air and then heated to 100 °C for 4 – 5 min.

The gentamycins C_{1a} (hR_f 35 – 40), C_2/C_{2a} (hR_f 40 – 45) and C_1 (hR_f 45 – 50) yielded intense yellow-colored zones on a pale yellow background.

In situ quantitation: The UV absorption was recorded in reflectance at $\lambda = 353$ nm. The detection limit was 40 ng substance per chromatogram zone.

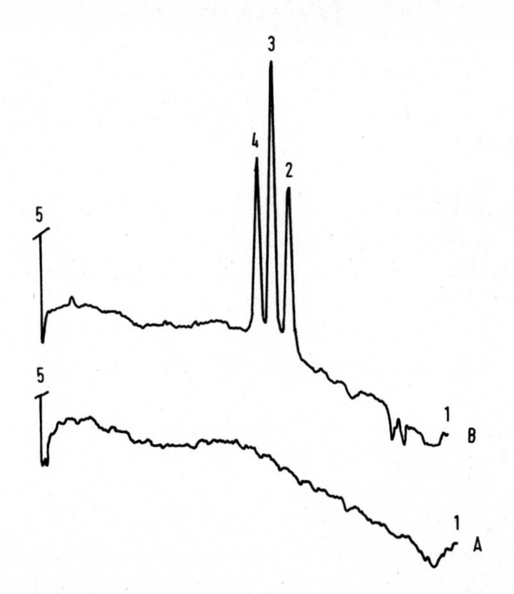

Fig. 1: Reflectance scan of a blank track (A) and of a gentamycin standard track (B) with 800 ng gentamycin mixture per starting zone. Start (1), gentamycin C_{1a} (2), gentamycin C_2/C_{2a} (3), gentamycin C_1 (4) and solvent front (5).

References

[1] Munier, R. L., Peigner, A., Thommegay, C.: *Chromatographia* **1970**, *3*, 205–210.
[2] Kiefer, U.: Thesis, Fachhochschule Gießen, Fachbereich Technisches Gesundheitswesen, 1984.
[3] Funk, W., Canstein, M. v., Couturier, Th., Heiligenthal, M., Kiefer, U. Schlierbach, S., Sommer, D.: Proceedings of the 3rd International Symposium on Instrumental HPTLC, Würzburg. Bad Dürkheim: IfC-Verlag 1985, p. 281–311.

Vanadium(V) – Sulfuric Acid Reagent
(Mandelin's Reagent) (Ammonium Monovanadate – Sulfuric Acid or Vanadium Pentoxide – Sulfuric Acid Reagent)

Reagent for:

- Carbohydrates and derivatives
  e.g. monosaccharides [1 – 4]
  oligosaccharides [1 – 3]
  sugar alcohols (hexitols) [1, 3, 4]
  alkylglycosides [1, 3]
  uronic acids [1, 3]
- Glycols [1, 4]
- Diethylene glycol [4]
- Reducing carboxylic acids [1]
  e.g. ascorbic acid, glycolic acid, oxalic acid, pyruvic acid
- Steroids [5]
- Antioxidants [5]
- Vitamins [5]
- Phenols [5, 6]
- Aromatic amines [5]
- Antihypertensives (β-blockers) [7, 9]
- Pyrazolidine derivatives [8]
- Laxatives [10]

NH_4VO_3
$M_r = 116.98$
Ammonium monovanadate

V_2O_5
$M_r = 181.88$
Vanadium pentoxide

H_2SO_4
$M_r = 98.08$
Sulfuric acid

Preparation of Reagent

Dipping solution Dissolve 0.6 g ammonium monovanadate (ammonium metavanadate) in 22.5 ml water and carefully add 2.5 ml conc. sulfuric acid and 25 ml acetone.

Spray solution Ia Dissolve 1.2 g ammonium monovanadate in 95 ml water and carefully add 5 ml conc. sulfuric acid [1, 2].

Spray solution Ib Dissolve 18.2 g vanadium pentoxide in 300 ml aqueous sodium carbonate solution, ($c_{Na_2CO_3}$ = 1 mol/l), with heating; after cooling carefully add 460 ml sulfuric acid, ($c_{H_2SO_4}$ = 2.5 mol/l), and make up to 1 l with water. Remove excess CO_2 in the ultrasonic bath [1, 3, 4].

Spray solution II *β-Blockers:* Saturated solution of ammonium monovanadate in conc. sulfuric acid [7].

Storage The reagent solutions may be stored for an extended period.

Substances Ammonium monovanadate
Vanadium(V) oxide
Sulfuric acid
Sodium carbonate decahydrate
Acetone

Reaction

The yellow colored vanadyl(V) ion is transformed to the blue vanadyl(IV) ion by reaction with reducing agents.

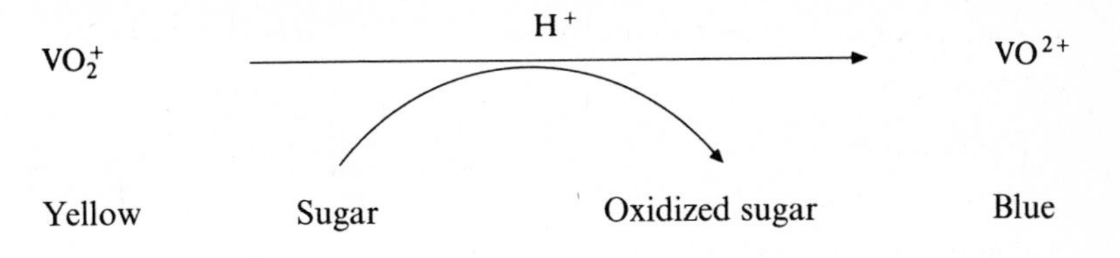

Method

The chromatogram is freed from mobile phase, immersed for 2 s in the dipping solution or sprayed evenly with spray solution Ia or Ib until it starts to appear transparent and then heated to 100–120 °C for 5 min. After cooling the chromatogram can be sprayed briefly once more to decolorize the background [4].

Spray solution II is employed for the detection of β-blockers [7]; here the plates are analyzed immediately after spraying.

In the case of carbohydrates blue chromatogram zones are produced on a yellow background that slowly fades [2]. Steroids, vitamins, antioxidants, phenols and aromatic amines yield, sometimes even at room temperature, variously colored chromatogram zones [5]. β-Blockers and laxatives also acquire various colors [7, 10]. The detection limits are in the nanogram to microgram range [5].

Note: The reagent can be employed on silica gel, kieselguhr, Si 50 000 and cellulose layers. At room temperature sugars and sugar derivatives react at different rates depending on the functional groups present [1], e.g. ketoses react more rapidly than aldoses. It is possible to differentiate substance types on this basis [1, 3].

Procedure Tested

β-Blockers [9]

Method	Ascending, one-dimensional development in a HPTLC chamber with chamber saturation.
Layer	HPTLC plates Silica gel 60 (MERCK) which had been prewashed by developing once with chloroform – methanol (50 + 50) and then heated to 110 °C for 30 min.
Mobile phase	Ethyl acetate – methanol – ammonia solution (25%) (40 + 5 + 5).
Migration distance	6 cm
Running time	15 min

Detection and result: The chromatogram was freed from mobile phase (20 min in a stream of warm air), immersed in the dipping solution for 2 s, dried briefly in a stream of warm air and heated to 120 °C for 10 min.

Atenolol (hR_f 25–30), bunitrolol (hR_f 40–45) and alprenolol (hR_f 50–55) appeared as light blue to white zones on a yellow background in visible light. They were visible as light blue fluorescent zones on a faint blue background in long-wavelength UV light ($\lambda = 365$ nm).

In situ quantitation: The fluorimetric determination was carried out in UV light (λ_{exc} = 313 nm, $\lambda_{fl} > 390$ nm; Fig. 1). The detection limits were 50 ng substance per chromatogram zone.

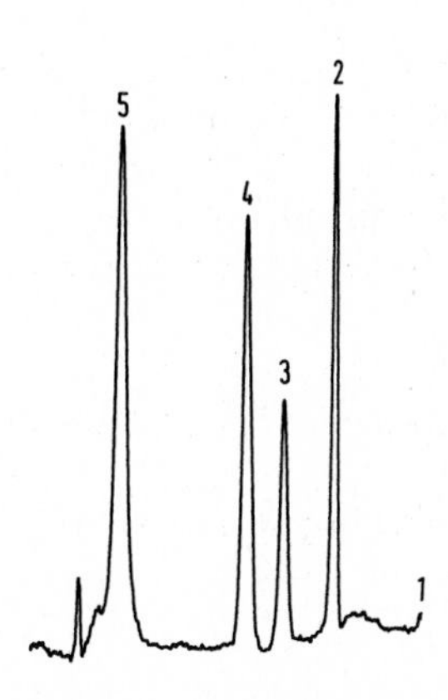

Fig. 1: Fluorescence scan of a chromatogram track with 1 µg substance per chromatogram zone. Start (1) atenolol (2), bunitrolol (3), alprenolol (4) and "dipping" front (5).

References

[1] Haldorsen, K. M.: *J. Chromatogr.* **1977,** *134,* 467–476: **1978,** *150,* 485–490.
[2] Alonso-Fernandez, J. R., Boveda, M. D., Parrado, C., Peña, J., Fraga, J. M.: *J. Chromatogr.* **1981,** *217,* 357–366.
[3] Friese, P.: *Fresenius Z. Anal. Chem.* **1980,** *301,* 389–397.
[4] Klaus, R., Fischer, W.: *Chromatographia* **1987,** *23,* 137–140.
[5] Malaiyandi, M., Barrette, J. P., Lanouette, M.: *J. Chromatogr.* **1974,** *101,* 155–162.
[6] Klaus, R., Fischer, W., Bayer, H.: *J. Chromatogr.* **1987,** *398,* 300–308.
[7] Jack, D. B., Dean, S., Kendall, M. J., Laugher, S.: *J. Chromatogr.* **1980,** *196,* 189–192.
[8] Wachowiak, R.: *Arzneim. Forsch.* **1979,** *29,* 599–602.
[9] Azarderakhsch, M.: Thesis, Fachhochschule Gießen, Fachbereich Technisches Gesundheitswesen 1988.
[10] Daldrup, T.: *Toxichem + Krimtech,* Mitteilungsblatt der Gesellschaft für toxikologische und forensische Chemie **1988,** *55,* 18–19.

Vanillin – Phosphoric Acid Reagent

Reagent for:

- Steroids, sterols [1–4]
  e.g. estrogens, anabolics [2, 3]
- Triterpenes [4, 5]
- Cucurbitacins
  (bitter principles) [4, 6]
- Digitalis glycosides [7, 8]
- Prostaglandins [9]
- Saponins [10, 11]

Vanillin	ortho-Phosphoric acid
$C_8H_8O_3$	H_3PO_4
$M_r = 152.15$	$M_r = 98.00$
Vanillin	*ortho*-Phosphoric acid

Preparation of Reagent

Dipping solution Dissolve 1 g vanillin (4-hydroxy-3-methoxybenzaldehyde) in 25 ml ethanol and add 25 ml water and 35 ml *ortho*-phosphoric acid.

Storage The solution may be kept at 4 °C for ca. 1 week.

Substances Vanillin
ortho-Phosphoric acid (85%)
Ethanol

Reaction

The reaction probably occurs according to the following scheme:

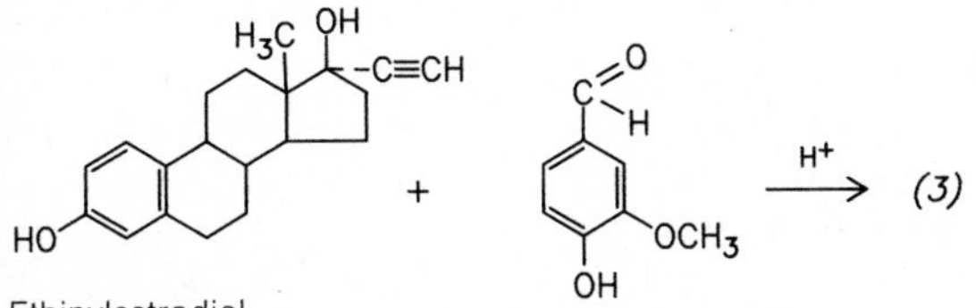

Ethinylestradiol

(1) (2)

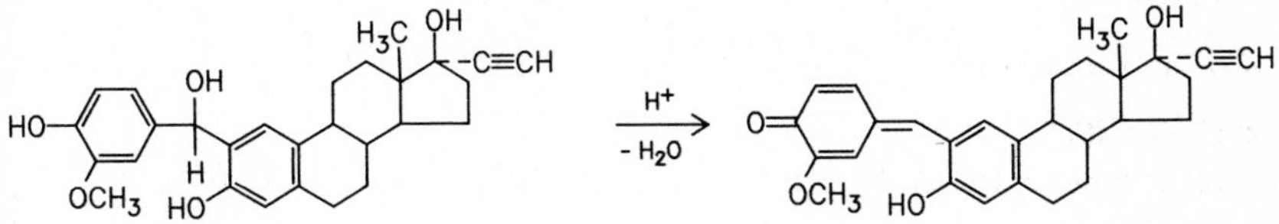

$\xrightarrow{H^+}$ (3)

(3) $\xrightarrow[- H_2O]{H^+}$ (4)

Method

The dried chromatograms (15 min in a stream of warm air) are briefly immersed in the reagent solution or homogeneously sprayed with it and then heated to 120–160 °C for 5–15 min. Colored zones are produced on a pale background; in the case of digitalis glycosides, cucurbitacins and sterols these fluoresce in long-wavelength UV light ($\lambda = 365$ nm) [4, 7].

Note: The reagent can be employed on silica gel, kieselguhr, Si 50 000 and RP layers. Hydrochloric or sulfuric acid can be employed in place of phosphoric acid (q.v.). The detection limits for steroids and digitalis glycosides are several nanograms per chromatogram zone.

Procedure Tested

Estrogens, Anabolics [3]

Method Ascending, one-dimensional development in a twin-trough chamber with 2 ml 25% ammonia solution in the chamber free from mobile phase. After application of the sample and before development the plate was preconditioned for 60 min over the mobile phase.

Layer HPTLC plates Silica gel 60 (MERCK), which had been prewashed three times with chloroform – methanol (50 + 50) and then dried at 110 °C for 30 min.

Mobile phase Chloroform – methanol (19 + 1).

Migration distance ca. 5 cm

Running time ca. 10 min

Detection and result: The chromatogram was dried in the air, immersed in the reagent solution for 5 s and then heated to 120 °C for 5 min. *cis*-Diethylstilbestrol (hR_f 15–20) and *trans*-diethylstilbestrol (hR_f 40–45) turned red; while ethynylestradiol (hR_f 50–55) appeared blue.

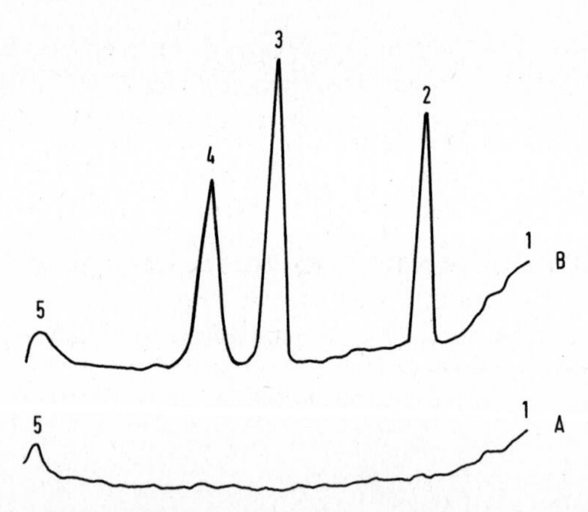

Fig. 1: Absorption scan of a blank track (A) and of a mixture of estrogen anabolics with 100 ng of each substance per chromatogram zone (B). Start (1), *cis*-diethylstilbestrol (2), *trans*-diethylstilbestrol (3), 17-α-ethynylestradiol (4), solvent front (5).

In situ quantitation: The absorption photometric analysis was made at $\lambda = 540$ nm (ethynylestradiol) and $\lambda = 605$ nm (diethylstilbestrol). The detection limit for ethynylestradiol was 12 ng and that for diethylstilbestrol 3 ng per chromatogram zone.

References

[1] Metz, H.: *Naturwissenschaften* **1961,** *48,* 569-570.
[2] Jarc, H., Rüttner, O., Krocza, W.: *Fleischwirtschaft* **1976,** *9,* 1326-1328.
[3] Funk, W., Canstein, M. v., Couturier, Th., Heiligenthal, M., Kiefer, U., Schlierbach, S., Sommer, D.: Proceedings of the 3rd International Symposium on Instrumental HPTLC, Würzburg, Bad Dürkheim: IfC-Verlag 1985, p. 281-311.
[4] Bauer, R., Wagner, H.: *Dtsch. Apoth. Ztg.* **1983,** *123,* 1313-1321.
[5] Jankov, L. K., Ivanov, T. P.: *Planta Med.* **1970,** *18,* 232-242.
[6] Bauer, R., Berganza, L. H., Seligmann, O., Wagner, H.: *Phytochemistry* **1985,** *24,* 1587-1591.
[7] Winsauer, K., Buchberger, W.: *Chromatographia* **1981,** *14,* 623-625.
[8] Hauser, W., Kartnig, Th., Verdino, G.: *Sci. Pharm.* **1968,** *36,* 237-241.
[9] Wallach, D. P., Daniels, E. G.: *Biochim. Biophys. Acta* **1971,** *231,* 445-457.
[10] Kartnig, Th., Wegschaider, O.: *Planta Med.* **1972,** *21,* 144-149.
[11] Kartnig, Th., Ri, C. Y.: *Planta Med.* **1973,** *23,* 269-271; 379-380.

Vanillin – Potassium Hydroxide Reagent

Reagent for:

- Amines, amino acids e.g. aminoglycoside antibiotics [1, 2]

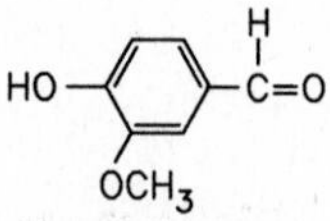

$C_8H_8O_3$	KOH
$M_r = 152.15$	$M_r = 56.11$
Vanillin	Potassium hydroxide

Preparation of Reagent

Dipping solution I Dissolve 1 g vanillin (4-hydroxy-3-methoxy-benzaldehyde) in 50 ml 2-propanol.

Dipping solution II Make 1 ml 1 mol/l potassium hydroxide solution up to 100 ml with ethanol.

Storage The solutions may be stored for several days in the refrigerator.

Substances Vanillin
2-Propanol
Potassium hydroxide solution (1 mol/l)
Ethanol

Reaction

Vanillin reacts with primary amines in weakly basic media to form fluorescent or colored SCHIFF's bases whereby colored phenolates are also produced at the same time.

$$HO-C_6H_3(OCH_3)-CHO + H_2N-R \xrightarrow[-H_2O]{} HO-C_6H_3(OCH_3)-CH=N-R$$

Vanillin Primary amines SCHIFF's bases

Method

The chromatograms are freed from mobile phase (10 min stream of warm air), immersed for 1 s in solution I, dried for 10 min at 110 °C, then immersed in solution II and finally heated once again for 10 min to 110 °C.
Variously colored zones are produced; these can frequently be excited to fluorescence by long-wavelength UV light ($\lambda = 365$ nm) [3].

Note: The dipping solution can also be employed as a spray solution.

Ornithine, proline, hydroxyproline, pipecolic acid and sarcosine yield red zones, glycine greenish-brown and the other amino acids weakly brown ones [3]. The colors of the zones are different if an alcoholic solution of potassium carbonate is used for basification instead of dipping solution II.

Characteristic fluorescence often appears under long-wavelength UV light (λ = 365 nm) after drying for the first time (before the use of reagent solution II): e.g. ornithine yields a strong greenish-yellow fluorescence and lysine a weak one, while hydroxyproline appears light blue.
The reagent can be employed on silica gel, kieselguhr or cellulose layers. Amino phases and polyamide layers are unsuitable.

Procedure Tested

Gentamycin C Complex [1, 2]

Method Ascending, one-dimensional development at 10–12 °C in a twin-trough chamber with 5 ml ammonia solution (25%) in the second part of the chamber; chamber saturation for 15 min.

Layer HPTLC plates Silica gel 60 (MERCK), which had been prewashed by developing three times with chloroform – methanol (1 + 1) and heated to 110 °C for 30 min after each development.

Mobile phase Chloroform – ammonia solution (25%) – ethanol (10 + 10 + 9); the *lower organic phase* was employed.

Migration distance ca. 5 cm

Running time ca. 20 min

Detection and result: The chromatogram was freed from mobile phase (45 min, stream of cold air), immersed in reagent solution I for 1 s, dried at 110 °C for 10 min, immersed in solution II and heated to 110 °C for a further 10 min. Gentamycins C_{1a} (hR_f 35–40), C_2/C_{2a} (hR_f 40–45), C_1 (hR_f 45–50) produce yellow zones on a pale yellow background.

In situ quantitation: The absorption photometric determination was carried out in long-wavelength UV light ($\lambda = 392$ nm). The detection limit was 40 ng gentamycin per chromatogram zone (Fig. 1).

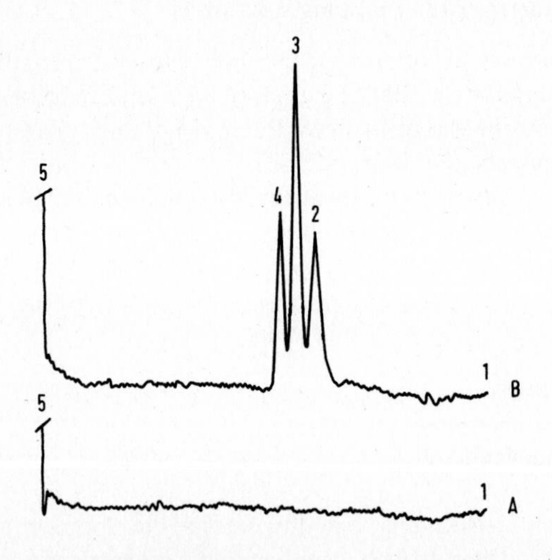

Fig. 1: Absorption scan of a blank track (A) and of a gentamycin standard track (B) with 500 ng gentamycin C_1–C_{2a} mixture per starting zone. Start (1), gentamycin C_{1a} (2), gentamycin C_2/C_{2a} (3), gentamycin C_1 (4) and solvent front (5).

References

[1] Kiefer, U.: Thesis, Fachhochschule Gießen, Fachbereich Technisches Gesundheitswesen, 1984.
[2] Funk, W., Canstein, M. v., Couturier, Th., Heiligenthal, M., Kiefer, U., Schlierbach, S., Sommer, D.: Proceedings of the 3rd International Symposium on Instrumental HPTLC, Würzburg. Bad Dürkheim: IfC-Verlag, 1985, p. 281–311.
[3] Curzon, G., Giltrow, J.: *Nature* (London) **1953**, *172*, 356–357: **1954**, *173*, 314–315.

Zirconium(IV) Oxide Chloride Reagent

Reagent for:

- Flavonoids [1 – 3]
- Mycotoxins
  e.g. sterigmatocystine [2]
- Steroids
  e.g. estrogens [4, 7]
- Purines, pyrimidines [4]
- Cardiac glycosides [5]
- Lipids
  e.g. fatty acids, phospholipids, cholesterol,
  cholesteryl esters [4]
  triglycerides [4, 6]
- Sugars [4]
- Prostaglandins [4]

$ZrOCl_2 \cdot 8H_2O$
$M_r = 322.25$

Preparation of Reagent

Dipping solution Dissolve 1 g zirconium(IV) oxide chloride octahydrate (zirconyl chloride) in 50 ml methanol.

Spray solution Dissolve 2 to 25 g zirconium(IV) oxide chloride octahydrate in 100 ml methanol [1], ethanol – water (55 + 15) [5] or water [4].

Storage The reagent solutions may be stored for an extended period.

Substances Zirconium(IV) oxide chloride octahydrate
Methanol
Ethanol

Reaction

The reaction mechanism has not yet been elucidated [4].

Method

The chromatograms are freed from mobile phase, immersed in the dipping solution for 1 s or sprayed evenly with it and normally heated to 100 – 120 °C for 10 min [2, 5, 6], or sometimes to 150 – 180 °C for 5 to 60 min [4].
Mainly yellow-green to bluish fluorescent chromatogram zones are formed on a dark background under long-wavelength UV light ($\lambda = 365$ nm).

Note: Flavonoids react with the reagent even at room temperature [1]; mycotoxins, steroids, purines, pyrimidines, cardiac glycosides and lipids only react on heating [2, 4 – 6]. Zirconyl sulfate can be used to replace the zirconyl chloride in the reagent; this is reported to result in an increase in the sensitivity to certain groups of substances (e.g. cholesteryl esters, triglycerides) [4].

The chromatogram can be sprayed with liquid paraffin – *n*-hexane (2 + 1) to increase the fluorescence intensity [2]. The detection limits per chromatogram zone are 0.5 – 1 ng for cardiac glycosides [5], triglycerides [6] and sterigmatocystine [2] and 25 pg for estrogens [7].

The reagent, which can also be employed to impregnate the layer before chromatography, is best suited for silica gel layers [4]; it can, however, also be employed on aluminium oxide, kieselguhr, Si 50000, cellulose and polyamide layers [4].

Procedure Tested

Estriol, Estradiol, Estrone [7]

Method Ascending, one-dimensional development in a trough chamber with chamber saturation.

Layer HPTLC plates Silica gel 60 (MERCK), which had been prewashed by developing once with chloroform – methanol (1 + 1) and then activated at 110 °C for 30 min before sample application.

Mobile phase Toluene – ethanol (9 + 1).

Migration distance 7 cm

Running time 15 min

Detection and result: The chromatogram was freed from mobile phase in a stream of cold air, conditioned for 15 min in ammonia vapor (placed in a twin-trough chamber whose second part contained 25% ammonia solution). It was then immediately immersed in the reagent solution for 1 s and heated to 180 °C for 15 min.

Estriol (hR_f 10 – 15), estradiol (hR_f 30 – 35) and estrone hR_f 40 – 45) appeared under long-wavelength UV light ($\lambda = 365$ nm) as light blue fluorescing chromatogram zones on a dark background.

In situ quantitation: Fluorimetric analysis was carried out by excitation at $\lambda_{exc} = 313$ nm and detection at $\lambda_{fl} > 390$ nm. The detection limits were 25 pg substance per chromatogram zone (Fig. 1).

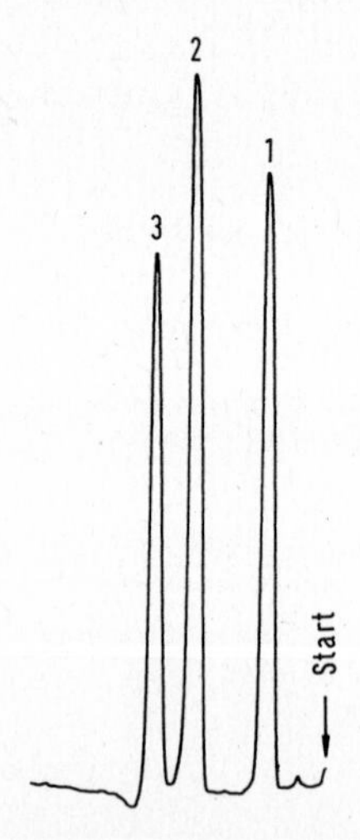

Fig. 1: Fluorescence scan of a chromatogram of ca. 50 ng each of estriol (1), estradiol (2) and estrone (3) per chromatogram zone.

References

[1] Poethke, W., Schwarz, C., Gerlach, H.: *Planta Med.* **1970**, *19*, 177–188.
[2] Gertz, C., Böschemeyer, L.: *Z. Lebensm. Unters. Forsch.* **1980**, *171*, 335–340.
[3] Thieme, H., Khogali, A.: *Pharmazie* **1975**, *30*, 736–743.
[4] Segura, R., Navarro, X.: *J. Chromatogr.* **1981**, *217*, 329–340.
[5] Hagiwara, T., Shigeoka, S., Uehara, S., Miyatake, N. Akiyama, K.: *J. High Resolut. Chromatogr. Chromatogr. Commun.* **1984**, *7*, 161–164.
[6] Nägele, U., Hägele, E. O., Sauer, G., Wiedemann, E., Lehmann, P., Wahlefeld, A., Gruber, W.: *J. Clin. Chem. Biochem.* **1984**, *22*, 165–174.
[7] Netz, S., Funk, W.: Private communication, Fachhochschule Gießen, Fachbereich Technisches Gesundheitswesen, 1987.

Name Reagents
Reagent Acronyms

Name Reagents

Reagent Acronyms

Index

B

N

Q